Characterizing Stellar and Exoplanetary Environments

Astrophysics and Space Science Library

More information about this series at
http://www.springer.com/series/5664

Helmut Lammer • Maxim Khodachenko
Editors

Characterizing Stellar and Exoplanetary Environments

Springer

Editors
Helmut Lammer
Maxim Khodachenko
Space Research Institute
Austrian Academy of Sciences
Graz
Austria

ISSN 0067-0057 ISSN 2214-7985 (electronic)
ISBN 978-3-319-09748-0 ISBN 978-3-319-09749-7 (eBook)
DOI 10.1007/978-3-319-09749-7
Springer Cham Heidelberg New York Dordrecht London

Library of Congress Control Number: 2014953599

Cover illustration: A model simulation showing an exoplanet interacting with the wind of it's host star

Printed on acid-free paper

Springer is part of Springer Science+Business Media (www.springer.com)

Preface

Exoplanetology is one of the fastest-growing fields in present day astrophysics and space science. Nineteen years after the discovery of 51 Peg b, the first Jupiter-type gas giant outside our Solar System, more than 1,100 exoplanets have been detected. Although most of the exoplanets have been discovered with the radial velocity method, recently an increasing the number of exoplanets with sizes from super-Earths to sub-Neptunes and Jupiter-type gas giants have been observed by using the transit detection technique. The discovery of these planets occurred because of the efforts of several international ground-based transit search projects as well as the capabilities of the CoRoT (COnvection, ROtation and planetary Transits) and Kepler space observatories. The detection of exoplanets at orbital distances $\leq$0.05 AU raises questions regarding their atmosphere structure, their interactions with the extreme stellar radiation and plasma environment, the role of possible magnetospheres for atmospheric protection, destructive tidal forces between the host star and the planets, the formation of plasma tori, comet-like escaping planetary plasma tails, as well as the stability of their upper atmospheres against thermal and nonthermal mass loss processes. The spectral and temporal behavior of exoplanet host stars is relevant to models related to the atmospheric chemistry and evolution of planetary atmospheres. Because of this relationship, recent observational and theoretical efforts aimed at a better understanding of the ultraviolet spectra of dwarf stars will be discussed.

The chapters and sections of this book address the analysis of observational findings during Hubble Space Telescope (HST) observations of transiting exoplanets in the ultraviolet (UV) and the application of advanced numerical models for characterizing planetary upper atmosphere structures and stellar environments. In addition to HST observations, the discovery of atoms and molecules by NASA's Spitzer telescope in the atmospheres of hot Jupiters during their secondary eclipses in the infrared (IR) will also be discussed. Observations of that kind are helpful in characterizing the temperature structure of the lower thermosphere with hydrodynamic and empirical upper atmosphere models. The observation and characterization of the upper atmosphere-magnetosphere-plasma environment of hydrogen-rich exoplanets in orbital locations less than 0.1 AU can also be used for the understanding of non-hydrostatic upper atmospheric conditions, the identification of magnetic obstacles and the validation of complex numerical models.

The authors of the various interdisciplinary, but connected, articles have been members of a team led by us with the title *Characterizing Stellar and Exoplanetary Environments via Observations and Advanced Modelling Techniques* supported by the Swiss-based International Space Science Institute (ISSI) in Bern which studied these processes during the past two years. The results summarized by these researchers will show that the study of exoplanets under extreme stellar radiation and plasma conditions will also help the planetary community to understand how terrestrial planets, including early Venus, Earth, and Mars, and their atmospheres evolved during the host star's active early phase. These observational and theoretical investigations and discoveries are timely and important in the context of the next generation of space telescopes, such as James Webb Space Telescope (JWST), which will have the capability of acquiring spectra in the far-infrared and mid-infrared. ESA's CHaracterizing ExOPlanets Satellite (CHEOPS) mission, and Europe's next Generation Planet Finder (PLATO 2.0), as well as NASA's Transiting Exoplanet Survey Satellite (TESS) mission, the Russian-led international UV space observatory World Space Observatory-UV (WSO-UV) and the planned European astrometry mission Nearby Earth Astrometric Telescope (NEAD).

Graz, Austria
October 2014

Helmut Lammer
Maxim L. Khodachenko

Acknowledgments

The Editors and the authors of this volume gratefully acknowledge the support of the International Space Institute (ISSI) and the friendly atmosphere created by the ISSI staff in Bern, Switzerland, for their support and hospitality during various meetings of the ISSI team *Characterizing stellar- and exoplanetary environments*, where this book project have been discussed and planned. We also acknowledge the Austrian Research Foundation FWF NFN project S11601-N16 'Pathways to Habitability: From Disks to Active Stars, Planets and Life', as well as the related FWF NFN subprojects, S116 604-N16 'Radiation & Wind Evolution from the T Tauri Phase to ZAMS and Beyond', S116 606-N16 'Magnetospheric Electrodynamics of Exoplanets', and S116607-N16 'Particle/Radiative Interactions with Upper Atmospheres of Planetary Bodies Under Extreme Stellar Conditions'.

Contents

List of Authors

Igor I. Alexeev Skobeltsyn Institute of Nuclear Physics (MSU SINP), Lomonosov Moscow State University, Moscow, Russian Federation

Elena S. Belenkaya Skobeltsyn Institute of Nuclear Physics (MSU SINP), Lomonosov Moscow State University, Moscow, Russian Federation

Dmitry V. Bisikalo Russian Academy of Sciences, Institute of Astronomy, Moscow, Russian Federation

Anders Erikson Institute for Planetary Research, DLR, Berlin, Germany

Nikolai V. Erkaev Russian Academy of Sciences, Institute for Computational Modelling, and Siberian Federal University, Krasnoyarsk, Russian Federation

Luca Fossati Argelander-Institut für Astronomie der Universität Bonn, Bonn, Germany

Malcolm Fridlund Institute for Planetary Research, DLR, Berlin, Germany

Leiden Observatory, Leiden, The Netherlands

Stephan Geier European Southern Observatory, Garching bei München, Germany

Ana-Ines Gómez de Castro Faculty of Mathematics, Universidad Complutense de Madrid (AEGORA), Madrid, Spain

Jean-Mathias Grießmeier LPC2E – Université d'Orléans/CNRS, France and Station de Radioastronomie de Nançay, Observatoire de Paris – CNRS/INSU, USR 704 – University of Orléans, Nançay, France

Maria S. Grygoryan Skobeltsyn Institute of Nuclear Physics (MSU SINP), Lomonosov Moscow State University, Moscow, Russian Federation

Manuel Güdel Department of Astrophysics, University of Vienna, Vienna, Austria.

Eike W. Guenther Thüringer Landessternwarte Tautenburg, Tautenburg, Germany

Carole A. Haswell Department of Physical Sciences, The Open University, Milton Keynes, UK

Mats Holmström Swedish Institute of Space Physics, Kiruna, Sweden

Dmitry E. Ionov Russian Academy of Sciences, Institute of Astronomy, Moscow, Russian Federation

Colin P. Johnstone Department of Astrophysics, University of Vienna, Vienna, Austria

Vladimir V. Kalegaev Skobeltsyn Institute of Nuclear Physics (MSU SINP), Lomonosov Moscow State University, Moscow, Russian Federation

Pavel V. Kaygorodov Institute of Astronomy RAS, Moscow, Russian Federation

Maxim L. Khodachenko Austrian Academy of Sciences, Space Research Institute, Graz, Austria

Skobeltsyn Institute of Nuclear Physics (MSU SINP), Lomonosov Moscow State University, Moscow, Russian Federation.

Kristina G. Kislyakova Austrian Academy of Sciences, Space Research Institute, Graz, Austria

Helmut Lammer Austrian Academy of Sciences, Space Research Institute, Graz, Austria

Joe Llama Lowell Observatory, Flagstaff, Arizona, USA

Jeffrey L. Linsky JILA, University of Colorado and NIST, Boulder, CO, USA

Theresa Lüftinger Department of Astrophysics, University of Vienna, Vienna, Austria

Heike Rauer Institute for Planetary Research, Berlin, Germany

Mikhai E. Sachkov Russian Academy of Sciences, Institute of Astronomy, Moscow, Russian Federation

Valery I. Shematovich Russian Academy of Sciences, Institute of Astronomy, Moscow, Russian Federation

Boris M. Shustov Russian Academy of Sciences, Institute of Astronomy, Moscow, Russian Federation.

Aline A. Vidotto SUPA, School of Physics and Astronomy, University of St Andrews, St Andrews, UK

Observatoire de Genève, Université de Genève, Versoix, Switzerland

Brian E. Wood Naval Research Laboratory, Washington, DC, USA

Acronyms

ACIS	Advanced CCD Imaging Spectrometer
ACS	Advanced Camera for Surveys
ADI	Angular Differential Imaging
AFTA	Astrophysics Focused Telescope Assets
AGB	Asymptotic Giant Branch
AGN	Active Galactic Nuclei
ALMA	Atacama Large Millimeter/submillimeter Array
ASPERA	Analyzer of Space Plasma and Energetic Atoms instrument
ASTEP	Antarctic Search for Transiting Exoplanets
AU	Astronomical Unit
CARMENES	Calar Alto high-Resolution search for M dwarfs with Exoearths with Near-infrared and optical Echelle Spectrographs
CFHT	Canada-France-Hawaii Telescope
CHEOPS	CHaracterising ExOPlanets Satellite
CME	Coronal Mass Ejection
CoRoT	COnvection, ROtation and planetary Transits space observatory
COS	Cosmic Origins Spectrograph
CRIRES	CRyogenic high-resolution InfraRed Echelle Spectrograph
CTTS	Classical T Tauri star
DI	Doppler Imaging
E-ELT	European-Extremely Large Telescope
EMD	Emission Measure Distributions
ESA	European Space Agency
ESO	European Southern Observatory
ESPaDOnS	Echelle SpectroPolarimetric Device for the Observation of Stars
ESPRESSO	Echelle SPectrograph for Rocky Exoplanet and Stable Spectroscopic Observations
EUV	Extreme Ultraviolet radiation

EUVE	Extreme Ultraviolet Explorer spacecraft
FTS	Fourier-Transform Spectrograph
FUSE	Far Ultraviolet Spectroscopic Explorer spacecraft
FUV	Far Ultraviolet radiation
GAIA	Global Astrometric Interferometer for Astrophysics
GALEX	Galaxy Evolution Explorer
GCR	Galactic Cosmic Rays
GMRT	Giant Metrewave Radio Telescope
GMT	Giant Magellan Telescope
HARPS	High Accuracy Radial Velocity Planet Searcher
HAT	Hungarian Automated Telescope
HCP	Head-on Collision Point
HR	Hertzsprung–Russell
HST	Hubble Space Telescope
HZ	Habitable Zone
IBEX	Interstellar Boundary Explorer spacecraft
IR	Infra-Red
ISM	Interstellar Medium
ISSI	International Space Science Institute
IUE	International Ultraviolet Explorer
JVLA	Jansky Very Large Array
JWST	James Webb Space Telescope
KELT	Kilodegree Extremely Little Telescope
LOFAR	Low Frequency Array
MAST	Mikulski Archive for Space Telescopes
MAVEN	Mars Atmosphere and Volatile Evolution spacecraft
METIS	Mid-infrared E-ELT Imager and Spectrograph
MHD	Magnetohydrodynamic
MIR	Mid Infra-Red
MOST	Microvariability and Oscillations of STars telescope
NASA	National Aeronautics and Space Administration
NEAT	Nearby Earth Astrometric Telescope
NGTS	Next-Generation Transit Survey
NIR	Near Infra-Red
NUV	Near Ultraviolet radiation
PFI	Planet Formation Imager
PLATO	PLAnetary Transits and Oscillations of stars
PN	Planetary Nebulae
ROSAT	Röntgensatellit
SEE	Solar EUV Experiment
SIRS	Solar Irradiance Reference Spectrum
SOPHIE	Spectrographe pour l'Observation des Phénomènes des Intérieurs stellaires et des Exoplanètes
SPHERE	Spectro-Polarimetric High-contrast Exoplanet Research
SPI	Star-Planet Interaction

STEREO	Solar TErrestrial RElations Observatory
STIS	Space Telescope Imaging Spectrograph
STScI	Space Telescope Science Institute
TESS	Transiting Exoplanet Survey Satellite
THEIA	Telescope for Habitable Exoplanets and Interstellar/Intergalactic Astronomy
TIMED	Thermosphere Ionosphere Mesosphere Energetics and Dynamics spacecraft
TMT	Thirty Meter Telescope
TNG	Telescopio Nazionale Galileo
TPF	Terrestrial Planet Finder
TrES	Tillinghast Reflector Echelle Spectrograph
UTR-2	Ukrainian T-shaped Radio telescope-2
UV	Ultraviolet radiation
VLT	Very Large Telescope
WASP	Wide Angle Search for Planets
WFIRST	Wide-Field Infrared Survey Telescope
WSO–UV	World Space Observatory–UltraViolet
WTTS	Weak-line T Tauri Stars
X-ray	Roentgen radiation
XMM-Newton	X-ray Multi-Mirror telescope
XUV	Soft-X-ray and EUV radiation
ZDI	Zeeman Doppler Imaging
ZIMPOL	Zurich IMaging POLarimeter

Part I
Exoplanet Host Star Radiation and Plasma Environment

The processes that lead to the evolution of planetary atmospheres, and hence habitability, cannot be considered separately from the evolution of the host star's radiation, plasma and magnetic environment during the lifetime of a planet's host star. In the following articles the activity of exoplanet host star UV, EUV and X-ray fluxes, stellar winds and their magnetic structure are discussed starting from their arrival at the Zero-Age-Main-Sequence.

Chapter 1
Exoplanet Host Star Radiation and Plasma Environment

Jeffrey L. Linsky and Manuel Güdel

Abstract Radiation from host stars controls the planetary energy budget, photochemistry in planetary atmospheres, and mass loss from the outer layers of these atmospheres. Stellar optical and infrared radiation, the major source of energy for the lower atmosphere and planetary surfaces, increases slowly as stars evolve from the Zero-Age-Main-Sequence. Ultraviolet radiation, including the Lyman-α emission line that dominates the UV spectrum of M dwarf stars, controls photochemical reactions of important molecules, including H_2O, CO_2, and CH_4. Extreme ultraviolet and X-radiation from host stars ionizes and heats the outer layers of planetary atmospheres driving mass loss that is rapid for close-in Jupiter-like planets. The strength of the stellar UV, EUV, and X-radiation depends on stellar activity, which decays with time as stellar rotation decreases. As a result, the evolution of an exoplanet's atmosphere depends on the evolution of its host star. We summarize the available techniques for measuring or estimating the X-ray, EUV, and UV radiation of host stars with different spectral types and ages.

1.1 Introduction: Relevance of Short Wavelength Radiation to Planetary Atmospheres

The discovery of many extrasolar planets (exoplanets) by radial velocity, transit, and imaging techniques (see also Chap. 13, Fridlund et al. (2014), Chap. 14, Shustov et al. (2014), and Chap. 15, Guenther (2014)) has stimulated observational and theoretical studies to characterize their atmospheric chemistry and physical properties (see Chap. 4 Fossati et al. 2014) as well as to investigate whether these exoplanets could sustain life forms (e.g., Kasting and Catling 2003; Seager and Deming 2010). As density decreases with height in exoplanet atmospheres,

J.L. Linsky (✉)
JILA, University of Colorado and NIST, 440 UCB Boulder, CO 80309-0440, USA
e-mail: jlinsky@jila.colorado.edu

M. Güdel
Department of Astrophysics, University of Vienna, Türkenschanzstr. 17, A-1180 Vienna, Austria
e-mail: manuel.guedel@univie.ac.at

H. Lammer, M. Khodachenko (eds.), *Characterizing Stellar and Exoplanetary Environments*, Astrophysics and Space Science Library 411,
DOI 10.1007/978-3-319-09749-7_1

photolysis (photodissociation of molecules and photoionization of atoms) will eventually dominate over thermal equilibrium chemistry. This change typically occurs at heights where the atmospheric pressure is less than 1 mbar. Photochemistry models have now been computed for terrestrial planets and super-Earths (Segura et al. 2005; Kaltenegger et al. 2011; Hu et al. 2012), hot-Neptunes (Line et al. 2011), and hot-Jupiters (Kopparapu et al. 2012; Moses et al. 2013; Line et al. 2010). Far Ultraviolet (FUV) radiation at wavelengths below 170 nm and, in particular, the very bright Lyman-α emission line (121.6 nm), control the photodissociation of such important molecules as H_2O, CH_4, and CO_2, which can increase the mixing ratio of oxygen (Tian et al. 2014). Ozone (O_3) has been called a potential biosignature in super-Earth atmospheres (Segura et al. 2005, 2010; Grenfell et al. 2012), but it is important to assess the extent to which photolysis of O_2 and subsequent chemical reactions rather than biological processes control its abundance (Tian et al. 2014). Future photochemical models based on realistic host star UV emission, including intrinsic Lyman-α fluxes, are needed to address questions of the reliability of proposed biosignatures and atmospheric chemical abundances. Recent models, such as those cited above, show that the C/O ratio, quenching reactions, thermal structure, and diffusion also play important roles in determining mixing ratios for important molecules in exoplanet atmospheres, but the short wavelength radiation of the host star is critically important.

Stellar ultraviolet spectra consist of emission lines and continua formed over a wide range of temperatures at different heights in the atmosphere. In this paper we call the 170–320 nm wavelength region the near ultraviolet (NUV), the 91.2–170 nm region the far ultraviolet (FUV), the 10–91.2 nm region the extreme ultraviolet (EUV), and below 10 nm the X-ray region. We chose 170 nm as the division between the NUV and FUV because photospheric emission usually dominates above this wavelength, and chromospheric emission dominates below this wavelength. Also, the 170–320 nm region is important for the photochemistry of O_2 and O_3, while the photochemistry of such important molecules as H_2O, CO, CO_2, and CH_4 is controlled primarily by radiation with wavelengths below 170 nm. Since the photoionization of hydrogen occurs at wavelengths below 91.2 nm, it is natural to define the EUV range as 10–91.2 nm. Figure 1.1 shows the quiet Sun's X-ray and EUV spectrum with some of the important spectral lines and continua indicated.

The reflectivity of optical coatings plays an important role in determining which instruments are capable of observing in these spectral regions. The reflectivity of MgF_2-coated optics used in the Space Telescope Imaging Spectrograph (STIS) and the Cosmic Origins Spectrograph (COS) instruments on board of the Hubble Space Telescope (HST) decreases rapidly at wavelengths below 117 nm. These instruments are the workhorses for NUV and FUV spectroscopy. At shorter wavelengths, one must use either normal-incidence spectrographs that have LiF- or SiC-coated optics used in the Far Ultraviolet Spectrograph Explorer (FUSE) satellite, or grazing-incidence optics used in the Extreme Ultraviolet Explorer (EUVE) satellite, and X-ray observatories. In both cases, the throughput was much smaller than for the HST spectrographs.

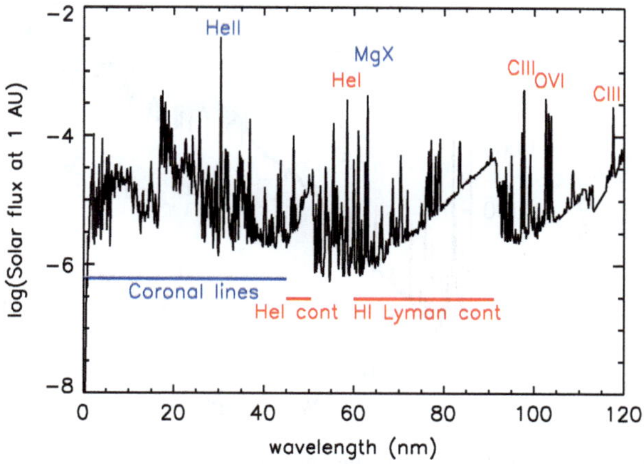

Fig. 1.1 Solar Irradiance Reference Spectrum (SIRS) obtained at solar minimum (2008 March–April). Flux units are $Wm^{-2}nm^{-1}$ at 1 AU. Important emission lines and continua are identified (After Linsky et al. 2014)

1.2 UV Radiation

The International Ultraviolet Explorer (IUE) provided the first high-resolution ultraviolet spectra of F–M dwarf stars, and the spectrographs on HST, especially the high throughput Cosmic Origins Spectrograph (COS), have provided spectra of fainter stars including M dwarfs. These data can be downloaded from the Mikulski Archive for Space Telescopes (MAST).[1] Ayres (2010) created the StarCAT[2] archive of individually calibrated *HST* stellar spectra covering the 117–320 nm wavelength range. This is a valuable archive for F, G, and K dwarfs stars, but contains only a few M dwarfs.

To establish an archive of UV spectra of M dwarf host stars,[3] France et al. (2012, 2013) obtained COS spectra of GJ 436 (M3.5 V), GJ 667C (M1.5 V), GJ 581 (M5.0 V), GJ 832 (M1.5), GJ 876 (M5.0 V), and GJ 1214 (M4.5). Comparison of the GJ 876 and quiet Sun spectra (see Fig. 1.2), as seen from each stars's habitable zone, shows an important result. The NUV spectrum of the relatively inactive GJ 876 is 1,000 times fainter than that of the Sun as a consequence of the M dwarf's much cooler photospheric temperatures where the NUV spectrum is formed and the exponential dependence on temperature of short-wavelength radiation at these low temperatures. On the other hand, the FUV spectrum of this M dwarf, especially

[1]http://mast.stsci.edu

[2]http://casa.colorado.edu/~ayres/StarCAT/

[3]spectra are available at http://cos.colorado.edu/~kevinf/muscles.html

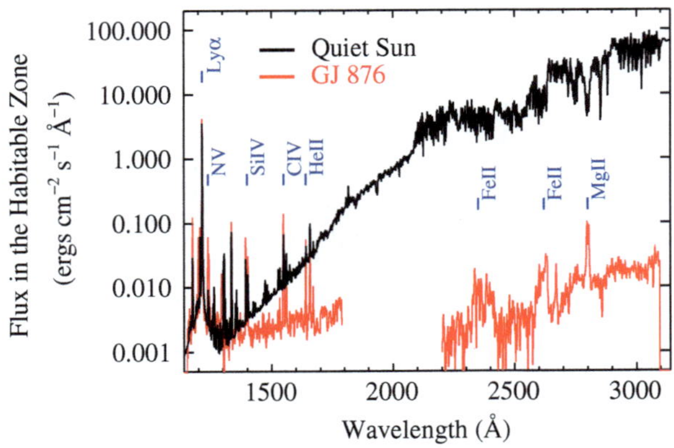

Fig. 1.2 Comparison of the FUV and NUV flux from the Sun and the M dwarf GJ 876 as viewed from the habitable zones of each star (1.0 AU for the Sun and 0.21 AU for GJ 876). The NUV radiation (170–320 nm) is emitted by the stellar photosphere and the FUV radiation (117–170 nm) by the stellar chromosphere (After France et al. 2012)

at wavelengths below 150 nm, is comparable in flux to the solar spectrum. This is because the FUV spectrum is formed in the chromosphere, where the emission depends on the magnetic heating rate rather than on the stellar effective temperature. Since the NUV radiation from GJ 876 and the other observed M dwarf host stars is faint, while the FUV radiation seen by an exoplanet in the habitable zone is as bright as the Sun seen from the Earth, the FUV spectra of M dwarf host stars must be included in photochemical calculations of exoplanet atmospheres. For M dwarfs, the combination of strong FUV radiation, which photodissociates CO_2 and H_2O to form oxygen, and weak NUV radiation, which photodissociates O_2 and O_3, provides an abiotic path for producing significant oxygen in an exoplanet's atmosphere. On this basis, France et al. (2013) and Tian et al. (2014) argue that the detection of oxygen or ozone in an exoplanet's atmosphere would not be sufficient to constitute a valid biomarker. This point was previously discussed by Canuto et al. (1982, 1983).

The brightest emission line in the solar spectrum by a large factor is Lyman-α ($\lambda = 121.6$ nm). This is also the brightest emission line for F–M dwarf stars, but neutral hydrogen absorption in the interstellar medium removes most of the intrinsic stellar flux. Wood et al. (2005) developed a technique for reconstructing the intrinsic Lyman-α line profile using information on the column density and velocity of interstellar hydrogen obtained from the deuterium Lyman-α line and other interstellar absorption lines. France et al. (2013) developed an alternative technique to solve for both the interstellar absorption and the intrinsic Lyman-α profile. As shown in Fig. 1.3, this method provides credible Lyman-α fluxes when the kinematic properties of the ISM along the line of sight to the star are relatively simple (Redfield and Linsky 2008). Even when there is no observed

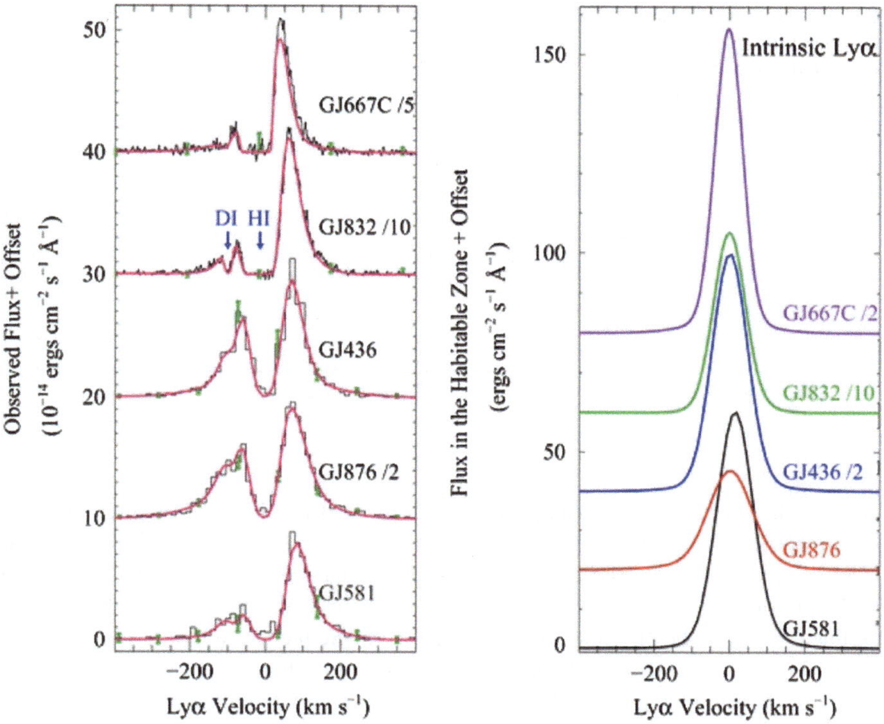

Fig. 1.3 Comparison of observed (*left*) and reconstructed (*right*) Lyman-α line profiles of five M dwarf host stars. The spectra are offset by 10 (*left*) or 20 (*right*) ergs cm^{-2} s^{-1} (After France et al. 2013)

Lyman-α profile to reconstruct, Linsky et al. (2013) showed that one can estimate the intrinsic Lyman-α line flux with somewhat larger uncertainty from the fluxes of other emission lines that are unaffected by interstellar absorption or even from the stellar spectral type and some measure of activity such as the stellar rotation rate.

The Lyman-α emission line flux plays a critically important role for photochemistry in the outer layers of exoplanet atmospheres because the line is very bright and the photodissociation cross-sections of H_2O, CO_2, CH_4, C_2H_2, and other important molecules in exoplanet atmospheres are high in the FUV (Ribas et al. 2010). In the habitable zone of the Sun, the Lyman-α flux is 30 % of the total 115–121 + 122–179 nm flux, but in the habitable zone of the M dwarf GJ 876 it is 2.3 times the flux in the rest of the FUV and nearly as large as the entire FUV + NUV flux (France et al. 2012).

The ultraviolet emission lines and continua (Linsky et al. 2012) of solar-type (early G) dwarf stars decrease as they age on the main sequence with the decrease in

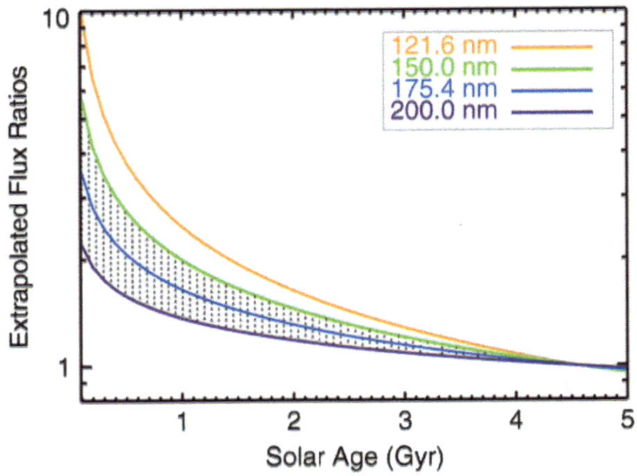

Fig. 1.4 Normalized flux ratios at different wavelengths compared to the present Sun as a function of solar age (After Claire et al. 2012)

stellar rotation rate, dynamo amplification of magnetic fields, and magnetic heating of chromospheres. Ayres (1997), Ribas et al. (2005, 2010), and recently Claire et al. (2012) have used the observed UV emission of solar-type stars with different ages and rotation rates to determine scaling laws for the UV emission as a function of stellar age (τ in Gyr) from the Zero-Age-Main-Sequence and rotation rate. For example, Claire et al. (2012) determined power law fits to the 150–215 nm flux in 10 nm bands of the form $F = \alpha\tau^{\beta}$, where α and β both decrease with age. Figure 1.4 shows the dependence of flux on stellar age and wavelength obtained from their analysis of six solar analog stars with ages 0.1–6.7 Gyr. Ribas et al. (2005) noted that the relative flux of the Lyman-α line compared to the EUV and X-ray flux increases with age for solar-type stars as the fluxes of higher-temperature emission lines decrease with age faster than Lyman-α. Although one would expect that similar scaling laws should apply to stars cooler than the Sun, there are no published studies presently available that develop such scaling laws for NUV and FUV emission.

The 91.2–117 nm spectral region contains the Lyman-β and higher emission lines of the hydrogen Lyman series and a number of bright transition region lines including the C II 103.6 nm doublet, C III 97.7 nm and the 117 nm multiplet, and the O VI 103.4 nm doublet. The FUSE spacecraft obtained stellar spectra in this spectral range (Moos et al. 2000; Sahnow et al. 2000). Redfield et al. (2002) measured emission line fluxes for seven A7 V to M0 V stars, but did not include the bright Lyman-β line and higher lines of the Lyman series that are contaminated by geocoronal emission seen through the large entrance aperture of FUSE. Recently, Linsky et al. (2014) estimated the Lyman series emission for these stars based on the flux ratio of the Lyman lines to Lyman-α in solar spectra.

1.3 EUV Radiation

The 10–91.2 nm spectral region (see Fig. 1.1) contains three free-bound continuua in emission: the Lyman continuum (∼60–91.2 nm), He I continuum (∼45–50.4 nm), and He II continuum (∼15–22.8 nm). Superimposed on these continua are the resonance lines of He I (58.4 nm) and He II (30.4 nm) and a host of transition-region and coronal emission lines. EUV radiation at wavelengths below the hydrogen photoionization edge (91.2 nm) ionizes and heats the outer atmospheres of close-in exoplanets, providing the energy for significant mass loss (Lecavelier des Etangs 2007; Murray-Clay et al. 2009; Ehrenreich and Désert 2011; Lammer et al. 2013). However, interstellar absorption prevents measurement of 40–91.2 nm radiation for even the nearest stars, except from the Sun.

Below 40 nm, the decreasing opacity of interstellar hydrogen and helium permits spectral observations of nearby stars. The EUVE spacecraft contained three spectrographs covering the 7–76 nm spectral region with 0.05–0.2 nm spectral resolution (Welsh et al. 1990; Bowyer and Malina 1991). Craig et al. (1997) and Monsignori Fossi et al. (1996) showed examples of EUVE spectra of F–M dwarf stars, and Sanz-Forcada et al. (2003) provided emission lines fluxes. Linsky et al. (2014) computed EUV fluxes in the 10–20, 20–30, and 30–40 nm wavelength intervals for 15 F5–M5 dwarf stars based on EUVE data downloaded from MAST data archive.

The available but limited number of EUV stellar spectra can be supplemented from two sources. There are high-resolution solar irradiance spectra (spectra of the Sun as a stellar point source) covering the entire EUV region. A good example is the Solar Irradiance Reference Spectrum (SIRS, Fig. 1.1) obtained when the Sun had a very low activity level (Woods et al. 2009; Chamberlin et al. 2009). There are also new semiempirical solar models (Fontenla et al. 2014) constructed to match the spectral irradiance of patches of the Sun for which increasing EUV fluxes are produced by increasing magnetic heating rates. These are 1-D non-LTE models of the thermal structure from the chromosphere to the corona, including the ionization of 21 elements and both continuum and line emission. The new models are updates of the Fontenla et al. (2009, 2011) models. Since the new models refer to the same star but different levels of magnetic heating, these models should be representative of at least solar-type stars with different activity levels.

Several authors have used the available spectra to estimate the EUV emission of stars as a function of age and rotation rate. Using HST, FUSE, EUVE, and X-ray spectra of six solar analog stars with ages 0.1–6.7 Gyr in the narrow spectral range G0–G2 and the present day Sun, Ribas et al. (2005) showed that these spectra can be fit with power laws of the form $F = \alpha\tau^{-\beta}$. Here F is the flux in a wavelength interval, τ is the stellar age in Gyr, and α and β depend on the wavelength interval. They computed α and β values that best fit the data for the 10–36, 36–92 (by interpolation), and 92–118 nm wavelength intervals. The value of β decreases with increasing wavelength, indicating a slower decrease in flux with stellar age. For example, the flux of solar mass stars decreases between ages 0.1 Gyr (EK Dra) and 6.7 Gyr (β Hyi) by a factor of 20,000 at 1 nm, a factor of 200 at 10 nm, but only a factor of 50 at 100 nm. Claire et al. (2012) extended this method for estimating the X-ray-to-infrared emission of the Sun over time with recent data (see Fig. 1.4). They computed high-resolution spectra of the Sun over time, but did not extend their analysis to stars with effective temperatures much different from the Sun.

To predict the emission from stars with different effective temperatures than the Sun, Sanz-Forcada et al. (2003, 2011) derived emission measures as a function of temperature, $EM(T) = \int n_e n_H dV$, from the analysis of stellar coronal emission lines in X-ray spectra. They computed EUV spectra from the emission measures of many stars, including the exoplanet host stars ϵ Eri (K2 V), GJ 436 (M3.5 V), GJ 876 (M5.0 V), HD 189733 (K1-2 V), and HD 209458 (G0 V). They found that the EUV luminosities (10–92 nm) of stars with spectral types F7–M3 can be fit with a simple log-log relation, $\log L_{\rm EUV} = (29.12 \pm 0.11) - (1.24 \pm 0.15) \log \tau$, where τ is the stellar age in Gyr. This formula can be used to estimate the total EUV emission of exoplanet host stars, but the emission measures at $\log T < 6.0$ are based on only a few emission lines, and the EUV spectra do not include the Lyman continuum, which is important between 70 and 92 nm, or chromospheric emission lines including the Lyman series. The predicted EUV emission of many stars is available from the X-exoplanet website.[4]

Recently, Linsky et al. (2014) developed a different approach for predicting stellar EUV emission. Models of the solar chromosphere and transition region for regions with different amounts of magnetic heating (Fontenla et al. 2011) have very similar thermal structures in which the effect of increased heating is to displace the temperature-height relation to lower heights where the densities are larger. The effect is to increase the flux in all emission lines and continua formed in the chromosphere and transition region, although emission lines formed at higher temperatures increase with the heating rate faster than those formed at lower temperatures. This implies that the ratio of EUV flux to that of a given emission line, for example Lyman-α, should vary smoothly with the emission line flux and stellar "activity".

Linsky et al. (2014) tested this method by plotting the flux ratio f(EUV)/f(Lyman-α) vs. f(Lyman-α) for the quiet Sun and 15 stars observed by EUVE in the 10–20, 20–30, and 30–40 nm wavelength bands (an example is Fig. 1.5) and the quiet Sun and five stars observed by FUSE in the 91.2–117 nm band (see Fig. 1.6). The flux ratios of F5–K7 stars can be fit by a least-squares relation, log [f($\Delta\lambda$)/f(Lyman-α)] = a + b log[f(Lyman-α)], and the flux ratios for the M stars with a similar relation. The emission predicted by the Fontenla et al. (2014) solar models ranging from the lowest emission model for dark regions of the solar atmosphere (model 1300) to model 1308 representing a very bright solar plage region can be fit by the same formula with similar parameters to those obtained from the observations. For the

[4] http://sdc.cab.inta-csic.es/xexoplanets/jsp/exoplanetsform.jsp

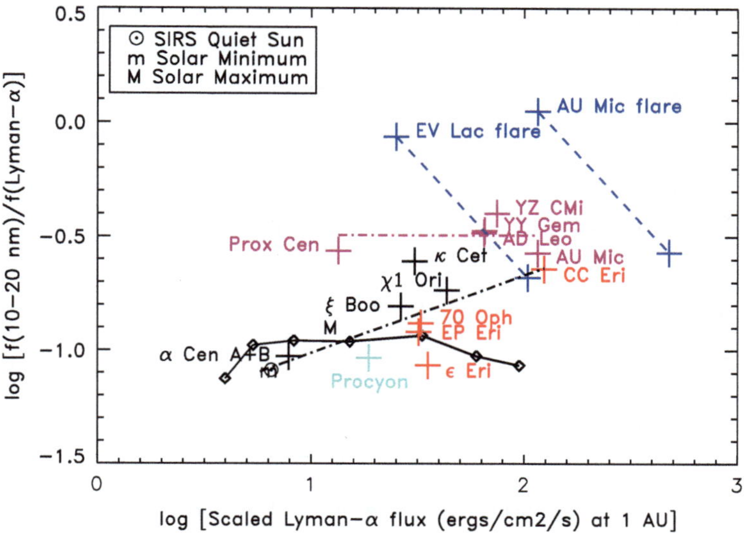

Fig. 1.5 Ratios of the intrinsic flux between 10 and 20 nm (corrected for interstellar absorption) divided by the intrinsic Lyman-α flux vs. the intrinsic Lyman-α flux at 1 AU. The *solid line-connected diamonds* are the flux ratios in this passband for the (Fontenla et al. 2014) semiempirical models 1300 to 1308. Flux ratios for one F star (*cyan color*), four G stars (*black*), four K stars (*red*), and six M stars (*plum*) based on *EUVE* spectra are shown as ± 15 error bar symbols. The *dash-dot (black) line* is the least-squares fit to the solar and F, G, and K star data. The *plum dash-dot line* is a least-squares fit to the M star data excluding the EV Lac flare and AU Mic flare data. Flux ratios for EV Lac and AU Mic during flares (*blue*) are plotted two ways. The *upper left symbols* are ratios of EUV fluxes to quiescent Lyman-α fluxes. *Dashed lines* extending to the *lower right* indicate the ratios for increasing Lyman-α flux. The symbols at the *lower end of the dashed lines* are ratios obtained using the most likely values of the Lyman-α fluxes fluxes during flares. The "m" and "M" symbols are the solar minimum and maximum data obtained with the SEE instrument on the *TIMED* spacecraft (Woods et al. 2005). The Sun symbol is the ratio for the SIRS quiet Sun data set (After Linsky et al. 2014)

unobservable 40–91.2 nm spectral region, Linsky et al. (2014) used formulae based only on the solar models. To infer the EUV flux from these formulae, one must have reconstructed Lyman-α fluxes appropriate for the same level of stellar activity. When Lyman-α flux measurements are not available, they may be estimated from other emission lines or stellar rotation rates (Linsky et al. 2013).

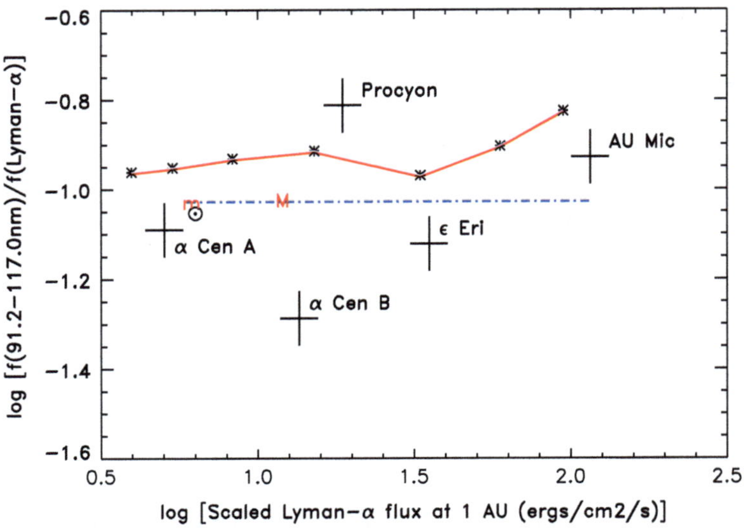

Fig. 1.6 Ratios of the total flux between 91.2 and 117.0 nm divided by the Lyman-α flux. *Solid-line*-connected asterisks are the flux ratios in this passband for the Fontenla et al. (2014) semiempirical models 1300 to 1308. Flux ratios for five stars based on FUSE spectra and estimated Lyman series fluxes are shown as $\pm 15\,\%$ error-bar symbols. The *Sun symbol* is the ratio for the SIRS quiet Sun data set. The *dash-dot line* is the least-squares fit to the stellar and SIRS data. The "m" and "M" symbols are the solar minimum and maximum data obtained with the *SEE* instrument on the TIMED spacecraft (Woods et al. 2005) (After Linsky et al. 2014)

1.4 X-Radiation

At wavelengths below 10 nm, X-rays suffer much less interstellar absorption than EUV radiation because the absorption cross section increases approximately as λ^3, making observation and interpretation of X-ray radiation properties relatively straightforward. X-ray luminosities, L_X, have been surveyed with the early X-ray satellites (e.g., Einstein, ROSAT) for all classes of cool stars including pre-main sequence stars in star-forming regions out to several hundred parsec. These low-spectral resolution data were interpreted with synthetic spectra typically consisting of two dominant isothermal components with different temperatures and emission measures, subject to a common interstellar absorption component. The large observatories of the latest generation, Chandra and XMM-Newton, carry high-resolution grating spectrometers providing direct access to spectral-line fluxes and therefore, through line-flux ratios, to element abundances, emission measure distributions in temperature, and, in a few cases, to coronal densities.

In cool stars, X-rays are produced mostly in the extended outer atmospheres ("coronae") held together by closed magnetic fields in and between active regions. Early surveys by, e.g., Pallavicini et al. (1981), recognized a steep dependence of L_X on stellar rotation and therefore stellar age, roughly of the form $L_X \propto P^{-2}$, where P is the rotation period, or $L_X \propto (v \sin i)^2$, where $v \sin i$ is the projected rotational velocity, and i is the inclination of the stellar axis with respect to the line of sight. At very rapid rotation rates (i.e., mostly in very young stars), X-ray activity saturates, i.e., becomes a function of the stellar bolometric luminosity for all cool stars; in that case, $L_X \approx 10^{-3} L_{bol}$, where the critical saturation period itself depends on L_{bol}, $P_{sat} \approx 1.2(L_{bol}/L_\odot)^{-1/2}$ (days) (Pizzolato et al. 2003).

Studies of the activity-age relation with respect to X-rays have been presented for stellar clusters with known ages τ, typically for $\tau < 1$ Gyr (e.g., Stern et al. 1995; Patten and Simon 1996). On the pre-main sequence and the young main-sequence (up to ages around 100 Myr), most stars rotate sufficiently rapidly to keep their X-ray luminosity in the saturation regime, even though the latter depends on the depth of the stellar convection zone, which varies greatly during the pre-main sequence evolution and is a function of spectral type (Preibisch et al. 2005). Depending on the initial rotation period as stars arrive on the main-sequence and on the wind-induced spin-down rate, cool stars eventually leave the X-ray saturation regime. The subsequent radiative evolution is much faster in G than M dwarfs. Attempts to unify the X-ray decay law for various spectral classes were described by Scalo et al. (2007), Guinan and Engle (2009), and Mamajek and Hillenbrand (2008). At an age of $\approx$700 Myr, M dwarfs still mostly reside at the saturation level, while L_X of solar analogs has declined by almost two orders off magnitude (Stern et al. 1995). The break point at which saturation is reached thus moves to progressively later spectral types as a cluster ages. As the stellar rotation period converges to a nearly unique value for a given spectral type and age (Soderblom et al. 1993), so does L_X.

Evolutionary trends have also been studied based on individual field stars with known ages, including stellar ages much beyond 1 Gyr and up to the solar age. For solar analogs, Güdel et al. (1997) studied a sample of "Sun in Time" targets and found, as with longer-wavelength radiation, a power-law dependence $L_X = \alpha\tau^{-\beta}$; specifically, $L_X \approx (3 \pm 1) \times 10^{28} \tau_9^{-1.5\pm0.3}$ [erg s^{-1}] (combined with results from Maggio et al. (1987); τ_9 is the age in Gyr). The X-ray trend is thus steeper than for EUV and UV, indicating a more rapid decay for higher-energy radiation (Ribas et al. 2005; Fig. 1.7). This decay trend implies that the luminosity range during a main-sequence lifetime is the largest for X-rays, amounting to about three orders of magnitude (where the present-day Sun is at the bottom level of activity for a solar analog; Fig 1.8).

The decay trends can be summarized for spectral lines with a maximum-formation temperature of T_{max} ($4 \leq \log T_{max} \leq 7$); the luminosity for a line forming at T_{max} decays, on average, as

$$L(T_{max}, t) = \alpha t_9^{-\beta}, \tag{1.1}$$

$$\beta = 0.32 \log T_{max} - 0.46 \tag{1.2}$$

(Güdel 2007); similar relations hold for the continuum, where T_{max} corresponds to $hc/(\lambda k)$, h is the Planck constant, k the Boltzmann constant, c the speed of light, and λ the continuum wavelength.

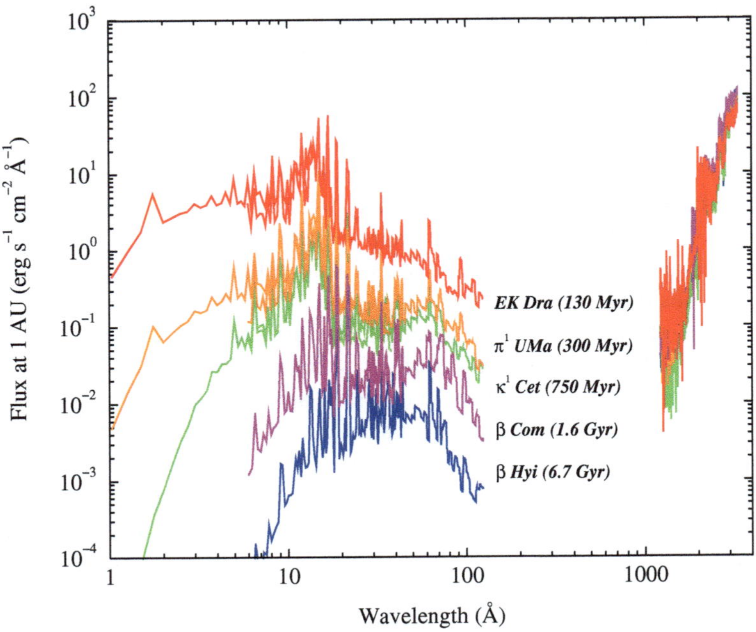

Fig. 1.7 Irradiances at 1 AU from solar analogs with different ages (after Guinan and Ribas 2002). Note the large decrease of magnetically induced radiation with age in the X-ray regime

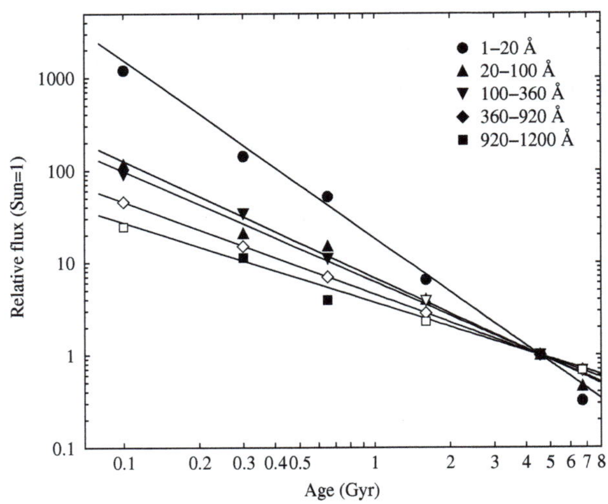

Fig. 1.8 Power-law decay of spectral output of a solar analog during its main-sequence (age 0.05–10 Gyr) life, normalized to the present-day solar flux. The shortest-wavelength flux decays the fastest (After Ribas et al. 2005)

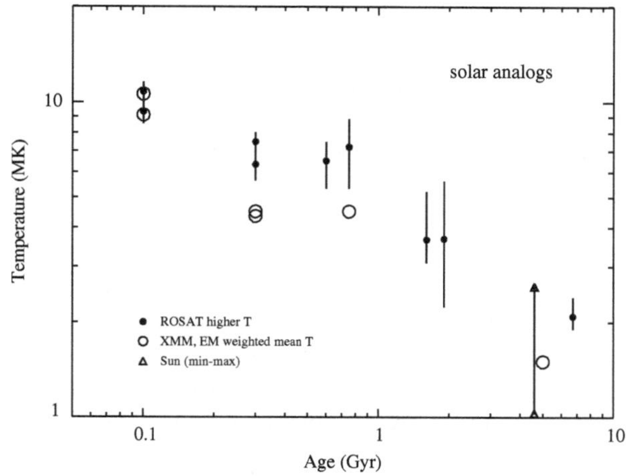

Fig. 1.9 Decay of coronal temperature with age for solar-type stars (After Güdel 2004)

Early two-component spectral interpretations of low-resolution X-ray spectra clearly showed that the hardness of the X-ray emission correlates with L_X (Schrijver et al. 1984; Schmitt et al. 1990) and, therefore, decays in time (Güdel et al. 1997), which has important implications for the penetration depth of X-rays in planetary atmospheres and pre-main sequence stellar disks. As a function of $\bar{T}$, the average coronal temperature in MK, Telleschi et al. (2005) found for solar analogs $L_X \approx 1.6 \times 10^{26} \bar{T}^{4.05}$ which, combined with the $L_X - \tau$ relation above, results in $\bar{T} \approx 3.6\tau_9^{-0.37}$ (MK), implying a decay in coronal temperature from about 10 MK to about 2 MK in the course of the main-sequence life of a solar analog (Fig. 1.9), with the consequent reduction in X-ray hardness.

The cause for the evolution of hardness cannot simply be related to higher coverage of young stars with magnetically active regions. Several lines of evidence suggest that continuous, stochastic flaring may be responsible for hotter coronae in younger, more X-ray luminous stars. Flares follow a correlation between peak temperature and peak luminosity similar to the above $T - L_X$ relation for average stellar values, and the rate of flares exceeding a given flare-energy threshold correlates with L_X as well (Audard et al. 2000). A more active corona therefore produces large flares at a higher rate, heating the corona to higher average levels. This picture is supported by X-ray timing analysis (Kashyap et al. 2002; Güdel et al. 2003; Arzner and Güdel 2004), the shape of stellar X-ray emission measure distributions (EMD) and their similarity to the EMDs of time-integrated flares (Güdel et al. 2003), and spectroscopically measured coronal densities (see Güdel 2004). If flares are indeed responsible for the highly elevated X-ray levels in more active stars, then an important amount of even harder emission, namely hard X-rays/gamma rays, as well as a much higher flux of high-energy particles should be present around active stars compared to the present-day Sun (e.g., Feigelson et al. 2002).

Conclusions

The FUV radiation from a host star controls the photochemistry of important molecules in an exoplanet's upper atmosphere, with important consequences on the question of whether detection of O_2 and O_3 indicates the presence of life forms. Similarly, the EUV radiation from a host star ionizes and heats an exoplanet's upper atmosphere, thereby driving mass loss. Observing programs with UV spectrographs on the *HST* now provide FUV and NUV fluxes of nearby exoplanets, including M dwarfs, and scaling laws for the dependence of these fluxes on stellar mass, age, and rotation rate. While EUV fluxes are available for only a few stars, it is now feasible to estimate EUV fluxes on the basis of scaling laws relative to observable emission lines and solar models. X-ray observations of the many stars observed with ROSAT, XMM-Newton, and Chandra provide a reliable basis for scaling laws to predict the X-ray fluxes of host stars and their evolution from the pre-main sequence.

Acknowledgements The authors acknowledge the support by the International Space Science Institute (ISSI) in Bern, Switzerland and the ISSI team *Characterizing stellar- and exoplanetary environments*. The authors have used the MAST Data Archive at the Space Telescope Science Institute. STScI is operated by the Association of Universities for Research in Astronomy, Inc., under NASA contract NAS 5-26555. M. Güdel acknowledges support from the Austrian Research Foundation FWF NFN project S11601-N16 'Pathways to Habitability: From Disks to Active Stars, Planets and Life', as well as the related FWF NFN subproject,S116 604-N16 'Radiation & Wind Evolution from T-Tauri Phase to ZAMS and Beyond'.

References

Audard, M., Güdel, M., Drake, J. J., Kashyap, V. (2000). *Astrophysical Journal, 541*, 396

Arzner, K., & Güdel, M. (2004). *Astrophysical Journal, 602*, 363

Ayres, T. R. (1997). *Journal of Geophysical Research, 102*(E1), 1641

Ayres, T. R. (2010). *Astrophysical Journal Supplement Series, 187*, 149

Bowyer, S., & Malina, R. F. (Eds.) (1991). *Extreme ultraviolet astronomy* (p. 333). New York: Pergamon.

Canuto, V. M., Levine, J. S., Augustsson, T. R., & Imhoff, C. L. (1982). *Nature, 296*, 816.

Canuto, V, M., Levine, J. S., Augustsson, T. R., Imhoff, C. L., & Giampapa, M. S. (1983). *Nature, 305*, 281.

Chamberlin, P. C., Woods, T. N., Crotser, D. A., Eparvier, F. G., Hock, R. A., & Woodraska, D. M. (2009). *Journal of Geophysical Research, 36*, 5102.

Claire, M. W., Sheets, J., Cohen, M., Ribas, I., Meadows, V. S., & Catling, D. C. (2012). *Astrophysical Journal Supplement Series, 757*, 95.

Craig, N., Abbott, M., Finley, D., et al. (1997). *Astrophysical Journal Supplement Series, 113*, 131.

Ehrenreich, D., & Désert, J.-M. (2011). *Astronomy and Astrophysics, 529*, L136.

Feigelson, E. D., Garmire, G. P., & Pravdo, S. H. (2002). *Astrophysical Journal, 572*, 335.

Fridlund, M., Rauer, H., & Erikson, A., (2014). H. Lammer & M. L. Khodachenko (Eds.), *Characterizing stellar and exoplanetary environments* (pp. 253). Heidelberg/New York: Springer.

Fontenla, J. M., Curdt, W., Haberreiter, M., Harder, J., & Tian, H. (2009). *Astrophysical Journal, 707*, 482.

Fontenla, J. M., Harder, J., Livingston, W., Snow, M., & Woods, T. (2011). *Journal of Geophysical Research, 116*, D20108.

Fontenla, J. M., Landi, E., Snow, M., & Woods, T. (2014). *Solar Physics, 289*, 515

Fossati, L., Haswell, C. A., Linsky, J. L., & Kislyakova, K. G. (2014). H. Lammer & M. L. Khodachenko (Eds.), *Characterizing stellar and exoplanetary environments* (pp. 59). Heidelberg/New York: Springer.

France, K., Linsky, J. L., Tian, F., Froning, C. S., & Roberge, A. (2012). *Astrophysical Journal, 750*, L32.

France, K., Froning, C. S., Linsky, J. L., Roberge, A., Stocke, J. T., Yian, F., Bushinsky, R., Désert, J.-M., Mauas, P., Vieytes, M., Walkowitz, L. M. (2013). *Astrophysical Journal, 763*, 149.

Grenfell, J. L., Gebauer, S., Godolt, M., Palczynski, K., Rauer, H., Stock, J., von Paris, P., Lehmann, R., & Selsis, F. (2012). *Astrobiology, 13*, 415.

Guenther, E. (2014). H. Lammer, & M. L. Khodachenko (Eds.), *Characterizing stellar and exoplanetary environments* (pp. 289). Heidelberg/New York: Springer.

Güdel, M. (2004). *Astronomy and Astrophysics Review, 12*, 71.

Güdel, M. (2007). *Living Reviews in Solar Physics, 4*(3). Online http://solarphysics.livingreviews.org/Articles/lrsp-2007-3/

Güdel, M., Guinan, E. F., & Skinner, S. L. (1997). *Astrophysical Journal, 483*, 947.

Güdel, M., Audard, M., Kashyap, V. L., Drake, J. J., & Guinan, E. F. (2003). *Astrophysical Journal, 582*, 423.

Guinan, E. F., & Ribas, I. (2002). B. Montesinos, A. Giménez, & E. F. Guinan (Eds.), *The evolving sun and its influence on planetary environments* (p. 85). San Francisco: ASP.

Guinan, E. F., & Engle, S. G. (2009). The ages of stars. In E. E. Mamajek, D. R. Soderblom, & R. F. G. Wyse (Eds.), *IAU symposium*, Rio de Janeiro (Vol. 258, pp. 395).

Hu, R., Seager, S., & Bains, W. (2012). *Astrophysical Journal, 761*, 166.

Kashyap, V., Drake, J. J., Güdel, M., & Audard, M. (2002). *Astrophysical Journal, 580*, 1118.

Kaltenegger, L., Segura, A., & Mohanty, S. (2011). *Astrophysical Journal, 733*, 35.

Kasting, J. F., & Catling, D. (2003). *Annual Review of Astronomy and Astrophysics, 41*, 429.

Kopparapu, R. K., Kasting, J. E., & Zahnle, K. J. (2012). *Astrophysical Journal, 745*, 77.

Lammer, H., Erkaev, N. V., Odert, P., Kislyakova, K. G., Leitzinger, M., & Khodachenko, M. L. (2013). *Monthly Notices of the Royal Astronomical Society, 430*, 1247.

Lecavelier des Etangs, A. (2007). *Astronomy and Astrophysics, 461*, 1185.

Line, M. R., Liang, M. C., & Yung, Y. L. (2010). *Astrophysical Journal, 717*, 496.

Line, M. R., Vasisht, G., Chen, P., Angerhausen, D., & Yung, Y. L. (2011). *Astrophysical Journal, 738*, 32.

Linsky, J. L., Bushinsky, R., Ayres, T., Fontenla, J., & France, K. (2012). *Astrophysical Journal, 745*, 25.

Linsky, J. L., France K., & Ayres, T. R. (2013). *Astrophysical Journal, 766*, 69.

Linsky, J. L., Fontenla, J., & France, K. (2014). *Astrophysical Journal, 780*, 61.

Maggio, A., Sciortino, S., Vaiana, G. S., Majer, P., & Bookbinder, et al. (1987). *Astrophysical Journal, 315*, 687.

Mamajek, E. E., & Hillenbrand, L. A. (2008). *Astrophysical Journal, 687*, 1264.

Monsignori Fossi, B. C., Landini, M., Del Zanna, G., & Bowyer, S. (1996). *Astrophysical Journal, 466*, 427.

Moos, H. W. et al. (2000). *Astrophysical Journal, 538*, L1.

Moses, J. I., Madhusudhan, N., Visscher, C., & Freedman, R. S. (2013). *Astrophysical Journal, 763*, 25.

Murray-Clay, R. A., Chiang, E. I., & Murray, N. (2009). *Astrophysical Journal, 693*, 23.

Pallavicini, R., Golub, L., Rosner, R., Vaiana, G. S., Ayres, T., & Linsky, J. L. (1981). *Astrophysical Journal, 248*, 279.

Patten, B. M., & Simon, T. (1996). *Astrophysical Journal Supplement Series, 106*, 489.

Pizzolato, N., Maggio, A., Micela, G., Sciortino, S., & Ventura, P. (2003). *Astronomy and Astrophysics, 397*, 147.
Preibisch, T., Kim, Y. C., Favata, F., Feigelson, E. D., & Flaccomio, E., et al. (2005). *Astrophysical Journal, Supplement Series 160*, 401.
Redfield, S., Linsky, J. L., & Ake, T. R., et al. (2002). *Astrophysical Journal, 581*, 626.
Redfield, S., & Linsky, J. L. (2008). *Astrophysical Journal, 673*, 283.
Ribas, I., Guinan, E. F., Güdel, M., & Audard, M. (2005). *Astrophysical Journal, 622*, 680.
Ribas, I., Porto de Mello, G. F., Ferreira, L. D., Hébrard, E., Selsis, F., Catalán, S., Garcés, A., do Nascimento Jr., J. D., & de Medeiros, J. R. (2010). *Astrophysical Journal, 714*, 384.
Sahnow, D. J. et al. (2000). *Astrophysical Journal, 538*, L7.
Sanz-Forcada, J., Brickhouse, N. S., Dupree, A. K. (2003). *Astrophysical Journal Supplement Series, 145*, 147.
Sanz-Forcada, J., Micela, G., Ribas, I., Pollock, A. M. T., Eiroa, C., Velasco, A., Solano, E., & Garcia-Alvarez, D. (2011). *Astronomy and Astrophysics, 532*, A6.
Scalo, J., Kaltenegger, L., Segura, A. G., Fridlund M., & Ribas, I., et al. (2007). *Astrobiology, 7*, 85.
Schmitt, J. H. M. M., Collura, A., Sciortino, S., Vaiana, G. S., Harnden, F. R. Jr., & Rosner, R. (1990). *Astrophysical Journal, 365*, 704.
Schrijver, C. J., Mewe, R., & Walter, F. M. (1984), *Astronomy and Astrophysics, 138*, 258.
Seager, S., & Deming, D. (2010). *Annual Review of Astronomy and Astrophysics, 48*, 631.
Segura, A., Kasting, J. F., Meadows, V., Cohen, M., Scalo, J., Crisp, D., Butler, R. A. H., & Tinetti, G. (2005). *Astrobiology, 5*, 706.
Segura, A., Walkowicz, L. M., Meadows, V., Kasting, J., & Hawley, S. (2010). *Astrobiology, 10*, 751.
Shustov, B. M., Sachkov, M. E., Bisikalo, D., & Gómez de Castro, A.-I. (2014). H. Lammer, & M. L. Khodachenko (Eds.), *Characterizing stellar and exoplanetary environments* (pp. 275). Heidelberg/New York: Springer.
Soderblom, D. R., Stauffer, J. R., MacGregor, K. B., & Jones, B. F. (1993). *Astrophysical Journal, 409*, 624.
Stern, R. A., Schmitt, J. H. M. M., & Kahabka, P. T. (1995). *Astrophysical Journal, 448*, 683.
Telleschi, A., Güdel, M., Briggs, K., Audard, M., Ness, J.-U., & Skinner, S. L. (2005). *Astrophysical Journal, 622*, 653.
Tian, F., France, K., Linsky, J. L., Mauas, P. J. D., & Vieytes, M. C. (2014). *Earth and Planetary Science Letters, 385*, 22.
Welsh, B., Vallerga, J. V., Jelinsky, P., Vedder, P.W., Bowyer, S., & Malina, R. F. (1990). *Optical Engineering, 29*(7), 752.
Wood, B. E., Redfield, S., Linsky, J. L., Müller, H.-R., & Zank, G. P. (2005). *Astrophysical Journal Supplement Series, 159*, 118.
Woods, T. N., Eparvier, F. G., Bailey, S. M., et al. (2005). *Journal of Geophysical Research, 110*, A01312.
Woods, T. N., Chamberlin, P. C., & Harder, J. W. et al. (2009). *Journal of Geophysical Research, 36*, L01101.

Chapter 2
Stellar Winds in Time

Brian E. Wood, Jeffrey L. Linsky, and Manuel Güdel

Abstract Exposure to stellar winds can have significant long term consequences for planetary atmospheres. Estimating the effects of these winds requires knowledge of how they evolve with time. Determining this empirically requires the ability to study the winds of stars of various ages and activity levels, but this is not easy to do as the coronal winds of solar-like stars are very hard to detect. Relevant observations are here reviewed, as well as more theoretical methods of addressing the problem.

2.1 Introduction: The Wind-Corona Connection

Besides bathing planets in electromagnetic radiation, stars also blast planets with high-speed winds. For cool main sequence stars like the Sun, the winds originate in hot ($T \sim 10^6$ K) coronae that surround the stars, heated by the release of magnetic energy generated by the dynamo in the stellar interior (Ossendrijver 2003; Charbonneau 2010). Planets in our Solar System are currently exposed to a solar wind with a mass loss rate of $\dot{M}_\odot = 2 \times 10^{-14}\ M_\odot\ \mathrm{year}^{-1}$ (Feldman et al. 1977). Mars is widely believed to be the planet that has been most affected by this wind. There is substantial evidence that Mars had a much thicker atmosphere in the distant past (Carr 1996; Jakosky and Phillips 2001), and erosion by the solar wind is a leading candidate for the cause of its loss (Luhmann et al. 1992; Perez de Tejada 1992; Jakosky et al. 1994; Kass and Yung 1995; Lammer et al. 2003; Terada et al. 2009; Brain et al. 2010). A recently launched NASA mission called Mars Atmosphere and Volatile Evolution spacecraft (MAVEN) will be entirely devoted to

B.E. Wood (✉)
Naval Research Laboratory, Washington, DC 20375, USA
e-mail: brian.wood@nrl.navy.mil

J.L. Linsky
JILA, University of Colorado, Boulder, CO 80309-0440, USA
e-mail: jlinsky@jila.colorado.edu

M. Güdel
Department of Astrophysics, University of Vienna, Türkenschanzstr. 17, A-1180 Vienna, Austria
e-mail: manuel.guedel@univie.ac.at

H. Lammer, M. Khodachenko (eds.), *Characterizing Stellar and Exoplanetary Environments*, Astrophysics and Space Science Library 411,
DOI 10.1007/978-3-319-09749-7_2

studying the effects of the solar wind on atmospheric loss when it arrives at Mars in late 2014. The effects of winds on exoplanetary atmospheres and magnetospheres is also of interest, especially for hot Jupiters that orbit very close to their stars, and which are therefore exposed to particle fluxes orders of magnitude higher than the Earth or Mars (e.g., Grießmeier et al. 2004; Khodachenko et al. 2012). This will be described in more detail in Chap. 5 (Bisikalo et al. 2014), Chap. 7 (Kislyakova et al. 2014), Chap. 8 (Vidotto et al. 2014), Chap. 10 (Alexeev et al. 2014), and Chap. 12 (Belenkaya et al. 2014) of this book.

A complete understanding of how winds affect planets must start with assessments of how stellar winds evolve with time. Merely assuming that stars of all ages and spectral types have winds identical to the current solar wind is unsatisfactory. The previous chapter describes how stellar coronae change significantly with time. Therefore, it is natural to expect that the winds that emanate from these coronae could evolve significantly as well.

Young stars rotate rapidly and have very active coronae that emit copious X-ray and EUV radiation (see also Chap. 1 Linsky and Güdel 2014). This coronal emission declines dramatically as stars age and their rotation rates slow down (e.g., Güdel et al. 1997; Ribas et al. 2005). Intuitively, wind strength might be expected to evolve in concert with the coronal emission, since winds originate in stellar coronae. Thus, young stars might be expected to have more massive winds. There is undoubtedly more material heated to coronal temperatures for young, active stars, and therefore more material available to be accelerated into a wind. However, the solar example provides evidence that the wind/corona connection may not be so simple. Coronal X-ray emission from the Sun varies by a factor of 5–10 over the course of the solar activity cycle (Judge et al. 2003), but the solar mass loss rate does not vary much at all on these timescales (Cohen 2011). This demonstrates that a more active corona does not automatically lead to a stronger coronal wind.

On the other hand, active stars might also have substantial mass loss associated with sporadic coronal mass ejections (CMEs). Strong flares on the Sun are usually accompanied by dramatic, fast CMEs. These CMEs do not account for most of the solar mass loss, but flare rates and energies are known to increase dramatically with stellar activity. Young, active stars have stronger and more frequent flares, meaning that the mass loss associated with flare-associated CMEs could be much higher.

Recently Drake et al. (2013) have computed the CME mass loss rate expected for young, active stars by extrapolating a known correlation between solar flare energy and CME mass to more active stars with more frequent and energetic flares. The result is shown in Fig. 2.1, which implies that the most active stars with $\log L_X \sim 3 \times 10^{30}$ ergs s^{-1} could have mass loss rates of $2 \times 10^4 \dot{M}_\odot$ due to CMEs alone. Limits on the percentage of a star's bolometric energy that can reasonably be expected to be spent on flares and CMEs ($\sim 10^{-3}$ L_{bol}) lead Drake et al. (2013) to suggest that a more reasonable ceiling is probably closer to $2500\dot{M}_\odot$. Nevertheless, Fig. 2.1 provides another argument for young, active stars having massive winds.

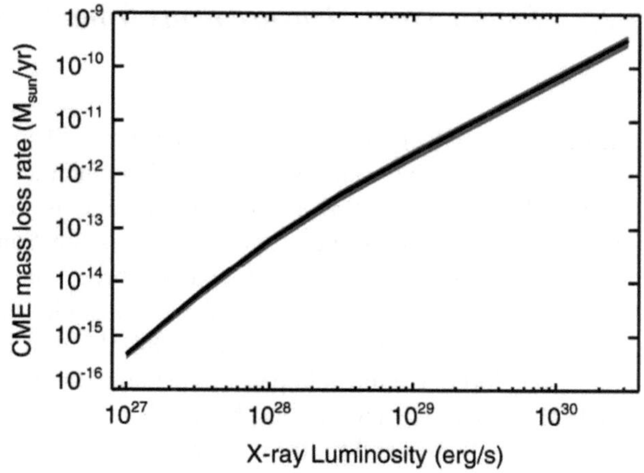

Fig. 2.1 Mass loss rate due to CMEs as a function of coronal X-ray luminosity for solar-like stars, computed by assuming that the solar-flare/CME-mass correlation can be extrapolated to younger, more active stars (After Drake et al. 2013)

2.2 Observational Constraints on Stellar Winds

2.2.1 Upper Limits from Direct Detection Techniques

Ultimately, the only way to truly establish how winds evolve with time is through observation, specifically by measuring the stellar winds of stars of various ages and activity levels. Unfortunately, this is easier said than done, because while coronal X-ray and UV emission are readily observed from stars, winds are very hard to detect. There are two direct methods of coronal wind detection that have been attempted: free-free radio emission and charge-exchange induced X-ray emission.

In the radio, observations have so far only provided upper limits for mass-loss rates (Brown et al. 1990; Lim et al. 1996; Gaidos et al. 2000). The radio arrays that they used were not sensitive enough to detect a solar-like wind around even a very nearby star, and the upper limits quoted for most non-detections are 2–3 orders of magnitude stronger than the solar wind. However, observations with new arrays such as the Atacama Large Millimeter/submillimeter Array (ALMA) or the Jansky Very Large Array (JVLA) could provide detections or at least lower upper limits. The possibility of using X-rays to search for winds has become apparent as it has been realized that most of the soft X-ray background is from our own solar wind, which emits X-rays when highly charged solar wind particles charge exchange with inflowing neutral atoms from the interstellar medium (ISM) (Lallement 2004; Koutroumpa et al. 2009). However, though potentially more sensitive than the radio technique, initial attempts to detect circumstellar wind-induced X-ray emission around nearby stars have not been successful (Wargelin and Drake 2002; Wargelin et al. 2008).

2.2.2 Stellar Wind Measurements from Astrospheric Absorption

The only clear detections of coronal stellar winds like that of the Sun are not of the winds themselves, but rather detections of the interactions between the winds and the ISM (Wood 2004, 2006). The interaction regions are referred to as astrospheres, analogous to the "heliosphere" that surrounds the Sun (Zank 1999). The global heliospheric structure is characterized by three boundaries: (1) the termination shock, where the solar wind is shocked to subsonic speeds, which Voyager 1 and Voyager 2 crossed at distances of 94 and 84 AU from the Sun, respectively (Stone et al. 2005, 2008), (2) the heliopause, separating the plasma flows of the solar wind and ISM, which Voyager 1 may have recently crossed at a distance of 121 AU (Gurnett et al. 2013), and (3) the bow shock, where a supersonic ISM flow would be decelerated to subsonic speeds, although recent measurements from the Interstellar Boundary Explorer (IBEX) imply that the flow may not be supersonic and therefore a bow shock may not exist (McComas et al. 2012; Zank et al. 2013; Zieger et al. 2013). Similar astrospheric structures would naturally be expected to exist around other stars with solar-like coronal winds.

The interstellar medium (ISM) immediately surrounding the Sun is only partially ionized. In the wind/ISM collision, the neutral atoms in the ISM do not interact as strongly as the ions, but they still take part through charge exchange. Modeling neutrals in the heliosphere is not easy because the charge exchange sends them entirely out of thermal and ionization equilibrium. Nevertheless, many modern heliospheric modeling codes have become sufficiently sophisticated to properly model the neutrals, starting with work by Baranov and Malama (1993, 1995) and Zank et al. (1996). These models predict that the heliosphere will be permeated by various populations of hot hydrogen atoms, defined by the region of the heliosphere in which charge exchange is occurring. Particularly important is H I created just beyond the heliopause, where the interstellar matter is decelerated, compressed, and heated relative to the undisturbed ISM. This region in the outermost heliosphere has been called the hydrogen wall.

The hydrogen wall can produce a detectable absorption signature in UV spectra of the H I Lyman-α lines of nearby stars from the Hubble Space Telescope (HST), if absorption from the ISM itself is not so broad as to obscure the heliospheric absorption. Furthermore, the observed lines of sight not only pass through our heliosphere, but they also pass through the astrospheres of the observed stars. Thus, it is also possible to detect astrospheric Lyman-α absorption, and thereby indirectly detect solar-like stellar winds. The first detection of hydrogen wall absorption was in HST observations of the two stars in the very nearby binary α Cen (G2 V + K1 V). Figure 2.2 shows the Lyman-α spectrum of α Cen B (Linsky and Wood 1996). The upper solid line is an estimate of the intrinsic Lyman-α emission line profile from the star. Intervening H I gas between HST and the star absorbs much of this Lyman-α emission, resulting in the very broad absorption line centered at about 1215.61 Å in the figure. Much narrower and weaker absorption is also seen from neutral deuterium (D I) at 1215.27 Å. Most of the intervening H I and D I between us and the star is interstellar, but the ISM cannot account for all of the H I absorption.

When the H I absorption line is forced to have a central velocity and temperature consistent with the central velocity and width of the D I Lyman-α absorption, the ISM H I absorption ends up too narrow to fit the data. This indicates that there is excess H I absorption on both sides of the absorption that cannot be interstellar. The excess on the red side of the line is due to heliospheric absorption. The reason for the redshift is due to deceleration and deflection of ISM neutrals as they approach the heliopause, which results in a redshift from our perspective inside the heliosphere. Conversely, from our perspective outside the astrospheres, astrospheric absorption should be blueshifted.

Thus, the blue side excess absorption in Fig. 2.2 is astrospheric, as first demonstrated by Gayley et al. (1997). Note that the same excess is also observed towards both α Cen A and α Cen B, as expected since both members of the binary are close enough that they will lie within the same astrosphere, meaning that the astrospheric absorption seen towards both stars is indicative of the combined winds of both stars. Many HST Lyman-α observations of solar-like stars have been analyzed to identify those with detectable heliospheric and/or astrospheric absorption (Dring et al. 1997; Wood et al. 1996, 2000a, 2005b).

Even though all observed lines of sight will pass through the heliosphere and the astrosphere of the observed star, the absorption signatures of these structures are not always detectable. One potential cause for non-detections is a high ISM H I column density, leading to broad ISM Lyman-α absorption, which can obscure the heliospheric and astrospheric absorption. Another factor is the orientation of the line of sight with respect to the upwind direction of the ISM flow in the stellar (or solar)

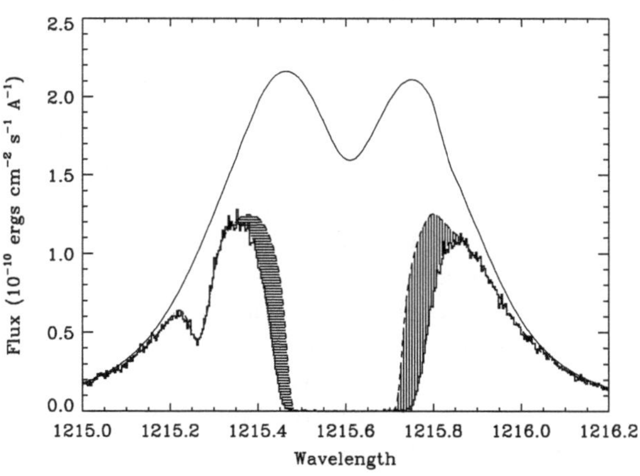

Fig. 2.2 HST Lyman-α spectrum of α Cen B, showing broad H I absorption at 1215.6 Å and D I absorption at 1215.25 Å. The *upper solid line* is the assumed stellar emission profile and the *dashed line* is the ISM absorption alone. The excess absorption is due to heliospheric H I (*vertical lines*) and astrospheric H I (*horizontal lines*) (After Linsky and Wood 1996)

rest frame. This is particularly apparent for heliospheric absorption, where many lines of sight have been observed with different angles from the upwind direction in the solar rest frame, demonstrating that the absorption is easiest to detect in upwind directions, consistent with model predictions (Wood et al. 2005b). A final major factor that applies solely to astrospheric detectability is the nature of the ISM surrounding the star. Although neutrals are present in the ISM around the Sun, this is actually not typical for stars within 100 pc. The Sun is within a region called the Local Bubble, in which most of the ISM is fully ionized (Lallement et al. 2003; Welsh et al. 2010, 2013). An astrosphere surrounded by a fully ionized ISM will contain no neutral H to produce Lyman-α absorption.

There are currently only 7 lines of sight with clear detections of heliospheric hydrogen wall absorption, all within 75° of the upwind direction of the ISM flow seen by the Sun (Wood et al. 2005b). The hydrogen wall neutrals are formed by charge exchange outside the heliopause, but it should be mentioned that in very downwind directions absorption from neutrals created by charge exchange in the inner heliosheath in between the termination shock and heliopause can become detectable. This broader but much shallower absorption signature has been clearly detected for 4 lines of sight within 20° of the downwind direction (Wood et al. 2007, 2014a) with an additional questionable detection towards Sirius (Izmodenov et al. 1999; Hébrard et al. 1999).

All 14 detections of astrospheric Lyman-α absorption are of the hydrogen wall variety, and ten are for main sequence stars like the Sun. Among these ten stars, the best analog for a young Sun is π^1 UMa, a 500 Myr old G1.5 V star (Wood et al. 2005b, 2014b). Figure 2.3 shows the Lyman-α spectrum for π^1 UMa, zooming in on the blue side of the line where the astrospheric absorption is detected. The amount of absorption will correlate with the strength of the stellar wind, but extracting a stellar mass loss rate from the Lyman-α data requires the assistance of hydrodynamic models of the astrosphere, such as the one in Fig. 2.4 for π^1 UMa, computed assuming a mass loss rate of $\dot{M} = 0.5\dot{M}_{\odot}$.

Astrospheric models like that in Fig. 2.4 are extrapolated from a heliospheric model that successfully reproduces heliospheric absorption, specifically a multi-fluid model described by Wood et al. (2000b). These models assume the same ISM characteristics as the heliospheric model, with the exception of the ISM flow speed in the stellar rest frame. Computing this speed requires knowledge of the star's unique space motion vector and the flow vector of the surrounding ISM. Stellar proper motions and radial velocities are very well known for the nearby stars with detected astrospheric absorption, so that is not an issue. The ISM flow vector varies somewhat from place to place, resulting in multiple ISM absorption components towards some stars, suggestive of multiple small, warm clouds in the solar neighborhood (Redfield and Linsky 2008). However, the velocity components are generally not widely separated, implying that the velocity vector of the cloud surrounding the Sun is a reasonable estimate of the ISM vector appropriate for other nearby stars. For π^1 UMa, for example, we estimate that the star sees an ISM wind speed of 34 km s^{-1}, compared to 23.8 km s^{-1} for the Sun (Redfield and Linsky

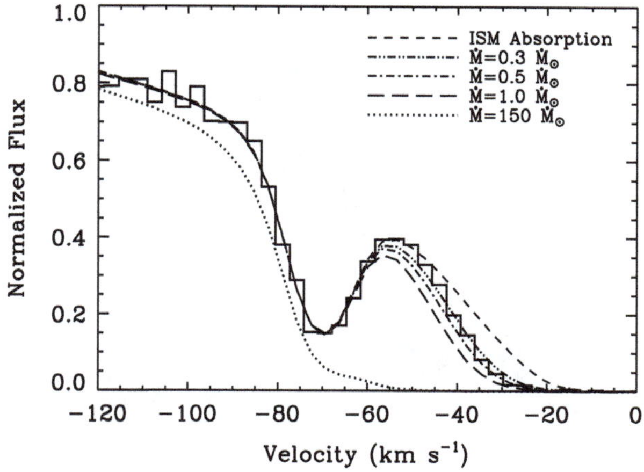

Fig. 2.3 The blue side of the Lyman-α absorption line of π^1 UMa, plotted on a heliocentric velocity scale. The absorption at -70 km s^{-1} is from D I. Since the ISM absorption cannot explain all of the H I absorption, the excess is assumed to be from the stellar astrosphere. The astrospheric absorption signature is compared with absorption predictions from four hydrodynamic models of the astrosphere, assuming four different mass-loss rates for π^1 UMa, after the astrospheric absorption is added to that of the ISM (After Wood et al. 2014b)

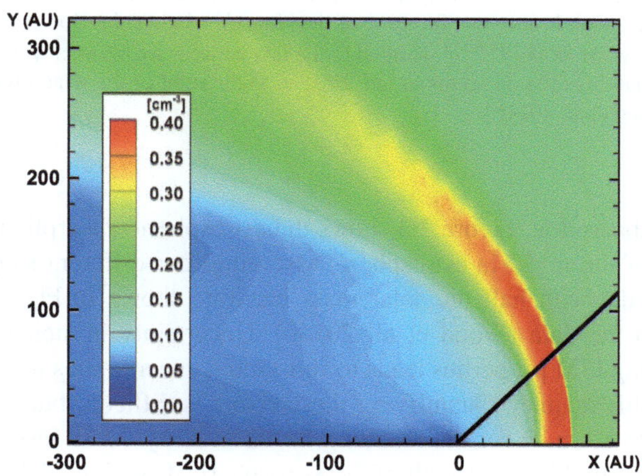

Fig. 2.4 The H I distribution of a hydrodynamic model of the π^1 UMa astrosphere, assuming $\dot{M} = 0.5\dot{M}_\odot$, which leads to the best fit to the data in Fig. 2.3. The star is at the origin, and the ISM is flowing from the right in this figure. The hydrogen wall is the parabolic shaped high-density region stretching around the star. The *black line* indicates our line of sight to the star (After Wood et al. 2014b)

2008), with our line of sight to the star lying 43° from the upwind direction, as shown in Fig. 2.4.

The astrospheric models are computed assuming different stellar wind densities, corresponding to different mass loss rates, and the Lyman-α absorption predicted by these models is compared with the data to see which best matches the observed

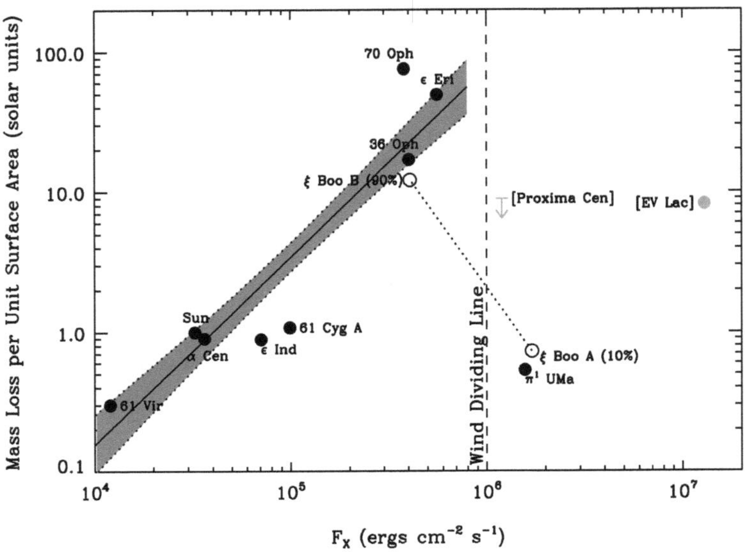

Fig. 2.5 A plot of mass loss rate (per unit surface area) versus X-ray surface flux for all main sequence stars with measured winds. Most of these are solar-like G and K stars, but the two with square-bracketed labels are M dwarfs. Separate points are plotted for the two members of the ξ Boo binary, assuming ξ Boo B accounts for 90 % of the binary's wind, and ξ Boo A only accounts for 10 %. A power law, $\dot{M} \propto F_X^{1.34\pm0.18}$, is fitted to the less active stars where a wind/corona relation seems to exist, but this relation seems to fail for stars to the right of the wind dividing line in the figure (After Wood et al. 2014b)

astrospheric absorption. Figure 2.3 shows the astrospheric absorption predicted by four models of the π^1 UMa astrosphere, assuming four different stellar mass loss rates. The model with half the solar mass loss rate shown in Fig. 2.4 is deemed the best fit to the data (Wood et al. 2014b). Uncertainty in stellar wind speeds and surrounding ISM properties leads to substantial uncertainties in mass loss rates measured in this manner. Quantifying these errors is difficult, but a factor of two is a reasonable estimate for the resulting $\dot{M}$ uncertainty. Mass loss rate estimates have been made in this way for all of the astrospheric detections (Wood et al. 2002, 2005a). Focusing on main sequence stars, Fig. 2.5 shows mass loss rates (per unit surface area) plotted versus coronal X-ray surface flux (Wood et al. 2014b). For the low-activity stars, mass loss increases with activity in a manner consistent with the $\dot{M} \propto F_X^{1.34\pm0.18}$ power law relation shown in the figure. For the ξ Boo binary, in which (like α Cen) the two members of the binary will share the same astrosphere, Fig. 2.5 indicates how the binary's combined wind strength of $\dot{M} = 5\dot{M}_\odot$ is most consistent with the other measurements if 90 % of the wind is ascribed to ξ Boo B, and only 10 % to ξ Boo A.

The connection between coronal activity and stellar winds appears to be a complex one. The Sun does not show a correlation of $\dot{M}$ with F_X, with $\dot{M}$ varying irregularly by less than a factor of two while the X-ray flux varies by a factor of 10 over the solar cycle (Wang 2010; Cohen 2011), possibly due to the suppression of mass flux by closed magnetic field regions where the density and X-ray emission are large. The $\dot{M} \propto F_X^{1.34\pm0.18}$ relation may result from the steep dependence of mass flux on X-ray luminosity in CMEs, which Drake et al. (2013) predict to be an important mass-loss mechanism in active stars but not the Sun. However, above an activity level corresponding to log $F_X = 10^6$ erg cm^{-2} s^{-1} this relation seems to fail, a boundary marked by a "Wind Dividing Line" in Fig. 2.5. Highly active stars above this limit appear to have surprisingly weak winds. This is suggested not only by the two solar-like G stars above the limit, ξ Boo A and π^1 UMa, but also for the two active M dwarfs above the limit, which have very modest mass loss rates. For Proxima Cen (M5.5 V), we only have an upper limit of $\dot{M} < 0.2\dot{M}_\odot$ (Wood et al. 2001), while for EV Lac $\dot{M} = 1\dot{M}_\odot$ (Wood et al. 2005a).

The apparent failure of the wind/corona correlation at the wind dividing line may indicate a fundamental change in magnetic field topology at that stellar activity level. Such a change is also suggested by observational evidence that very active stars usually have stable, long-lived polar starspots (Schrijver and Title 2001; Strassmeier 2002), in contrast to the solar example where sunspots are only observed at low latitudes. Perhaps the polar spots are indicative of a particularly strong dipolar magnetic field that envelopes the entire star and inhibits stellar wind flow, thereby explaining why very active stars have surprisingly weak winds. Very active stars may also be enveloped with strong toroidal fields as well (Donati and Landstreet 2009).

Given that young stars are more active than old stars (e.g., Ribas et al. 2005), the correlation between mass loss and activity indicated in Fig. 2.5 implies an anticorrelation of mass loss with age. The following relation between stellar X-ray flux and age for solar-like stars has been discovered by Ayres (1997): $F_X \propto t^{-1.74\pm0.34}$. Combining this with the power law relation from Fig. 2.4 yields the following relation between X-ray flux and age: $\dot{M} \propto t^{-2.33\pm0.55}$ (Wood et al. 2005a). Figure 2.6 shows what this relation suggests for the history of the solar wind, and for the history of winds from any solar-like star for that matter. The truncation of the power law relation in Fig. 2.5 near $F_X = 10^6$ erg cm^{-2} s^{-1} leads to the mass-loss/age relation in Fig. 2.6 being truncated as well at about $t = 0.7$ Gyr, slightly older than the age of the Hyades cluster. The plotted location of π^1 UMa in Fig. 2.6 indicates what the solar wind may have been like at times earlier than $t = 0.7$ Gyr. Despite the Wind Dividing Line, the stellar wind measurements suggest that winds of younger stars are generally stronger than the solar wind (at least for $t > 0.7$ Gyr), making it more likely that winds have significant erosive effects on planetary atmospheres over time. Analyses of lunar surface soils have also suggested a stronger solar wind in the past, though quantifying the effect is difficult from such data (e.g., Geiss 1974).

Before concluding, it should be said that inferences about the history of the solar wind from stellar astrospheric observations would certainly benefit from more data. The mass-loss/activity and mass-loss/age relations are based on only a handful of astrospheric detections. More detections would be especially desirable at high

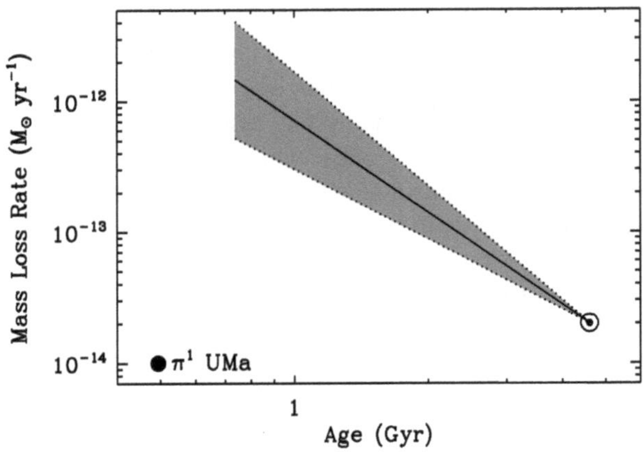

Fig. 2.6 The mass loss history of the Sun inferred from the power law relation in Fig. 2.5. The truncation of the relation in Fig. 2.4 means that the mass-loss/age relation is truncated as well. The low mass loss measurement for π^1 UMa suggests that the wind weakens at $t \approx 0.7$ Gyr as one goes back in time

activity levels to better determine what the solar wind was like when the Sun was very young and active. The Space Telescope Imaging Spectrograph (STIS) on HST is the only current instrument capable of observing the Lyman-α spectra used to detect the astrospheres.

2.2.3 T Tauri Star Winds

The previous subsection discussed our empirical knowledge about the winds of stars on the main sequence, which are expected to be analogous to the coronal wind of the Sun. However, very early in their history planets could also be affected by stellar winds that may be of a somewhat different character, due to the presence of a protoplanetary disk, and accretion of that disk onto the star.

Pre-Main-Sequence stars are well known to be extremely active in all stages of their evolution (Class I protostars, classical and weak-line T Tauri stars); a telltale signature is their high level of X-ray luminosity, corresponding to a saturation regime with $L_X/L_{bol} = 10^{-4} - 10^{-3}$ (e.g., Preibisch et al. 2005; Güdel et al. 2007), very high plasma temperatures (e.g., Telleschi et al. 2007), and frequent strong flares (Wolk et al. 2005). These features are taken as evidence that basically the same type of magnetic coronal structure is present and similar magnetic energy release mechanisms are at work in these objects as in main-sequence stars. By analogy, then, wind mass loss from the stellar surface may follow the trends seen in main-sequence stars. If indeed wind mass loss is suppressed in the most active stars, then we expect this to apply to Pre-Main-Sequence stars as well.

What is the observational evidence? The situation is complicated by the presence of several types of winds in disk-surrounded Pre-Main-Sequence stars; in view of the many open challenges to interpret the observations, we only briefly touch on some suggestive evidence. One manifestation of gas flows are the optically revealed bi-polar jets seen in young stellar objects; jets flow at velocities of several hundred km s^{-1} but are thought to be related to the presence of accretion in a protostellar or protoplanetary disks, evidenced by a correlation between accretion rate and mass-loss rate (Hartigan et al. 1995). Jets are probably launched magnetohydrodynamically at the star-disk interface or from the inner disk (Königl and Pudritz 2000; Shu et al. 2000). A second component, spectroscopically identified as a low-velocity wind with velocities of a few km s^{-1}, is likely related to X-ray or ultraviolet-induced photoevaporation of the disk surface at several AU to tens of AU from the central star (Alexander et al. 2004; Ercolano et al. 2008), generating a wide-angle wind that may carve out cavities in protostellar envelopes (Arce et al. 2013).

The question then is how a coronal wind escapes from the star, how it interacts with the jets and photoevaporative winds further out, and whether it energetically competes with those latter flows. The role of accretion in launching any of these wind components still also needs clarification. For example, Kwan et al. (2007), proposed that the emission profile of the He I λ10830 line arises from some sort of stellar wind in most observed Pre-Main-Sequence stars, although the presence of these features exclusively in the presence of disks suggests an accretion-powered wind. The most likely origins of these winds are the stellar polar regions above the magnetic funnel flows (Kwan et al. 2007). Direct evidence for ionized stellar winds was also suggested in far-ultraviolet observations by Dupree et al. (2005). P Cygni-type line profiles, asymmetries and absorption indicated the presence of $\approx$300,000 K winds with velocities of order 400 km s^{-1} in the classical T Tauri stars TW Hydrae and T Tauri. The mass-loss rates would then reach levels of 10^{-12} to $10^{-11} M_\odot$ year^{-1}. However, a re-analysis of these data by Johns-Krull and Herczeg (2007) puts this interpretation in doubt; it sets an upper limit of only 10,000–30,000 K to the wind temperature and implies that the ionization state is due to photoionization, making these outflows different from solar-type coronal winds.

Magnetized stellar winds have also been suggested for such disk-surrounded Pre-Main-Sequence stars to explain their relatively slow rotation, as an alternative to the problematic magnetic disk-locking mechanism. For example, Matt and Pudritz (2005) postulated accretion-powered massive winds escaping into large solid angles with open fields anchored in the stellar polar regions, with $\dot{M}_w \approx 0.1\dot{M}_a$, where $\dot{M}_a$ is the stellar accretion rate. Wind mass-loss rates of order $10^{-10} M_\odot$ year^{-1} or even more should therefore be expected. More sophisticated models of this kind were elaborated in a series of studies by Matt et al. (2012). The accretion-related launching mechanisms of stellar winds have also been considered by Cranmer (2008) suggesting turbulence-driven polar winds energized by accretion. We refer to the contribution in Chap. 3 by Lüftinger et al. (2014) in this volume for further details.

2.3 Expectations from Theoretical Models

Given the difficulty of studying coronal winds observationally, it is worthwhile to assess what can be inferred about wind evolution from theory. For cool main sequence stars, proposed wind acceleration mechanism include thermal pressure-driven winds originally proposed by Parker (1958), magnetic wave-driven winds originally proposed by Webber and Davis (1967), and magnetocentrifugally-driven winds (e.g., Vidotto et al. 2011). Most recent models include magnetic effects for the wind acceleration mechanism. We summarize here three theoretical studies that explore the different wind acceleration mechanisms and predict mass-loss rates that can be compared with observations.

Holzwarth and Jardine (2007) extended the Webber and Davis (1967) model by using thermal wind parameters (temperature, density, and magnetic field strength) that depend functionally on the stellar rotation rate. They found that for slowly-rotating stars the main driving force is thermal pressure gradients in the corona, but for fast rotators the winds are accelerated mainly by magneto-centrifugal forces. Their model is consistent with the mass-loss rates measured by Wood et al. (2005a) for slowly- rotating stars like the Sun and the most rapidly-rotating active stars (e.g., ξ Boo A and EV Lac). However, their model fails to explain the moderately-rotating K dwarfs (ϵ Eri, 36 Oph, and 70 Oph), which have mass flux rates nearly 100 times solar but coronal densities inferred from X-ray luminosities that are not sufficiently large to explain the very high mass loss rates. Since they could find no consistent set of parameters to explain both the observed very high mass-loss rates for the moderately-rotating stars and the modest mass-loss rates of the rapidly-rotating stars, they concluded that the moderately-rotating K dwarfs are uncharacteristic for the winds of cool main-sequence stars. They conclude with a model in which the mass-loss rate of a $1M_{\odot}$ star decreases by one order of magnitude between 10^6 years and the present age of the Sun.

Cranmer and Saar (2011) developed a magnetic turbulence-driven wind model in which they follow the MHD turbulence energy flux from the subphotosphere convective zone through the corona where the energy emerges as wind expansion through open magnetic field lines. Their model is physically self-consistent. For cool dwarf stars, the driving mechanism for mass outflow is primarily thermal pressure gradients, although they do include Alfvén wave pressure that dominates for giant stars. For cool dwarf stars, photospheric magnetic field strengths are close to the equipartition value with the filling factors increasing rapidly with faster rotation (Cuntz et al. 1998). For solar mass stars, their standard model predicts mass flux rates about 100 times the present Sun at ages 10^7 to 10^8 years, but with an age dependence proportional to $t^{-1.1}$ compared with the empirical relation $\dot{M} \sim t^{-2.33\pm0.55}$ found by Wood et al. (2005a). It is suggested by Cranmer and Saar (2011) that a steeper age dependence of $\dot{M}$ would result from an alternative dependence of rotation rate on age. Their approach has the advantage that it is based on physical principals and empirical stellar parameters, but except for active M dwarfs like EV Lac, it provides mass flux estimates within an order of magnitude of empirical values. They suggest that a different driving mechanism, perhaps flares and coronal mass ejections, may explain the observed large mass loss rates.

For very rapidly-rotating stars, centrifugal forces likely play an important role in driving mass loss. A recent study by Vidotto et al. (2011) considered the M4 dwarf star V374 Peg with a rotational period of 0.44 days. They solved the magnetohydrodynamic (MHD) equations in the strong poloidal magnetic field obtained from Zeeman Doppler imaging. The rapid rotation and strong magnetic fields produce a wind with a high speed, 1,500–2,300 km/s, and extremely high mass-loss rates, 3–50 $\times 10^{-11}$ $M_{\odot}$ year. These mass-loss rates are 3–4 orders of magnitude larger than solar and 2–3 orders of magnitude larger than found for the M dwarf star EV Lac by the astrospheres technique. These mass-loss rates predict short time scales for rotational braking and thus may be inconsistent with the short rotational period of the star. Despite these problems, this study shows that centrifugal forces likely play an important role in driving winds of rapidly-rotating stars.

Although these theoretical models are roughly consistent with the observations, there remain serious problems in explaining the low mass-loss rates of the active stars. We look forward to the next generation of stellar wind models that will, hopefully, better explain the present and future observations.

Conclusion

There is much room for improvement with regards to our knowledge of how stellar winds vary with time. Astrospheric Lyman-α absorption currently provides the only method to detect and measure the coronal winds of cool main sequence stars, but there are still only a handful of detections. These data are generally consistent with younger, more active stars having stronger winds, up to $\dot{M} \sim 100\dot{M}_{\odot}$ for stars about 0.7 Gyr old, increasing the likelihood that winds have significant impacts on the evolution of planetary atmospheres. However, the recent π^1 UMa measurement suggests a different regime for the wind-corona connection at earlier ages, when stars may have surprisingly weak winds. More data are clearly required to explore these issues.

Unfortunately, the astrospheric diagnostic has substantial drawbacks that make the acquisition of additional relevant measurements uncertain. In the near future UV space observatories such as the World Space Observatory-UV (WSO-UV) (see Chap. 14 by (Shustov et al. 2014)) can be used for observations. At present HST is the only observational platform capable of these observations. Obtaining HST time for these purposes is not easy, particularly since the astrospheric detection likelihood for most stars is quite low (Wood et al. 2005b). What is worse is that the reason for most non-detections (being surrounded by an ISM without neutral H) means that

(continued)

non-detections do not even provide meaningful upper limits to $\dot{M}$ in most instances. Hopefully a more direct wind detection technique can ultimately be found without these drawbacks. Free-free radio emission from the ionized coronal winds would seem to be the most likely observational approach, though even the expanded VLA and the new ALMA array are likely to fall short of the sensitivity necessary to detect the modest winds successfully detected through astrospheric absorption. A far more sensitive radio telescope will ultimately be required to survey the stellar winds of nearby stars in the way that X-ray surveys have successfully studied the stellar coronae responsible for these winds.

Acknowledgements The authors acknowledge the support by the International Space Science Institute (ISSI) in Bern, Switzerland and the ISSI team *Characterizing stellar- and exoplanetary environments*. Support for this work was also provided by NASA through an award from the Space Telescope Science Institute (program GO-12596), which is operated by the Association of Universities for Research in Astronomy, Inc., under NASA contract NAS 5-26555. M. Güdel acknowledges support from the Austrian Research Foundation FWF NFN project S11601-N16 'Pathways to Habitability: From Disks to Active Stars, Planets and Life', as well as the related FWF NFN subprojects, S116 604-N16 'Radiation & Wind Evolution from T-Tauri Phase to ZAMS and Beyond'.

References

Alexander, R. D., Clarke, C. J., & Pringle, J. E. (2004). *Monthly Notices of the Royal Astronomical Society, 354*, 71.

Alexeev, I. I., Grygoryan, M. S., Belenkaya, E. S., Kalegaev, V. V., & Khodachenko, M. L. (2014). H. Lammer & M. L. Khodachenko (Eds.), *Characterizing stellar and exoplanetary environments* (pp. 189). Heidelberg/New York: Springer.

Arce, H. G., Mardones, D., Corder, S. A., Garay, G., Noriega-Crespo, A., Raga, A. C. (2013). *Astrophysical Journal, 774*, 39.

Ayres, T. R. (1997). *Journal of Geophysical Research, 102*, 1641.

Baranov, V. B., & Malama, Y. G. (1993). *Journal of Geophysical Research, 98*, 15157.

Baranov, V. B., & Malama, Y. G. (1995). *Journal of Geophysical Research, 100*, 14755.

Belenkaya, E. S., Khodachenko, M. L., & Alexeev, I. I. (2014). H. Lammer & M. L. Khodachenko (Eds.), *Characterizing stellar and exoplanetary environments* (pp. 239). Heidelberg/New York: Springer.

Bisikalo, D. V., Kaygorodov, P. V., Ionov, D. E., & Shematovich, V. I. (2014). H. Lammer, & M. L. Khodachenko (Eds.), *Characterizing stellar and exoplanetary environments* (pp. 81). Heidelberg/New York: Springer.

Brain, D., et al. (2010). *Icarus, 206*, 139.

Brown, A., Vealé, A., Judge, P., Bookbinder, J. A., & Hubeny, I. (1990). *Astrophysical Journal, 361*, 220.

Carr, M. H. (1996). *Water on mars*. New York: Oxford University Press

Charbonneau, P. (2010). *Living Reviews in Solar Physics, 7*, 3. http://www.livingreviews.org/lrsp-2010-3

Cohen, O. (2011). *Monthly Notices of the Royal Astronomical Society, 417*, 2592.

Cranmer, S. R. (2008). *Astrophysical Journal, 689*, 316.

Cranmer, S. R. & Saar, S. H. (2011). *Astrophysical Journal, 741*, 54

Cuntz, M., Ulmschneider, P., & Musielak, Z. E. (1998). *Astrophysical Journal, 493*, L117

Donati, J.-F., & Landstreet, J. D. (2009). *Annual Review of Astronomy and Astrophysics, 47*, 333.

Drake, J. J., Cohen, O., Yashiro, S., Gopalswamy, N. (2013). *Astrophysical Journal, 764*, 170.

Dring, A. R., et al. (1997). *Astrophysical Journal, 488*, 760.

Dupree, A. K., Brickhouse, N. S., Smith, G. H., & Strader, J. (2005). *Astrophysical Journal, 625*, L131.

Ercolano, B., Drake, J. J., Raymond, J. C., & Clarke, C. C. (2008). *Astrophysical Journal, 688*, 398.

Feldman, W. C., Asbridge, J. R., Bame, S. J., Gosling, J. T. (1977). O. R. White (Ed.) *The solar output and its variation* (p. 351). Boulder: Colorado Associated University Press.

Gaidos, E. J., Güdel, M., & Blake, G. A. (2000). *Geophysical Research Letters, 27*, 501.

Gayley, K. G., Zank, G. P., Pauls, H. L., Frisch, P. C., & Welty, D. E. (1997). *Astrophysical Journal, 487*, 259.

Geiss, J. (1974), D. R. Criswell & J. W. Freeman (Eds.), *Conference on lunar interactions: Interactions of the interplanetary plasma with the modern and ancient moon* (p. 110). Houston: Lunar Science Institute.

Grießmeier, J.-M., et al. (2004). *Astronomy and Astrophysics, 425*, 753.

Güdel, M., et al. (2007). *Astronomy and Astrophysics, 468*, 353.

Güdel, M., Guinan, E. F., Skinner, S. L. (1997). *Astrophysical Journal, 483*, 947.

Gurnett, D. A., Kurth, W. S., Burlaga, L. F., Ness, N. F. (2013). *Science, 341*, 1489.

Hartigan, P., Edwards, S., Ghandour, L. (1995). *Astrophysical Journal, 452*, 736.

Hébrard, G., Mallouris, C., Ferlet, R., Koester, D., Lemoine, M., Vidal-Madjar, A., & York, D. (1999) *Astronomy and Astrophysics, 350*, 643.

Holzwarth, V. & Jardine, M. (2007) *Astronomy and Astrophysics, 463*, 11.

Izmodenov, V. V., Lallement, R., Malama, Y. G. (1999). *Astronomy and Astrophysics, 342*, L13.

Jakosky, B. M., Pepin, R. O., Johnson, R. E., & Fox, J. L. (1994). *Icarus, 111*, 271.

Jakosky, B. M., & Phillips, R. J. (2001). *Nature, 412*, 237.

Johns-Krull, C. M., & Herczeg, G. J. (2007). *Astrophysical Journal, 655*, 345.

Judge, P. G., Solomon, S. C., & Ayres, T. R. (2003). *Astrophysical Journal, 593*, 534.

Kass, D. M., & Yung, Y. L. (1995) *Science, 268*, 697.

Khodachenko, M. L., et al. (2012). *Astrophysical Journal, 744*, 70.

Kislyakova, K. G., Holmström, M., Lammer, H., & Erkaev, N. V. (2014). H. Lammer, & M. L. Khodachenko (Eds.), *Characterizing stellar and exoplanetary environments* (pp. 137). Heidelberg/New York: Springer.

Königl, A., & Pudritz, R. E. (2000). V. Mannings, A. P. Boss, & S. S. Russell (Eds.), *Protostars planets IV* (pp. 759). Tucson: University of Arizona Press.

Koutroumpa, D., Lallement, R., Raymond, J. C., & Kharchenko, V. (2009). *Astrophysical Journal, 696*, 1517.

Kwan, J., Edwards, S., & Fischer, W. (2007). *Astrophysical Journal, 657*, 897.

Lallement, R. (2004). *Astronomy and Astrophysics, 418*, 143.

Lallement, R., Welsh, B. Y., Vergely, J. L., Crifo, F., & Sfeir, D. (2003). *Astronomy and Astrophysics, 411*, 447.

Lammer, H., et al. (2003). *Icarus, 165*, 9.

Lim, J., White, S. M., & Slee, O. B. (1996). *Astrophysical Journal, 460*, 976.

Linsky, J. L., & Wood, B. E. (1996). *Astrophysical Journal, 463*, 254.

Linsky, J. L., Güdel, M., (2014). H. Lammer & M. L. Khodachenko (Eds.), *Characterizing stellar and exoplanetary environments* (pp. 3). Heidelberg/New York: Springer.

Luhmann, J. G., Johnson, R. E., & Zhang, M. H. G. (1992). *Geophysical Research Letters, 19*, 2151.

Lüftinger, T., Vidotto, A. A., & Johnstone, C. P. (2014). H. Lammer & M. L. Khodachenko (Eds.), *Characterizing stellar and exoplanetary environments* (pp. 37). Heidelberg/New York: Springer.

Matt, S., Pinzón, G., Greene, T. P., & Pudritz, R. E. (2012). *Astrophysical Journal, 745*, 101.
Matt, S., & Pudritz, R. E. (2005). *Astrophysical Journal, 632*, L135.
McComas, D. J., et al. (2012). *Science, 336*, 1291.
Ossendrijver, M. (2003). *A&ARv, 11*, 287.
Parker, E. N. (1958). *Astrophysical Journal, 128*, 664.
Perez de Tejada, H. (1992). *Journal of Geophysical Research, 97*, 3159.
Preibisch, T., et al. (2005). *Astrophysical Journal, 160*, 401.
Redfield, S., & Linsky, J. L. (2008). *Astrophysical Journal Supplement Series, 673*, 283.
Ribas, I., Guinan, E. F., Güdel, M., & Audard, M. (2005). *Astrophysical Journal, 622*, 680.
Schrijver, C. J., & Title, A. M. (2001). *Astrophysical Journal, 551*, 1099.
Shu, F. H., Najita, J. R., Shang, H., & Li, S.-Y. (2000). V. Mannings, A. P. Boss, & S. S. Russell (Eds.), *Protostars planets IV* (p. 789). Tucson: University of Arizona Press.
Stone, E. C., Cummings, A. C., McDonald, F. B., Heikkila, B. C., Lal, N., & Webber, W. R. (2005). *Science, 309*, 2017.
Stone, E. C., Cummings, A. C., McDonald, F. B., Heikkila, B. C., Lal, N., & Webber, W. R. (2008). *Nature, 454*, 71.
Strassmeier, K. G. (2002). *Astronomische Nachrichten, 323*, 309.
Shustov, B. M., Sachkov, M. E., Bisikalo, D., & Gómez de Castro, A.-I. (2014). H. Lammer & M. L. Khodachenko (Eds.), *Characterizing stellar and exoplanetary environments* (pp. 275). Heidelberg/New York: Springer.
Telleschi, A., Güdel, M., Briggs, K. R., Audard, M., & Palla F. (2007). *Astronomy and Astrophysics, 468*, 425.
Terada, N., Kulikov, Y. N., Lammer, H., Lichtenegger, H. I. M., Tanaka, T., Shinagawa, H., & Zhang, T. (2009). *Astrobiology, 9*, 55.
Vidotto, A. A., Jardine, M., Opher, M., Donati, J. F., & Gombosi, T. I. (2011), *Monthly Notices of the Royal Astronomical Society, 412*, 351.
Vidotto, A. A., Bisikalo, D. V., Fossati, L., & Llama, J. (2014). H. Lammer & M. L. Khodachenko (Eds.), *Characterizing stellar and exoplanetary environments* (pp. 153). Heidelberg/New York: Springer.
Wang, Y.-M. (2010). *Astrophysical Journal, 715*, L121.
Wargelin, B. J., & Drake, J. J. (2002). *Astrophysical Journal, 578*, 503.
Wargelin, B. J., Kashyap, V. L., Drake, J. J., García-Alvarez, D., & Ratzlaff, P. W. (2008). *Astrophysical Journal, 676*, 610.
Webber, E. J. & Davis, L. J. (1967). *Astrophysical Journal, 148*, 217.
Welsh, B. Y., Lallement, R., Vergely, J. L., & Raimond, S. (2010). *Astronomy and Astrophysics, 510*, A54.
Welsh, B. Y., Wheatley, J., Dickinson, N., & Barstow, M. A. (2013). *PASP, 125*, 644.
Wolk, S. J., Harnden, F. R., Jr., Flaccomio, E., Micela, G., Favata, F., Shang, H., & Feigelson, E. D. (2005). *Astrophysical Journal Supplement Series, 160*, 423.
Wood, B. E. (2004). *Living Reviews in Solar Physics, 1*, 2. http://www.livingreviews.org/lrsp-2004-2
Wood, B. E. (2006). *Space Science Reviews, 126*, 3.
Wood, B. E., Alexander, W. R., & Linsky, J. L. (1996). *Astrophysical Journal, 470*, 1157.
Wood, B. E., Izmodenov, V. V., Alexashov, D. B., Redfield, S., & Edelman, E. (2014a). *Astrophysical Journal, 780*, 108.
Wood, B. E., Izmodenov, V. V., Linsky, J. L., & Malama, Y. G. (2007). *Astrophysical Journal, 657*, 609.
Wood, B. E., Linsky, J. L., Müller, H.-R., & Zank, G. P. (2001). *Astrophysical Journal, 547*, L49.
Wood, B. E., Linsky, J. L., & Zank, G. P. (2000a). *Astrophysical Journal, 537*, 304.
Wood, B. E., Müller, H.-R., Redfield, S., & Edelman, E. (2014b). *Astrophysical Journal, 781*, L33.
Wood, B. E., Müller, H.-R., & Zank, G. P. (2000b). *Astrophysical Journal, 542*, 493.
Wood, B. E., Müller, H.-R., Zank, G. P., & Linsky, J. L. (2002). *Astrophysical Journal, 574*, 412.
Wood, B. E., Müller, H.-R., Zank, G. P., Linsky, J. L., & Redfield, S. (2005a). *Astrophysical Journal, 628*, L143.

Wood, B. E., Redfield, S., Linsky, J. L., Müller, H.-R., & Zank, G. P. (2005b). *Astrophysical Journal Supplement Series, 159*, 118.
Zank, G. P. (1999). *Space Science Reviews, 89*, 413.
Zank, G. P., Heerikhuisen, J., Wood, B. E., Pogorelov, N. V., Zirnstein, E., McComas, D. J. (2013). *Astrophysical Journal, 763*, 20.
Zank, G. P., Pauls, H. L., Williams, L. L., & Hall, D. T. (1996). *Journal of Geophysical Research, 101*, 21639.
Zieger, B., Opher, M., Schwadron, N. A., McComas, D. J., & Tóth, G. (2013). *Geophysical Research Letters, 40*, 2923.

Chapter 3
Magnetic Fields and Winds of Planet Hosting Stars

Theresa Lüftinger, Aline A. Vidotto, and Colin P. Johnstone

Abstract Stellar magnetism is a crucial driver of activity, ionization, photodissociation, chemistry and winds in stellar environments. It therefore has an important impact on the atmospheres and the magnetospheres of surrounding planets. Modelling of stellar magnetic fields and their winds is extremely challenging, both from the observational and the theoretical points of view, and only recent ground breaking advances in observational instrumentation, as well as a deeper theoretical understanding of magnetohydrodynamic processes in stars enable us to model stellar magnetic fields and winds – and the resulting influence on surrounding planets – in more and more detail. Here we review what is known about the magnetic fields of cool stars, covering relevant techniques such as Zeeman Doppler Imaging (ZDI), field extrapolation and wind simulations, as well as relevant observational results.

3.1 Introduction: Stellar Magnetic Fields

Magnetic fields play a key role in many physical processes important for stellar and planetary formation and evolution. They crucially influence the collapse of molecular clouds, their fragmentation into individual stars and planetary systems, the angular momentum evolution within the initial cloud, and the generation of winds, outflows and jets. Magnetic fields may also give rise to enhanced hydrodynamical instabilities and strongly affect a star's angular momentum evolution throughout its life. The resulting magnetic field and wind properties of a star have an important influence on the atmospheres, as described in Chap. 4 (Fossati et al. 2014), and

T. Lüftinger (✉) • C.P. Johnstone
Department of Astrophysics, University of Vienna, Türkenschanzstr. 17, A-1180 Vienna, Austria
e-mail: theresa.rank-lueftinger@univie.ac.at; colin.johnstone@univie.ac.at

A.A. Vidotto
SUPA, School of Physics and Astronomy, University of St Andrews, North Haugh, KY16 9SS, St Andrews, UK

Observatoire de Genève, Université de Genève, Ch. des Maillettes 51, Versoix, CH-1290, Switzerland
e-mail: Aline.Vidotto@unige.ch

H. Lammer, M. Khodachenko (eds.), *Characterizing Stellar and Exoplanetary Environments*, Astrophysics and Space Science Library 411,
DOI 10.1007/978-3-319-09749-7_3

Chap. 7 (Kislyakova et al. 2014) and the magnetospheres of surrounding planets, outlined in Chap. 8 (Vidotto et al. 2014a), and Chap. 11 (Grießmeier 2014), as is evident from the Sun-Earth relation and studied for more distant stars in e.g. (Catala et al. 2007; Shkolnik et al. 2008; Donati et al. 2008a; Fares et al. 2009).

For young, low mass stars still contracting towards the main sequence and surrounded by gaseous and dusty accretion discs, the presence of strong kG fields is reported in several studies such as that of Johns-Krull (2007). The first studies to reconstruct the photospheric magnetic field spatial distributions show that low mass protostars possess strong large-scale fields whose intensities and topologies seem to strongly depend on the stellar internal structure. As suggested by Hussain et al. (2009) and Gregory et al. (2012), and discussed in more detail in Sect. 3.4, the complexity of the (supposedly) dynamo generated magnetic fields is related to the size of the convection zone, with more complex fields found in classical T Tauri stars (CTTSs) with radiative cores, while less complex fields are observed for fully convective stars. On the main-sequence, more or less all cool, low-mass solar-type stars show magnetic fields comparable to that of our Sun.

3.2 Analyzing Stellar Magnetic Fields: Techniques

Various different techniques are currently used to detect and analyze stellar magnetic fields, being based on either high-resolution spectroscopy or photo- and spectropolarimetry. While high-resolution spectroscopy allows detailed studies of spectral line profile shapes, photo- or spectropolarimetry focuses on the analysis of the polarized light component that magnetic fields produce via the Zeeman effect. Here we mention two techniques important for the analysis of magnetic field strengths and their structure on the surfaces of stars.

3.2.1 Zeeman Broadening and Spectropolarimetry

In the presence of a magnetic field, a spectral line is split into several components, the so called π and σ components, through the Zeeman effect, whereby the distance between the components is also a function of the field strength. Analysis via the Zeeman effect is the best known and most widely used diagnostic technique for all types of magnetic fields, from the very weak fields of molecular clouds (μG range) to the very strong fields of white dwarfs.

For a magnetic field B (in kG),

$$\Delta\lambda_B = 4.67\lambda_0^2\,\bar{g}B \tag{3.1}$$

gives the average wavelength displacement $\Delta\lambda_B$ of a σ component from its zero field wavelength λ_0 (in μm), where $\bar{g}$ represents the effective Landé factor measuring the average magnetic sensitivity of a line. Thus, from the separation of the π and σ components of a line with known Landé factor $\bar{g}$, the intensity of the magnetic field averaged over the visible hemisphere of the star, B, can be measured. As a more detailed discussion of the Zeeman effect on atoms in a magnetic field is out of the aim and scope of this publication, we refer the reader to Donati and Landstreet (2009) for an elaborate summary.

The detection and analysis of the polarization properties of a Zeeman-split line provide another means of measuring magnetic fields. A major gain of this method is that polarization signatures give access to the orientation of the field properties: circular polarisation (from σ components) is sensitive to the line-of-sight (or longitudinal) component of the magnetic field, while linear polarization (from π and σ components) gives access to the perpendicular (or transverse) component of the magnetic field. In astrophysics, polarized radiation is usually described via the Stokes parameters I, Q, U, and V (introduced by G. G. Stokes in 1852). A definition of the Stokes parameters in terms of ideal filters for circular and linear polarization that is commonly used in optical astronomy was proposed by Shurcliff (1962). Ideal filters for linear polarization, the so-called polarizers, let the component of the electric field of a radiation beam perpendicular to the direction of propagation be transmitted, whereas the component of the electric field in the orthogonal direction cannot pass. Combining an ideal quarter-wave plate with an ideal linear polarizer, whose transmission axis is rotated counterclockwise by 45° (for positive circular polarization) and clockwise (for negative circular polarization), with respect to the fast-axis of the wave-plate, we obtain a filter for the circularly polarized component of the electromagnetic wave. The values of the Stokes parameters can then be described as:

$$I = kS, Q = k(S_0 - S_{90}), U = k(S_{45} - S_{135}), \text{ and } V = k(S_+ - S_-),$$

where S is the signal obtained without any filter, S_0, S_{45}, S_{90}, and S_{135} are the signals obtained with the transmission axis of a linear polarizer set to 0°, 45°, 90°, and 135°, respectively, and S_+ and S_- refer to a filter set interposed, allowing us to obtain positive and negative circular polarization, and k is the normalization constant. Further details on Stokes polarimetry can be found in Landi Degl'Innocenti and Landolfi (2004).

3.2.2 *Zeeman Doppler Imaging*

The development and application of the Doppler Imaging (DI) and Zeeman Doppler Imaging (ZDI) techniques (e.g., Kochukhov et al. 2004; Donati et al. 2006), makes it possible to invert a time-series of high-resolution Stokes parameter observations of stellar spectra into surface maps of parameters such as temperature, elemental abundance and magnetic field geometry (Lüftinger et al. 2010a,b).

Built around complex mathematical procedures, this technique has become one of the most powerful astrophysical remote sensing methods. From the mathematical point of view, during the inversion process, a total discrepancy function $\Psi = D + R$ is minimized, whereby D characterizes the discrepancy between the observed and theoretical phase-resolved spectra, and R is the regularization function.

This regularization function ensures stability of the complex optimization algorithm within ZDI and the simplest possible and unique solution independent from the initial guess and the surface discretisation. State of the art ZDI codes have been described in detail by Donati (2001), Donati et al. (2006), Piskunov and Kochukhov (2002), Kochukhov and Piskunov (2002), and Wade et al. (2001). These are based on elaborate spectrum synthesis, taking into account all relevant physics of polarized line formation. Recently, the code of Kochukhov and Piskunov (2002) was extended towards a new version for temperature and magnetic mapping of the surface structures in cool, active stars (Kochukhov and Piskunov 2009) including the treatment of molecular opacities (Rosén and Kochukhov 2012). It should be mentioned, however, that ZDI misses possibly present small-scale magnetic flux, whose polarization signatures cancel out as stated e.g. in Reiners and Basri (2009).

Although the addition of a small-scale field increases the stellar surface flux, it was found by Lang et al. (2014), that the large-scale open flux, which is important when studying the influence of stellar fields on surrounding planets and which governs the loss of mass and angular momentum in the wind, remains unaffected. Thus, the possible neglect of small-scale fields is unlikely to significantly influence the spin-down times and wind configurations calculated from magnetograms derived via ZDI (see also Sect. 3.5.1).

Due to the development of dedicated novel instrumentation, such as ESPaDOnS at the Canada-France-Hawaii Telescope (CFHT), NARVAL at the Télescope Bernard-Lyot (Pic du Midi, France), and HARPSpol of the European Southern Observatory (ESO), we now have our hands on excellent spectropolarimetric data to exploit the full amount of information available from time-resolved observations of Stokes profiles. In Sects. 3.3 and 3.4, we present recent results based on ZDI.

3.3 Rotation and Magnetism in Low Mass Main-Sequence Stars

3.3.1 The Sun

Most of what we know about the magnetic activity of low-mass stars is based on analogies with the Sun. The most obvious manifestation of the solar magnetic field is the presence of dark cool sunspots in the photosphere that can be seen in visible light. These spots are found in active regions contained within low latitude bands and are caused, as mentioned above, by strong kG magnetic fields (Hale 1908). Outside of sunspots, magnetic fields exist all over the solar photosphere at all times.

Most of this detected magnetic flux is contained within discreet structures, covering a small fraction of the solar surface and largely concentrated within intergranular lanes, with approximately uniform field strengths of ~1.5 kG (Stenflo 1973; Solanki 1993; Stenflo 2011).

In addition, a weaker and much more complex small-scale field covers the rest of the solar surface, though the nature of this field is still a matter of debate (for a review, see Solanki 2009). The photospheric magnetic field extends into the corona where it is responsible for heating the coronal plasma to MK temperatures. Since most of the small scale field structures that exist in the photosphere do not extend into the corona, the coronal field is much simpler than the photospheric field. High in the corona, the field becomes very simple and can often be approximately described by a dipole.

The number of sunspots, and therefore the amount of magnetic flux, on the solar surface varies over 11-year cycles. At the beginning of a cycle, there are no large active regions, and the field strength at the poles is at its maximum. The solar global magnetic field is at this time simple, axisymmetric, and dipolar. As the cycle progresses, the number of active regions increases, and the strength of the polar magnetic field decreases until approximately half-way through the cycle where the dipole component flips in polarity. At this point, the solar magnetic field is highly complex and non-dipolar. This change in the photospheric field leads to large effects on the coronal structure. The corona on global scales can be approximately broken down into regions of closed magnetic field and regions of open magnetic field. Closed field regions are where the magnetic field dominates over the coronal plasma and is therefore able to prevent it from expanding away from the Sun. Open field regions are where the magnetic field is not able to hold in the coronal plasma and is therefore dragged out with the expanding wind. At solar minimum, large regions of open field cover the poles, and at low-latitudes, the corona is almost entirely closed. At solar maximum, the poles are often covered in regions of closed field, and the surface is covered in complex distributions of open and closed field regions (Wang and Sheeley 1990b).

3.3.2 Solar Type Stars

Essentially all cool, low-mass solar-type stars show magnetic fields comparable to that of our Sun. They can exhibit enormous activity, also showing dark spots on their surfaces (e.g. Berdyugina 2005) appearing and disappearing on comparably short time scales, from days (linked to the star's rotation), to months (as they form and disappear again over the spot's lifetime) or years (changing in number and location with the stellar activity cycle). This leads to a spotty surface with often huge magnetic field and temperature gradients. The current understanding is that this activity, in analogy to the solar activity, is triggered by dynamo processes resulting from the conversion of convective and rotational mechanical energy into magnetic energy implying rotational shearing and cyclonic turbulence

(Parker 1958). During these processes, variable and complex magnetic fields are generated, whose properties are strongly correlated with the stellar rotation rate, mass, and age (e.g. Donati and Landstreet 2009). Although it is not fully understood in detail, the basic principles of the dynamo mechanism are well established (e.g. Dobler 2005). Dynamo processes in the case of the Sun and in many other cool stars are presumably concentrated in the tachocline, a thin interface layer at the base of the convection zone, where radial gradients in rotation rates are steepest (Charbonneau 2010). Due to the presence of convective envelopes up to a certain stellar mass and surface temperature, all cool, solar type stars are expected to show solar-type dynamo magnetism. As the convective envelopes gradually decrease with increasing spectral type, dynamo type magnetic fields and the related stellar activity are expected to decrease until they seem to disappear (or become immeasurable with present instrumentation) at a certain mass and temperature range for the majority of early F-, and A-type stars.

3.3.3 M Dwarfs

Planet-hosting M dwarfs have their habitable zones much closer to the central star than higher mass stars, but magnetic activity and the corresponding energetic radiation decline much slower in time than in solar type stars (Ribas et al. 2005). Magnetic field structures of M dwarfs, as has recently been found by Donati et al. (2008b) and Morin et al. (2008), seem to be rather manifold: fully convective mid-M dwarfs show evidence of simpler, large-scale magnetic fields (almost fully poloidal) than early (more massive) M dwarfs, which likely host a convective core and show more complex fields. In this part of the HR-diagram, as also observed for CTTSs (discussed in more detail in Sect. 3.4), it seems that fully convective stars have simple axisymmetric fields that are dominated by the dipolar component and as soon as the inner radiative zones start to be developed (with increasing mass), the fields become more complex.

However, in a later study on cooler M dwarfs of spectral types between M5 – M8, (Morin et al. 2010) find M stars of similar stellar parameters exhibiting drastically different magnetic topologies, from very simple fields to complex geometries. A possible explanation for this ambiguity is that for M dwarfs of this mass range, a bistable dynamo process is operating (Morin et al. 2011), which is responsible for the diverse magnetic field structures. This is likely to result in a diversity in the stellar environments among M dwarfs having otherwise very similar properties.

3.3.4 Rotation and Magnetism

When the first successful model for the solar wind was produced by Parker (1958), it was immediately clear that an ionized stellar wind emanating from a rotating magnetized star would lead to a slowing down of the star's rotation rate with time. This slowing down of the rotation rate is due to the transfer of angular momentum from the star by the wind, mostly in the form of stresses in the magnetic field (Weber and Davis 1967). This was later proven observationally by Skumanich (1972). Since then, a large number of studies have shown that while this result is approximately correct in certain circumstances, the situation is much more complex. Stars start out their main-sequence lives with a huge two-orders of magnitude spread in rotation rates, as can be seen from measured rotation periods in young clusters (e.g. Irwin et al. 2008, 2009; Hartman et al. 2010), which can be traced back to the early stages of their Pre-Main-Sequence lifetimes. As these stars age and spin-down, at most stellar masses, the rotation rates quickly converge at a rate that is strongly mass dependant. For solar mass stars, the rotation rates have almost completely converged within the first Gyr (Bouvier et al. 1997; Meibom et al. 2011; Gallet and Bouvier 2013).

The rotation rate of a star is closely related to the strength of the star's magnetic field with quickly rotating stars possessing significantly stronger magnetic fields than slowly rotating stars. This link was pointed out by Skumanich (1972), who showed that stellar Ca II H and K emission, which is known from the Sun to be a good tracer of magnetic field strength, decreases in the exact same way with age as rotation. A similar relation between rotation and X-ray luminosity has been observed (Pallavicini et al. 1981; Maggio et al. 1987; Wright et al. 2011). This has been confirmed by direct measurements of the surface averaged magnetic field strength fB (Saar 1996, 2001; Reiners 2012). In a recent study, Vidotto et al. (2014b) investigated how large-scale surface magnetic fields reconstructed via ZDI vary with age, rotation and X-ray emission and confirm the above mentioned relations for non-accreting dwarfs (0.1 to 2 $M_\odot$). As can be seen in the studies cited above, this relation saturates at high rotation rates. At rotation rates above the saturation threshold, magnetic field strengths are independent of rotation and remain at a constant level.

3.4 Low Mass Pre-Main-Sequence Stars

On the Pre-Main-Sequence, because stellar internal structure is a function of both mass and age, the situation is more complex than on the main-sequence and it can be interesting to compare what is found for such stars with the results for main-sequence stars. When a cloud collapses to form a central star, due to conservation of angular momentum, a circumstellar disc of gas and dust is formed around the star. The star itself starts off fully convective, and as it ages, if it has a mass above approximately 0.35 $R_\odot$, it develops a inner radiative zone. The age at which a Pre-Main-Sequence star develops a radiative zone is a strong function of stellar mass, with stars of half a solar mass remaining fully convective until ages of approximately 10 Myrs, and stars of two solar masses developing radiative cores at ages of approximately 0.5 Myr (Gregory et al. 2012). The circumstellar disc typically lasts a few Myr (Fedele et al. 2010) and can extend very close to the star, and in some cases can extend all the way to the stellar surface. As it is currently understood, due to the redistribution of angular momentum via viscosity in the disc, material moves inwards through the disc and accretes onto the star, either directly onto the stellar surface through a boundary layer, or through magnetospheric accretion, a process in which material is channeled by the stellar magnetic field into discreet accretion funnels where it falls at approximately free-fall speeds onto the star. In general, the low-mass Pre-Main-Sequence stars that are still accreting are CTTSs, and those that are not are weak-line T Tauri stars (WTTS).

On the Pre-Main-Sequence, stars contract and spin-up as they age. Naively, we expect that the accretion of high specific angular momentum material onto the star would lead to accreting stars spinning faster than we would otherwise expect. However, it is now firmly established observationally that while these stars still possess discs, their rotation rates remain approximately constant, requiring that they lose large amounts of angular momentum (Edwards et al. 1993; Rebull et al. 2004; Gallet and Bouvier 2013). Although the physical mechanisms responsible for this loss of angular momentum are poorly understood, it is clear that they involve star-disc magnetic interactions. Magnetic activity is a universal property of low-mass Pre-Main-Sequence stars. Measurements of the Zeeman effect on T Tauri stars have consistently shown surface averaged magnetic field strengths of several kG (Basri et al. 1992; Guenther et al. 1999; Johns-Krull 2007; Yang et al. 2008; Yang and Johns-Krull 2011).

Such strong magnetic fields inevitably lead to high levels of X-ray emission (Getman et al. 2005; Güdel et al. 2007). T Tauri stars show X-ray luminosities that are typically in the range of 10^{28}–10^{32} erg s^{-1}, and the emission is dominated by magnetically confined coronal plasma with temperatures typically $\sim$10 MK or higher, as opposed to the solar X-ray luminosity of $\sim 10^{27}$ erg s^{-1} and coronal temperatures typically below 2 MK (Judge et al. 2003).

In addition to the spatially unresolved magnetic field measurements, a large amount of information about the structures and strengths of the large-scale fields of Pre-Main-Sequence stars has become available in recent years. At the time of writing, magnetic maps have been produced for 11 CTTSs (e.g. Donati et al. 2007; Hussain et al. 2009; Donati et al. 2012, 2013) and for several other Pre-Main-Sequence stars (e.g. Dunstone et al. 2008; Marsden et al. 2011). These magnetic imaging studies have shown that CTTSs can possess a range of different large-scale magnetic field strengths and topologies, with large scale field strengths that can be as high as 6 kG (Donati et al. 2012). Most of the stars in the sample possess large-scale magnetic fields that are significantly non-dipolar, though with dipole components that are able to disrupt their discs far from the stellar surface leading to magnetospheric accretion. In general, the more complex magnetic fields are also the weakest (Johnstone et al. 2014). The magnetic maps and 3D coronal field structures for the two CTTSs AA Tau and V2247 Oph are shown in Fig. 3.1.

It appears that in general, the complexity and strength of the magnetic fields are strong functions of the internal stellar structure (Gregory et al. 2012). The general trend appears to be that stars that are fully convective possess simple axisymmetric

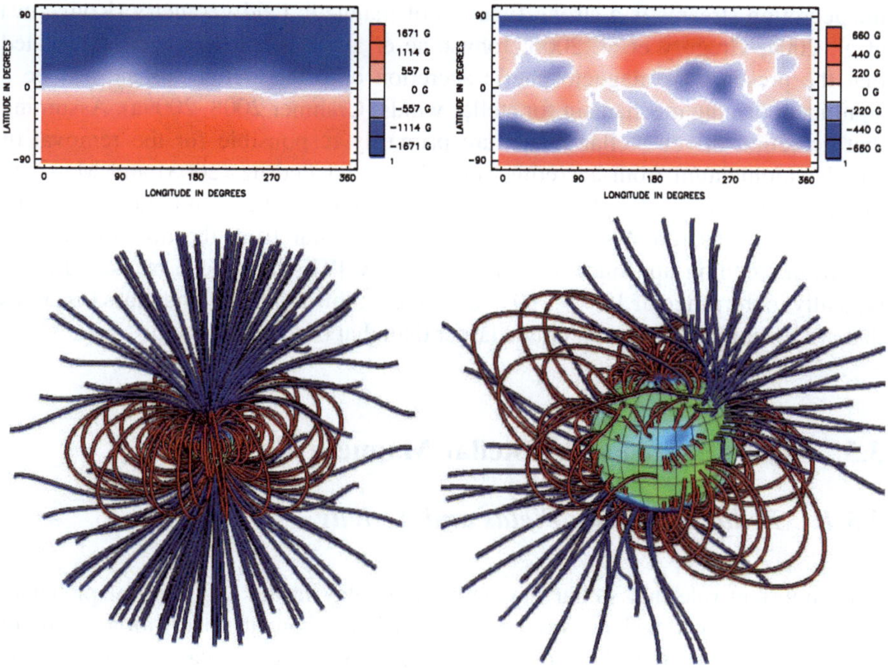

Fig. 3.1 *Upper panels*: Magnetic maps showing the surface distributions of the radial component of the fields of the classical T Tauri stars AA Tau (*left*) and V2247 Oph (*right*) from Donati et al. (2008a) and Donati et al. (2008b), respectively. *Lower panels*: Magnetic field extrapolations showing the coronal 3D field structures for these two stars with closed magnetic field lines (*red*) and open magnetic field lines (*blue*) (After Johnstone et al. 2014)

fields that are dominated by the dipole components, and when these stars develop inner radiative zones, they become more complex and the strength of the dipole component decreases. Initially, when their radiative zones are still small, their fields are complex and axisymmetric, and as their radiative zones grow, the fields become less axisymmetric. The exception is the low-mass V2247 Oph, which, based on the analogy with the M dwarfs discussed in Sect. 3.3.3, might be on the weak field branch of the bistable dynamo (Gregory et al. 2012).

The winds of Pre-Main-Sequence stars are currently poorly understood. As discussed above, these stars possess coronae that are much hotter and several orders of magnitude more luminous in X-rays than the solar corona. Therefore, although the links between coronal properties and wind properties are not well understood, it is reasonable to expect that such stars have winds that are much stronger than the solar wind. The situation is more complicated during the CTTS phase than on the main-sequence because of the presence of a circumstellar disc that is accreting gas onto the star. The radius of the inner edge of the disc is likely to be several stellar radii from the stellar surface and is determined primarily by the strength of the dipole component of the magnetic field (Johnstone et al. 2014). In addition, the shape of the inner edge of the disc and the trajectories of the accretion streams is a strong function of magnetic field geometry (Romanova et al. 2004; Gregory et al. 2006; Johnstone et al. 2014). It has been speculated that the presence of magnetospheric accretion onto the stellar surface can be a significant driving mechanism for stellar winds (Cranmer 2008, 2009a). Assuming that accretion driven stellar winds are primarily responsible for the removal of angular momentum from accreting stars, Matt and Pudritz (2005) and Matt and Pudritz (2008) found that the mass loss rates from the stellar wind must be at least $\sim$10 % of the accretion rate of material onto the star from the disc to stop the star from spinning up. Since accretion rates of CTTSs have been measured to be typically in the range of $10^{-10} - 10^{-7}\,M_{\odot}\,year^{-1}$, this requires wind mass loss rates that are several orders of magnitude larger than that of the solar wind.

3.5 Winds Launched by Stellar Magnetic Fields

3.5.1 Stellar Magnetic Fields and Activity

The important role that stellar magnetic fields play in the evolution of planetary systems is well accepted, but still poorly constrained from an observational point of view. Due to the present-day technology used in exoplanetary searches, most of the currently known exoplanets are found orbiting at extremely close distances to their host stars (<0.1 AU).

At such close distances, the interplanetary medium conditions experienced by the expolanets is likely to be remarkably distinct from the conditions experienced by Solar System planets, and this is very likely to play a significant role in planetary habitability (Vidotto et al. 2013).

Short-period planets orbiting their central star within about 10 stellar radii can cause enhanced activity within the star's upper atmosphere through magnetic and tidal star-planet interaction (Shkolnik et al. 2010). Close in hot-Jupiters (<0.1 AU) might lie within the Alfvén radius of their parent star and thus can magnetically directly interact with the surface of their host star. In addition, the tidal effect of a hot Jupiter can regulate the rotation rate (Pont 2009), and as a consequence the activity level of it's host star. It has been speculated by Cuntz et al. (2000) that star-planet interactions (SPIs) can lead to local instabilities in tidal bulges and, as a consequence, modify the properties of a local dynamo (see also Chap. 9, Guenther and Geier 2014).

Researchers such as Cebron et al. (2011a) and Cebron et al. (2011b) suggest, based on theoretical studies of the effects of tidally driven elliptical instabilities caused by a close-in hot Jupiter on its host-star, that these instabilities could even produce a dynamo.

Magnetic properties of planet host stars based on spectropolarimetric observations have been studied by Fares et al. (2013) who found a wide range of topologies in their sample of stars that cover stellar masses of 0.8–1.4 $M_{\odot}$. They find large-scale

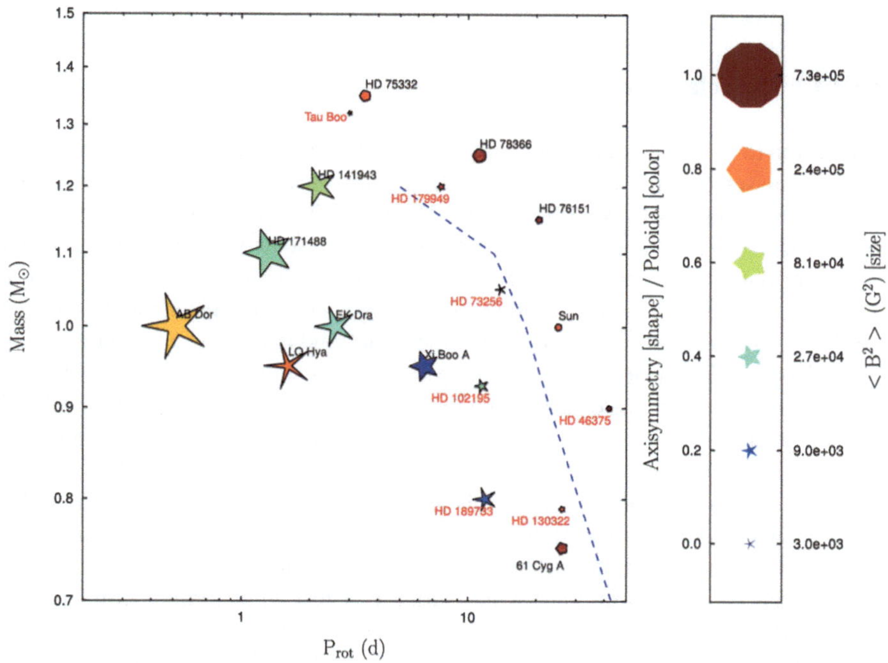

Fig. 3.2 A mass-rotation diagram of 18 reconstructed stellar magnetic fields, figure reproduced from Fares et al. (2013) by permission of Oxford University Press. Planet-hosting stars studied in their paper have their names indicated in *red*, while other stars without detected hot Jupiters have names indicated in *black* (data from Donati and Landstreet 2009). The *dashed line* shows where the Rossby number is 1.0 (calculated, using results of Landin et al. (2010)). Sizes of the symbols represent the field strength, their colour the contribution of the poloidal component to the field, and their shapes how axisymmetric the poloidal components are. For τ Boo, the field for one epoch of observation (mainly poloidal) is shown here. Hot Jupiter host stars do not seem to have different magnetic properties to the other stars

field strengths between 2 and 40 G and (except for two epochs of two of their sample stars), fields mostly dominated by their poloidal component. Comparing them to the field structures of stars without a detected close-in giant planet, they conclude that hot Jupiter host stars do not seem to show a magnetic behavior distinguishing them from comparable stars without a surrounding close-in planet. Figure 3.2, reproduced from their study, illustrates this comparison in a mass-rotation plane.

Current studies show that magnetic field topologies are often considerably more complex than a simple dipole or quadrupole. In addition, the long-term evolution of the star's magnetic fields also needs to be taken into account, as not only the Sun, but also other stars have been observed to show the presence of magnetic field cycles (Donati et al. 2008a; Fares et al. 2009; Morgenthaler et al. 2011). It is therefore essential, when studying the influence of host stars on their surrounding planets, to continue obtaining time-resolved spectropolarimetric observations, applying the ZDI technique, and to use the resulting maps of field configurations as an input for studies on wind modelling and SPI.

3.5.2 *Winds*

3.5.2.1 The Solar Wind

In addition to having a hot magnetically confined corona, as described in Sect. 3.3.1, the Sun possesses a hot ionized wind that streams outwards in all directions. The physical mechanisms responsible for the driving of the solar wind are not yet well understood (for a review, see Cranmer 2009b). The solar wind is in general not isotropic, but can be approximately divided into two distinct components. These are the slow wind and the fast wind and they have speeds of approximately 400 and 800 km s^{-1} respectively. The structure of the solar wind is very closely related to the structure of the Sun's large scale magnetic field. It was shown by Wang and Sheeley (1990a) and Arge and Pizzo (2000) that the speed of the solar wind is closely related to how much the magnetic field expands between the solar surface and the top of the closed corona. For a simple axisymmetric dipole magnetic field, such as the one seen on the Sun at solar minimum, this leads to fast wind coming from the poles and slow wind coming from the equator. For a complex non-axisymmetric magnetic field, such as the one seen on the Sun at solar maximum, this leads to complex distributions of the slow and fast wind, with the fastest wind coming from coronal holes, and the slowest wind coming from above closed field regions. Although the speed of the wind is variable, in general, the mass flux in the solar wind at 1 AU is approximately uniform, with the weak trend that the mass flux in the fast wind is slightly lower on average than the mass flux in the slow wind (Withbroe 1989; Wang 2010). The mass flux over the solar surface, on the other hand, is approximately proportional to the surface magnetic field strength (which is itself strong evidence that the solar wind is driven by the magnetic field) and the uniformity of the mass flux at 1 AU is a result of the clear relation between surface magnetic field strength and magnetic field expansion (Wang 2010). The example of the solar wind is a clear indication that a good understanding of the properties of stellar winds must be based on a good understanding of the strengths and geometries of stellar magnetic fields.

3.5.2.2 Winds of Stars

To constrain interactions between exoplanets and their host-star's magnetized winds and to characterize the interplanetary medium that surrounds exoplanets, it is important to adopt more realistic stellar wind models, which account for factors such as stellar rotation and the complex stellar magnetic field configurations of cool stars.

As previously discussed, the stellar magnetism of cool, low-mass stars can be significantly different from the solar one, both in topology and intensity. Thanks to ZDI reconstructions of stellar large-scale magnetic fields (Sect. 3.2.2), more realistic models of magnetized stellar winds, which incorporate observationally reconstructed magnetic maps into numerical simulations, have become available (Vidotto et al. 2011c, 2012, 2014c; Jardine et al. 2013; Llama et al. 2013).

In these models, the radial component of the magnetic field B_r reconstructed from observations is used as boundary condition and held fixed throughout the simulation run. As the simulations evolve in time, both the wind and magnetic field lines are allowed to interact with each other. The resultant solution, obtained self-consistently, is found when the system reaches steady state in the reference frame corotating with the star.

In a study on the influence of changing magnetic field topologies on wind properties of early-dM stars, which primarily show irregular asymmetric Alfvén surfaces, Vidotto et al. (2014c) find that fields, which are more non-axisymmetric, lead to more asymmetric mass fluxes without any preferred colatitudes contributing to the stellar mass loss. The asymmetric nature and the complexity of the magnetic fields likely also lead to a lack of symmetry in the shapes and distances of astropauses and can play an important role in the angular momentum evolution of a star. In addition, if an analogy to the Sun-Earth system holds true (the cosmic ray rate, that impacts the Earth is inversely correlated to the non-axissymmetric component of the total open solar magnetic flux; Wang et al. 2006), the lack of symmetry of a stellar field geometry can likely also affect an orbiting planet. Planets surrounding stars having largely non-axisymmetric fields might be the most shielded planets from galactic cosmic rays.

With the goal to characterize the environment surrounding their close-in planets, recently, the simulations of winds of two planet-hosting stars, namely τ Boo and HD 189733, were performed by Vidotto et al. (2012) and Llama et al. (2013), respectively. These simulations incorporated the observed maps derived from ZDI and are briefly detailed below.

3.5.2.3 τ Boo

One of the currently most outstanding planet hosting stars is τ Boo (spectral type F7V). This star not only hosts a giant planet orbiting very close in (located at 0.046 AU from the star) but also has a large scale magnetic field that has been shown to flip in polarity at the visible pole cyclically (Catala et al. 2007; Donati et al. 2008a; Fares et al. 2009, 2013). This to date makes it the only star other than the Sun for which full magnetic cycles have been reported. These observations suggest that τ Boo undergoes magnetic cycles similar to the Sun, but with a cycle period that is about one order of magnitude smaller than the solar one (about 2 years as opposed to 22 years for the solar magnetic cycle).

Vidotto et al. (2012) performed numerical simulations of the stellar wind of τ Boo. They used the surface magnetic maps reconstructed by Catala et al. (2007), Donati et al. (2008a), and Fares et al. (2009) as boundary conditions for stellar wind simulations. Figure 3.3 shows the self-consistent solution found for the magnetic field lines after the stellar wind solution relaxed in the grid. Note that the magnetic field lines become stressed, wrapping around the rotational axis of the star (pointing towards positive z). Color-coded is the reconstructed large-scale surface magnetic field intensity observed in January 2008 (Fares et al. 2009).

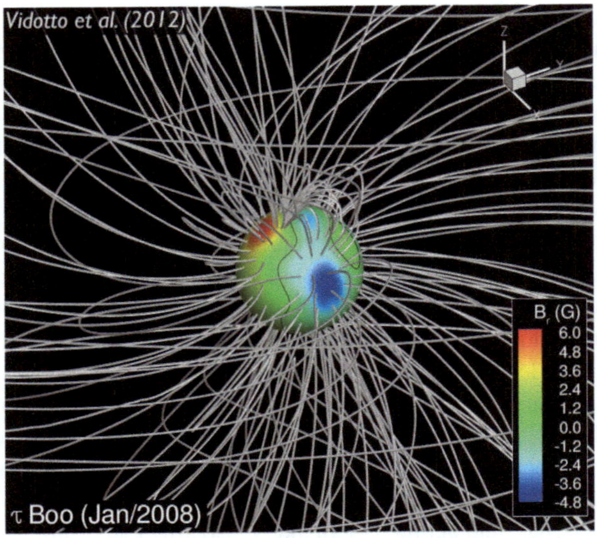

Fig. 3.3 Self-consistent solution found for the magnetic field lines of τ Boo, after the stellar wind solution relaxed in the grid. Color-coded is the reconstructed large-scale surface magnetic field intensity observed in January 2008 (Fares et al. 2009). This field is used as boundary conditions for the simulations. The simulations were published in Vidotto et al. (2012)

It has been found by Vidotto et al. (2012) that variations of the stellar magnetic field during the cycle directly influence the outflowing wind. By analogy with the Solar System, it is believed that supersonic stellar winds impacting on planets are able to form bow shocks that surround the magnetospheres of exoplanets. As the planet orbits through winds that vary both temporally and spatially, the strength and geometry of its bow shock changes. The rapid variation of the large-scale magnetic field of τ Boo implies that the environment surrounding the close-in planet should be varying quite rapidly.

3.5.2.4 HD 189733

Another detailed modelling of the wind of a planet-hosting star was recently done by Llama et al. (2013), who explored the effects the stellar wind has on the conditions surrounding the transiting planet HD 189733b. Using a similar technique as that used for the τ Boo system (see Sect. 3.5.2.3), Llama et al. (2013) investigated both the spatial and temporal variations of the stellar wind by incorporating the reconstructed magnetic maps of HD 189733 derived by Fares et al. (2010) at two separate epochs. Figure 3.4 shows the final configuration of the stellar magnetic field derived from the 3D wind simulations at these two different epochs. As can be seen, the stellar magnetic field configuration varied considerably from one epoch to the other. As a consequence, the stellar wind simulated at these epochs also showed some differences (Llama et al. 2013).

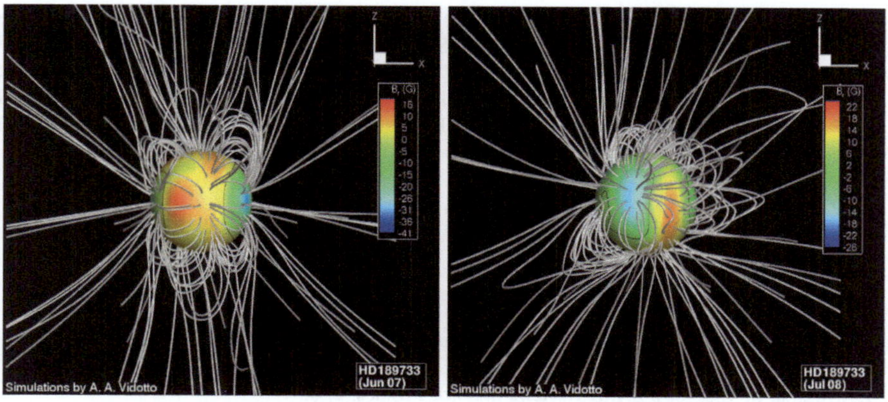

Fig. 3.4 Same as Fig. 3.3, but for HD 189733. The simulations were presented in Llama et al. (2013)

The results of a 3D stellar wind model has been used by Llama et al. (2013), firstly to determine the local stellar wind conditions throughout the orbital path of HD 189733b, and secondly to predict the geometry and density of the bow shock that forms around the magnetosphere of the planet. For that, they assumed the planet to have a magnetosphere similar to that of Jupiter. Assuming this bow shock is able to absorb stellar light in the near-UV, similar to suggestions developed for the hot-Jupiter WASP-12b (Vidotto et al. 2010, 2011a,b; Llama et al. 2011), simulated transit light curves in the optical and in the NUV have been performed by Llama et al. (2013). Depending on the nature of the stellar magnetic field, and hence its wind, Llama et al. (2013) found that both the transit duration and ingress time can vary when compared to optical light curves and that even consecutive near-UV transit light curves may vary significantly.

Conclusion

Tremendous progress has been made during the past years in the analysis of the magnetic field topologies of stars. Ongoing observational efforts, deriving magnetic field properties via Zeeman Doppler Imaging and advanced modelling efforts now allow us to properly take into account the interactions between the stellar magnetic field and the stellar wind plasma, and to realistically model the propagation of the wind from the stellar surface to the orbits of habitable planets. Thus, we now have finally reached a stage, where we can realistically assess the role that stellar magnetic fields and the related activity phenomena play in the habitability of exoplanetary environments – one of the most challenging and most exciting research areas of present astrophysics.

Acknowledgements The authors acknowledge the support by the International Space Science Institute (ISSI) in Bern, Switzerland and the ISSI team *Characterizing stellar- and exoplanetary environments*. T. Lüftinger and C. P. Johnstone acknowledge support by the FWF NFN project S11601-N16 Pathways to Habitability: From Disks to Active Stars, Planets and Life, and the related FWF NFN subproject S116 604-N16; AAV acknowledges support from a Royal Astronomical Society Fellowship.

References

Arge, C. N., & Pizzo, V. J. (2000). *Journal of Geophysical Research, 105*(A5), 10465–10480.

Basri, G., Marcy, G.W., & Valenti, J. A. (1992). *Astrophysical Journal, 390*, 622.

Berdyugina, S. V. (2005). *Living Reviews in Solar Physics, 2*, 8.

Bouvier, J., Forestini, M., & Allain, S. (1997). *Astronomy and Astrophysics, 326*, 1023.

Catala, C., Donati, J. F., Shkolnik, E., Bohlender, & D., Alecian, E. (2007). *Monthly Notices of the Royal Astronomical Society, 374*, L42.

Cebron, D., Le Bars, M., Moutou, C., Maubert, P., & Le Gal, P. (2011a). *EPSC-DPS Joint Meeting 2011*, Nantes, France (p. 1080).

Cebron, D., Moutou, C., Le Bars, M., Le Gal, P., & Fares, R. (2011b). *The European Physical Journal Web of Conferences, Observatoire de Haute-Provence, France 11*, 3003.

Charbonneau, P. (2010). *Living Reviews in Solar Physics, 7*, 3.

Cranmer, S. R. (2008). *Astrophysical Journal, 689*, 316.

Cranmer, S. R. (2009a). *Astrophysical Journal, 706*, 824.

Cranmer, S. R. (2009b). *Living Reviews in Solar Physics, 6*, 3.

Cuntz M., Saar S. H., & Musielak Z. E. (2000). *Astrophysical Journal, 533*, L151.

Dobler, W. (2005). *Astronomische Nachrichten, 326*, 254.

Donati, J.-F. (2001). *LNP, 573*, 207.

Donati, J.-F., Howarth, I. D., Jardine, M. M., Petit, P., Catala, C., Landstreet, J. D., Bouret, J.-C., Alecian, E., Barnes, J. R., Forveille, T., Paletou, F., & Manset, N. (2006). *Monthly Notices of the Royal Astronomical Society, 370*, 629.

Donati, J.-F., Jardine, M. M., Gregory, S. G., Petit, P., Bouvier, J., Dougados, C., M[́e]nard, F., Cameron, A. C, Harries, T. J., Jeffers, S. V., & Paletou, F. (2007). *Monthly Notices of the Royal Astronomical Society, 380*, 1297.

Donati, J. F., Moutou, C., Farès, R., Bohlender, D., Catala, C., Deleuil, M., Shkolnik, E., Collier Cameron, A., Jardine, M. M., & Walker, G. A. H. (2008a). *Monthly Notices of the Royal Astronomical Society, 385*, 1179.

Donati, J.-F., Morin, J., Petit, P., Delfosse, X., Forveille, T., Aurirè, M., Cabanac, R., Dintrans, B., Fares, R., Gastine, T., Jardine, M. M., Lignières, F., Paletou, F., Ramirez Velez, J. C., & Thèado, S. (2008b), *Monthly Notices of the Royal Astronomical Society, 390*, 545.

Donati, J.-F., & Landstreet, J. D. (2009). *ARA & A, 47*, 333.

Donati, J.-F., Gregory, S. G., Alencar, S. H. P., Hussain, G., Bouvier, J., Dougados, C., Jardine, M. M., M[́e]nard, F., & Romanova, M. M. (2012). *Monthly Notices of the Royal Astronomical Society, 425*, 2948.

Donati, J.-F., Gregory, S. G., Alencar, S. H. P., Hussain, G., Bouvier, J., Jardine, M. M., M[́e]nard, F., Dougados, C., & Romanova, M. M. (2013). *Monthly Notices of the Royal Astronomical Society, 436*, 881.

Dunstone, N. J., Hussain, G. A. J., Collier Cameron, A., Marsden, S. C., Jardine, M., Stempels, H. C., Ramirez Velez, J. C., & Donati, J.-F. (2008). *Monthly Notices of the Royal Astronomical Society, 387*, 481.

Edwards, S., Strom, S. E., Hartigan, P., Strom, K. M., Hillenbrand, L. A., Herbst,W., Attridge, J., Merrill, K. M., Probst, R., & Gatley, I. (1993). *Astrophysical Journal, 106*, 372.

Fares, R., Donati, J., Moutou, C., Bohlender, D., Catala, C., Deleuil, M., Shkolnik, E., Cameron, A. C., Jardine, M. M., & Walker, G. A. H. (2009). *Monthly Notices of the Royal Astronomical Society, 398*, 1383.
Fares, R., Donati, J., Moutou, C., Jardine, M. M., Grießmeier, J., Zarka, P., Shkolnik, E. L., Bohlender, D., Catala, C., & Cameron, A. C. (2010). *Monthly Notices of the Royal Astronomical Society, 406*, 409.
Fares, R., Moutou, C., Donati, J. F., Catala, C., Shkolnik, E. L., Jardine, M. M., Cameron, A. C., & Deleuil, M. (2013). *Monthly Notices of the Royal Astronomical Society, 435*, 1451.
Fedele, D., van den Ancker, M. E., Henning, Th., Jayawardhana, R., & Oliveira, J. M. (2010). *Astronomy and Astrophysics, 510*, 72.
Fossati, L., Haswell, C. A., Linsky, J. L., Kislyakova, K. G. (2014). H. Lammer & M. L. Khodachenko (Eds.), *Characterizing stellar and exoplanetary environments* (pp. 59). Heidelberg, New York: Springer.
Gallet, F., & Bouvier, J. (2013). *Astronomy and Astrophysics, 556*, 36.
Getman, K. V., Flaccomio, E., Broos, P. S., Grosso, N., Tsujimoto, M., Townsley, L., Garmire, G. P., Kastner, J., Li, J., Harnden, F. R., Jr.,Wolk, S., Murray, S. S., Lada, C. J., Muench, A. A., McCaughrean, M. J., Meeus, G., Damiani, F., Micela, G., Sciortino, S., Bally, J., Hillenbrand, L. A., Herbst, W., Preibisch, T., & Feigelson, E. D. (2005). *Astrophysical Journal, 160*, 319.
Guenther, E. W., Lehmann, H., Emerson, J. P., & Staude, J. (1999). *Astronomy and Astrophysics, 341*, 768.
Guenther, E., & Geier, S. (2014). H. Lammer & M. L. Khodachenko (Eds.), *Characterizing stellar and exoplanetary environments* (pp. 169). Heidelberg, New York: Springer.
Gregory, S. G., Jardine, M., Simpson, I., & Donati, J.-F. (2006). *Monthly Notices of the Royal Astronomical Society, 371*, 999.
Gregory, S. G., Donati, J.-F., Morin, J., Hussain, G. A. J., Mayne, N. J., Hillenbrand, L. A., Jardine, M., & CanWe (2012). *Astrophysical Journal, 755*, 97.
Grießmeier, J.-M. (2014). H. Lammer & M. L. Khodachenko (Eds.), *Characterizing stellar and exoplanetary environments* (pp. 213). Heidelberg, New York: Springer,
Hale, G. E. (1908). *Astrophysical Journal, 28*, 100.
Hartman, J. D., Bakos, G., Kovécs, G., & Noyes, R. W. (2010). *Monthly Notices of the Royal Astronomical Society, 408*, 475.
Hussain, G. A. J., Cameron, A. C., Jardine, M. M., Dunstone, N., Velez, J. R., Stempels, H. C., Donati, J.-F., Semel, M., Aulanier, G., Harries, T., Bouvier, J., Dougados, C., Ferreira, J., Carter, B. D., & Lawson, W. A. (2009). *Monthly Notices of the Royal Astronomical Society, 398*, 189.
Irwin, J., Hodgkin, S., Aigrain, S., Bouvier, J., Hebb, L., & Moraux, E. (2008). *Monthly Notices of the Royal Astronomical Society, 383*, 1588.
Irwin, J., Aigrain, S., Bouvier, J., Hebb, L., Hodgkin, S., Irwin, M., & Moraux, E. (2009). *Monthly Notices of the Royal Astronomical Society, 392*, 1456.
Jardine, M., Vidotto, A. A., van Ballegooijen, A., Donati, J. F., Morin, J., Fares, R., & Gombosi, T. I. (2013). *Monthly Notices of the Royal Astronomical Society, 431*, 528.
Johns-Krull, C. M. (2007). *Astrophysical Journal, 664*, 975.
Johnstone, C. P., Jardine, M., Gregory, S. G., Donati, J.-F., & Hussain, G. (2014). *Monthly Notices of the Royal Astronomical Society, 437*, 3202.
Judge, P. G., Solomon, S. C., & Ayres, T. R. (2003). *Astrophysical Journal, 593*, 534.
Kislyakova, K. G., Holmström, M., Lammer, H., Erkaev, N. V. (2014). H. Lammer & M. L. Khodachenko (Eds.), *Characterizing stellar and exoplanetary environments* (pp. 137). Heidelberg, New York: Springer.
Kochukhov, O., & Piskunov, N. (2002). *Astronomy and Astrophysics, 388*, 868.
Kochukhov, O. & Piskunov, N. 2009, *In Solar Polarization 5* (San Francisco: ASP), 539.
Kochukhov, O., Bagnulo, S., Wade, G. A., Sangalli, L., Piskunov, N., Landstreet, J. D., Petit, P., & Sigut, T. A. A. (2004). *Astronomy and Astrophysics, 414*, 613.

Landi de Innocenti, E., & Landolfi, M. (2004). *Polarisation in spectral lines*. Dordrecht/ Boston/London: Kluwer.
Landin, N. R., Mendes, L. T. S., & Vaz, L. P. R. (2010). *Astronomy and Astrophysics, 510*, id.A46.
Lang, P., Jardine, M., Morin, J., Donati, J.-F., Jeffers, S., Vidotto, A. A., & Fares, R. (2014). *Monthly Notices of the Royal Astronomical Society, 439*, 2122–2131.
Llama, J., Wood, K., Jardine, M., Vidotto, A. A., Helling, C., Fossati, L., & Haswell, C. A. (2011). *Monthly Notices of the Royal Astronomical Society, 416*, L41.
Llama, J., Vidotto, A. A., Jardine, M., Wood, K., Fares, R., & Gombosi, T. I. (2013). *Monthly Notices of the Royal Astronomical Society, 436*, 2179.
Lüftinger, T., Fröhlich, H.-E., Weiss, W., Petit, P., Aurière, M., Nesvacil, N., Gruberbauer, M., Shulyak, D., Alecian, E., Baglin, A., Baudin, F., Catala, C., Donati, J.-F., Kochukhov, O., Michel, E., Piskunov, N., Roudier, T., & Samadi, R. (2010a). *Astronomy and Astrophysics, 509A*, A43.
Lüftinger, T., Kochukhov, O., Ryabchikova, T., Piskunov, N., Weiss, W. W., & Ilyin, I. (2010b). *Astronomy and Astrophysics, 509A*, 71.
Maggio, A., Sciortino, S., Vaiana, G. S., Majer, P., Bookbinder, J., Golub, L., Harnden, F. R., Jr., & Rosner, R. (1987). *Astrophysical Journal, 315*, 687.
Marsden, S. C., Jardine, M. M., Ramìrez Vélez, J. C., Alecian, E., Brown, C. J., Carter, B. D., Donati, J.-F., Dunstone, N., Hart, R., Semel, M., & Waite, I. A. (2011). *Monthly Notices of the Royal Astronomical Society, 413*, 1922.
Matt, S., & Pudritz, R. E. (2005). *Astrophysical Journal, 632*, 135.
Matt, S., & Pudritz, R. E. (2008). *Astrophysical Journal, 681*, 391.
Meibom, S., Barnes, S. A., Latham, D. W., Batalha, N., Borucki, W. J., Koch, D. G., Basri, G., Walkowicz, L. M., Janes, K. A., Jenkins, J., Van Cleve, J., Haas, M. R., Bryson, S. T., Dupree, A. K., Furesz, G., Szentgyorgyi, A. H., Buchhave, L. A., Clarke, B. D., Twicken, J. D., & Quintana, E. V. (2011). *Astrophysical Journal, 733*, L9, 5pp.
Morgenthaler A., Petit P., Morin J., Aurière M., Dintrans B., Konstantinova-Antova R., & Marsden S. (2011). *Astronomische Nachrichten, 332*, 866.
Morin, J., Donati, J.-F., Forveille, T., Delfosse, X., Dobler, W., Petit, P., Jardine, M. M., Collier Cameron, A., Albert, L., Manset, N., Dintrans, B., Chabrier, G., & Valenti, J. A. (2008). *Monthly Notices of the Royal Astronomical Society, 384*, 77.
Morin, J., Donati, J.-F., Petit, P., Delfosse, X., Forveille, T., & Jardine, M. M. (2010). *Monthly Notices of the Royal Astronomical Society, 407*, 2269.
Morin, J., Dormy, E., Schrinner, M., & Donati, J.-F. (2011). *Monthly Notices of the Royal Astronomical Society, 418*, 133.
Pallavicini, R., Golub, L., Rosner, R., Vaiana, G. S., Ayres, T., & Linsky, J. L. (1981). *Astrophysical Journal, 248*, 279.
Parker, E. N. (1958). *Astrophysical Journal, 128*, 664.
Piskunov, N., & Kochukhov, O (2002). *Astronomy and Astrophysics, 381*, 736.
Pont, F. (2009). *Monthly Notices of the Royal Astronomical Society, 396*, 1789.
Rebull, L. M., Wolff, S. C., & Strom, S. E. (2004). *The Astronomical Journal, 127*, 1029.
Reiners A., & Basri G. (2009). *Astronomy and Astrophysics, 496*, 787.
Reiners, A. (2012). *Living Reviews in Solar Physics, 9*, 1.
Ribas, I., Guinan, E. F., Nudel, G., & Audard, M. (2005). *Astrophysical Journal, 622*, 680.
Romanova, M. M., Ustyugova, G. V., Koldoba, A. V., & Lovelace, R. V. E. (2004). *Astrophysical Journal, 610*, 920.
Rosén, L., & Kochukhov, O. (2012). *Astronomy and Astrophysics, 548*, A8, 12pp.
Saar, S. H. (1996). *International Astronomical Union Symposium, 176*, 237
Saar, S. H. (2001). *American Shetland Pony Club, 223*, 292.
Shkolnik, E., Bohlender, D. A., Walker, G. A. H., & Cameron, A. C. (2008). *Astrophysical Journal, 676*, 628.
Shkolnik, E., et al. (2010). EGU General Assembly 2010, held 2-7 May 2010 in Vienna, Austria, p. 13591
Shurcliff, W. A. (1962). *Journal of the Royal Astronomical Society of Canada, 56*, 269.

Skumanich, A. (1972). *Astrophysical Journal, 171*, 565.
Solanki, S. K. (1993). *Space Science Reviews, 63*, 1.
Solanki, S. K. (2009). *American Shetland Pony Club, 405*, 135.
Stenflo, J. O. (1973). *Solar Physics, 32*, 41.
Stenflo, J. O. (2011). *Astronomy and Astrophysics, 529*, 42.
Vidotto, A. A., Jardine, M., & Helling, C. (2010). *Astrophysical Journal, 722*, L168.
Vidotto, A. A., Jardine, M., & Helling, C. (2011a). *Monthly Notices of the Royal Astronomical Society, 411*, L46.
Vidotto, A. A., Jardine, M., & Helling, C. (2011b). *Monthly Notices of the Royal Astronomical Society, 414*, 1573.
Vidotto, A. A., Jardine, M., Opher, M., Donati, J., F., & Gombosi, T. I. (2011c). *Monthly Notices of the Royal Astronomical Society, 412*, 351.
Vidotto, A. A., Fares, R., Jardine, M., Donati, J. F., Opher, M., Moutou, C., Catala, C., & Gombosi, T. I. (2012). *Monthly Notices of the Royal Astronomical Society, 423*, 3285.
Vidotto, A. A., Jardine, M., Morin, J., Donati, J. F., Lang, P., & Russell, A. J. B. (2013). *Astronomy and Astrophysics, 555*, A67.
Vidotto, A. A., Bisikalo, D. V., Fossati, L., & Llama, J. (2014a). H. Lammer & M. L. Khodachenko (Eds.), *Characterizing stellar and exoplanetary environments* (pp. 153). Heidelberg/New York: Springer.
Vidotto, A. A., Jardine, M., Morin, J., Donati, J. F., Opher, M., & Gombosi, T. I. (2014b). *Monthly Notices of the Royal Astronomical Society, 438*, 1162.
Vidotto, A. A., Gregory, S. G., Jardine, M., Donati, J.-F., Petit, P., Morin, J., Folsom, C. P., Bouvier, J., Cameron, A. C., Hussain, G., Marsden, S., Waite, I. A., Fares, R., Jeffers, S., & do Nascimento, J. D., Jr. (2014c). MNRAS, 441, 2361–2374.
Wang, Y.-M., & Sheeley, N. R., Jr. (1990a). *Astrophysical Journal, 355*, 726.
Wang, Y.-M., & Sheeley, N. R., Jr. (1990b). *Astrophysical Journal, 365*, 372.
Wang, Y.-M., Sheeley, N. R., Jr., & Rouillard, A. P. (2006). *Astrophysical Journal, 644*, 638.
Wang, Y.-M. (2010). *Astrophysical Journal, 715*, 121.
Wade, G. A., Bagnulo, S., Kochukhov, O., Landstreet, J. D., Piskunov, N., & Stift, M. J. (2001). *Astronomy and Astrophysics, 374*, 265.
Weber, E. J., & Davis, L. Jr. (1967). *Astrophysical Journal, 148*, 217.
Withbroe, G. L. (1989). *Astrophysical Journal, 337*, 49.
Wright, N. J., Drake, J. J., Mamajek, E. E., & Henry, G. W. (2011). *Astrophysical Journal, 743*, 48.
Yang, H., Johns-Krull, C. M., & Valenti, J. A. (2008). *Astrophysical Journal, 136*, 2286.
Yang, H., & Johns-Krull, C. M. (2011). *Astrophysical Journal, 729*, 83.

Part II
Exoplanet Upper Atmospheres and Stellar Interaction: Observations and Modelling

Transits of exoplanets offer the opportunity not only to discern the size of a planet but also to study its atmosphere and plasma environment as well, which to date has not been possible for non-transiting planets. The radiation from exoplanet host stars is transmitted through its atmosphere during the transit, while the thermal radiation and the reflected light from the planet disappear and reappear during the secondary eclipse. By measuring the planet-to-star flux ratio as a function of wavelength, spectra of the planetary atmosphere and in some cases also the plasma environment around an exoplanet are obtained. The following sections discuss the latest knowledge of exoplanet atmosphere and host star interaction based on observations and various modeling efforts.

Chapter 4
Observations of Exoplanet Atmospheres and Surrounding Environments

Luca Fossati, Carole A. Haswell, Jeffrey L. Linsky, and Kristina G. Kislyakova

Abstract The study of exoplanets is arguably the most exciting and fastest-growing field in astrophysics. Given the youth of exoplanet science, the field is strongly driven by observations. Here we summarise current knowledge of the atmospheres and wider environments of the known exoplanets giving particular emphasis on the upper atmospheres and the surrounding environment, rather than on the deeper atmospheric layers.

4.1 Introduction: Exoplanet Atmospheres

In the two decades since the first exoplanet was discovered, many ingenious strategies have been used to measure their properties. Here we summarize the current knowledge of the atmospheres and wider environments of the known exoplanets. In fitting with the remainder of this book, our emphasis is on the upper atmospheres and the surrounding environment, rather than on the deeper atmospheric layers.

L. Fossati (✉)
Argelander-Institut für Astronomie der Universität Bonn, Auf dem Hügel 71, 53121, Bonn, Germany
e-mail: lfossati@astro.uni-bonn.de

C.A. Haswell
Department of Physical Sciences, The Open University, Walton Hall, Milton Keynes MK7 6AA, UK
e-mail: c.a.Haswell@open.ac.uk

J.L. Linsky
JILA, University of Colorado and NIST, 440UCB Boulder, CO 80309-0440, USA
e-mail: jlinsky@jilau1.colorado.edu

K.G. Kislyakova
Austrian Academy of Sciences, Space Research Institute, Schmiedlstrasse 6, A-8042 Graz, Austria
e-mail: kristina.kislyakova@oeaw.ac.at

H. Lammer, M. Khodachenko (eds.), *Characterizing Stellar and Exoplanetary Environments*, Astrophysics and Space Science Library 411,
DOI 10.1007/978-3-319-09749-7_4

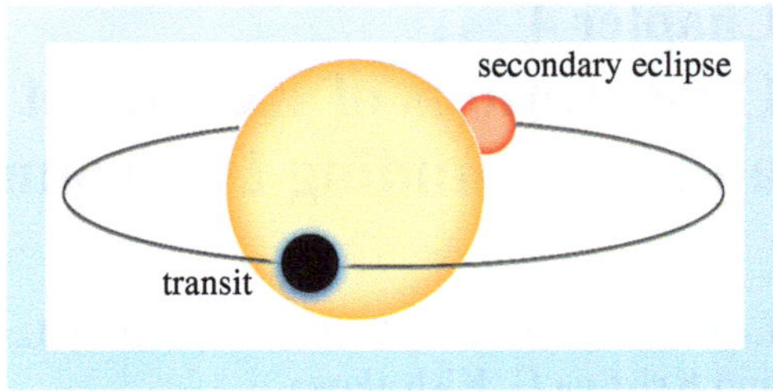

Fig. 4.1 A transit occurs when a planet crosses the face of its host star. During secondary eclipse the star occults the planet

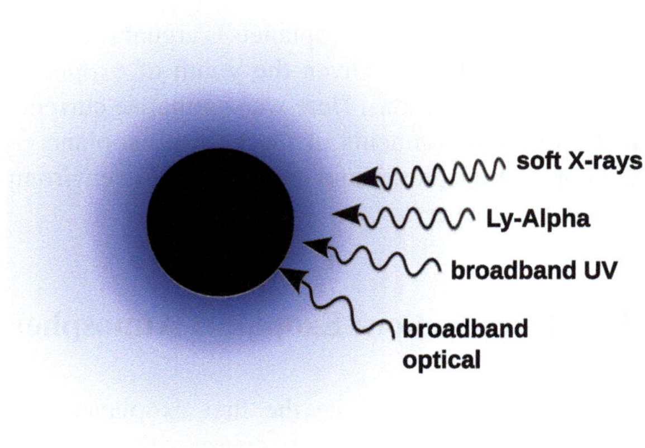

Fig. 4.2 Schematic representation of the absorption depths in a hot Jupiter atmosphere (After Poppenhaeger et al. 2013)

Figure 4.1 illustrates the opportunities to examine an exoplanet's atmosphere through transmission spectroscopy during transit, and to use a subtraction technique to determine the emitted spectrum of the planet during secondary eclipse. Since the first transiting planet was discovered (Charbonneau et al. 2000), observations of exoplanets' atmospheres have been attempted, with varying success, from X-ray (Pillitteri et al. 2010) to radio (Lecavelier des Etangs et al. 2013) wavelengths.

Exoplanets have been extensively studied using ground-based narrow- and broad-band transit photometry, primarily to precisely measure planets' radii and to roughly characterize the atmospheric structure and composition (Ciceri et al. 2013; Mancini et al. 2013a,b,c; Nikolov et al. 2013b).

As Fig. 4.2 shows, generally the scale-height at which the atmosphere becomes opaque increases with increasing photon energy. This basic fact underpinned the spectacular seminal paper (Vidal-Madjar et al. 2003) reporting an extended exosphere surrounding the hot Jupiter HD 209458 b which obscured ~15 % of the stellar Lyman-α flux during the planet's transit (Vidal-Madjar et al. 2003); note the bulk of the planet occults only 1.5 % of the stellar flux. The hot Jupiters transiting the bright host stars HD 209458 and HD 189733 dominated the transmission spectroscopy work until about 2010. Transmission spectroscopy offers the opportunity to identify strongly absorbing constituents of exoplanetary atmospheres, provided that they are present at a scale-height large enough to cause a measurable enhancement to the transit depth. While the scale-height is favorable for X-rays, low stellar photon count-rates and intrinsic variability in stellar X-ray emission are unfavorable. For measurement of the emitted light from exoplanets, a favorable contrast ratio rather than brightness of the host star has a dominating effect on detectability.

4.2 The Deepest Observed Layers of Hot Jupiter Atmospheres

4.2.1 Heat Transport in the Thermosphere

The Spitzer Space Telescope has made important observations of hot Jupiter thermospheres. One highlight is the detection of day-night temperature gradients (Knutson et al. 2007, 2009b, 2011, 2012). These observations provide tests of global circulation models for exoplanets, in particular for the tidally-locked hot Jupiters which differ significantly from the Solar System giant planets in their atmospheric dynamics. For a more thorough description of observations at infrared wavelengths see the review by Seager and Deming (2010).

4.2.2 The Dayside Emitted Spectrum

The infrared region of the spectrum provides the most favorable contrast ratio, and detections of the emitted light from exoplanets have been made using the Spitzer Space Telescope and from the ground, notably with the Very Large Telescope (VLT). These observations have explored planets' albedos, energy budgets and tidal histories, as well as detected features in the infrared dayside emission spectra which are attributed to molecular species like CO and water e.g., de Kok et al. (2013), Birkby et al. (2013), and Brogi et al. (2012, 2013), though the modelling of planets' emission spectra is complicated by the large parameter space making the fits of the models to the observations degenerate.

One of the most important discoveries was the detection of a temperature inversion (i.e., the temperature rises with height above the surface) in the atmosphere of some exoplanets (see e.g., Knutson et al. 2008, 2009a; Madhusudhan and Seager 2013). Almost all Solar System planets present a temperature inversion, mostly caused by the absorption of solar UV radiation by molecules such as CH_4 or O_3 (in the case of Earth). This phenomenon was a surprise in the case of hot Jupiters because of the expected absence of the absorbing molecules. The temperature inversions observed in some hot Jupiters are probably driven by absorption of stellar flux in high altitude absorbing layers, with UV irradiation possibly playing an important role (Hubeny et al. 2003; Fortney et al. 2008; Knutson et al. 2010). Possible absorbing molecules include TiO and VO (Hubeny et al. 2003; Fortney et al. 2008; Zahnle et al. 2009), but modelling performed by Spiegel et al. (2009) shows that photochemistry and sulphur compounds might play a more important role.

4.2.3 Clouds, Hazes and Aurorae

Just as in Earth's atmosphere, intervening clouds and hazes can obscure our view of the atmosphere beyond (Brown 2011). High altitude hazes have been observed in the two prototype hot Jupiters, HD 209458 b and HD 189733 b, with Rayleigh scattering attributed to H_2 molecules causing the planet radius to increase with decreasing wavelength due to the λ^{-4} dependence of the scattering cross-section[1] (e.g., Lecavelier des Etangs et al. 2008; Sing et al. 2011a; Jordan et al. 2013; Pont et al. 2013). Hazes have also been suggested in WASP-12 b, WASP-17 b, and WASP-19 b (Mandell et al. 2013). Conversely, in XO-2 b no evidence of high altitude clouds/hazes has been found (Sing et al. 2011b, 2012). Clouds and hazes play an important role in determining the planet's optical transmission spectrum, blocking our view of short-wavelength features at depths below ~1,500 km (Sing et al. 2008a). Aerosols and clouds are also crucial to determining which physical and chemical processes are most important in planetary atmospheres, see for example Nikolov et al. (2013a) for a short review.

The molecular hydrogen responsible for Rayleigh scattering in hot Jupiter atmospheres might also be expected to fluoresce as a result of the intense irradiation from the host star. The fluorescence is driven primarily by the intense Lyman-α emission line. France, et al. (2010) searched for molecular hydrogen emission in far-UV (FUV; 1200–1700 Å) spectra of HD 209458 b taken at quadrature, but they could only establish upper limits to any auroral or dayglow emission.

4.2.4 Alkali Metal Features

As predicted by Brown (2011), the Na I D resonance doublet in the optical spectrum is a prominent feature in the transmission spectra of HD 209458 b and HD 189733 b (Charbonneau et al. 2002; Sing et al. 2008b, 2011a). Sing et al. (2008a) showed that in HD 209458 b the atomic sodium extends up to altitudes of ~ 3500 km but the abundance of atomic sodium declines dramatically with increasing altitude from ~ 1500 km; this is attributed to condensation into sodium sulfide.

[1]Ballester et al. (2007) interpreted the deep transit of HD 209458b that they obtained with HST/STIS in the region across the Balmer jump as absorption caused by thermospheric/exospheric hot hydrogen, while later observations and theoretical works led to the conclusion this detection was due to Rayleigh scattering (Sing et al. 2011a; Christie et al. 2013; Vidal-Madjar et al. 2013).

The past 10 years have seen a dramatic increase in planet atmospheric studies at optical wavelengths, mostly to derive planets' atmospheric temperature-pressure profile and chemical composition (e.g., Deming et al. 2006 and Sing et al. 2008a). Several systems have been studied and the general conclusion is that the optical spectrum of a hot Jupiter atmosphere is usually dominated by alkali metals, in particular sodium and potassium (e.g., Charbonneau et al. 2002; Snellen et al. 2008; Sing et al. 2012; Colón et al. 2012; Zhou and Bayliss 2012).

Recent work has established that there are strong differences among the hot Jupiters observed so far. For example, HD 209458 b presents strong and broad sodium features and a lack of potassium (e.g., Charbonneau et al. 2002; Narita et al. 2005; Sing et al. 2008b; Snellen et al. 2008), while XO-2 b is the only hot Jupiter currently known where both sodium and potassium have been detected (Sing et al. 2011b, 2012).

4.2.5 *Balmer Lines*

Hα was detected in ground-based transmission spectroscopy of HD 189733b, showing the presence of hydrogen in the $n = 2$ state (Jensen et al. 2012). This detection obtained for HD 189733 b led to the conclusion that strong stellar UV irradiation reaches the base of the planet thermosphere and that there is a rather low density upper atmosphere which is not in thermodynamic equilibrium with the lower atmosphere.

4.3 Transmission Spectroscopy of Hot Jupiter Exospheres

4.3.1 *Far-UV Observations*

Vidal-Madjar et al.'s exciting discovery of the 15 % deep transit of HD 209458 b in Lyman-α (Vidal-Madjar et al. 2003; Lecavelier des Etangs et al. 2004) was interpreted as revealing that the planet's upper atmosphere overflows the Roche lobe and is therefore escaping. This so-called hydrodynamic blow-off occurs when the outflow velocity at the base of the exosphere is larger than the planet's escape velocity (Öpik 1963). Further FUV STIS transit observations revealed enhanced transit depths at the position of O I and C II lines suggesting that these species are entrained in the hydrodynamic blow-off (Fig. 4.3). A further analysis of the velocity distribution of the atmospheric absorption signature showed that the material lost by the planet is probably being blown away by the stellar wind, forming a cometary tail behind the planet (Schneiter et al. 2007; Ehrenreich et al. 2008; Bourrier and Lecavelier des Etangs 2013a).

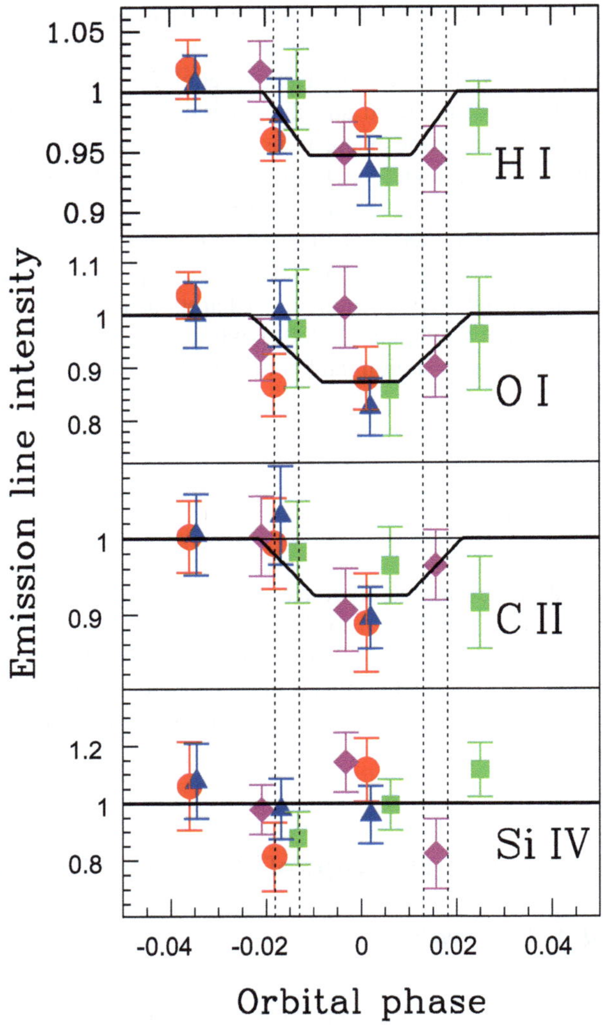

Fig. 4.3 Light curves of H I (i.e., Ly-α), O I, C II, and Si IV FUV spectral lines during transits of HD 209458b. The four observed transits are shown by *circles*, *squares*, *triangles*, and *diamonds*, respectively. The *vertical dotted lines* indicate the position of the first to fourth contact. The *thick line* shows the best fit to the observed transit data. The transit is observed in the H I, O I, and C II lines, while no significant absorption is detected in the Si IV lines (After Vidal-Madjar et al. 2004)

Linsky et al. (2010) obtained HST/COS high-resolution ultraviolet spectra of HD 209458 at times of transit, secondary eclipse, and quadrature of its hot Jupiter planet HD 209458 b. During transit, the emission line profiles of the C II λ 1334, 1335 Å and Si III λ 1206 Å lines show absorption features, but the Si IV λ 1393 Å line shows no absorption (see Fig. 4.4). The combination of the C II and Si III lines shows absorption centered at -10 and $+15\,\mathrm{km\,s^{-1}}$. They argued that the strong hydrodynamic blow-off brings up carbon and silicon from the lower atmosphere

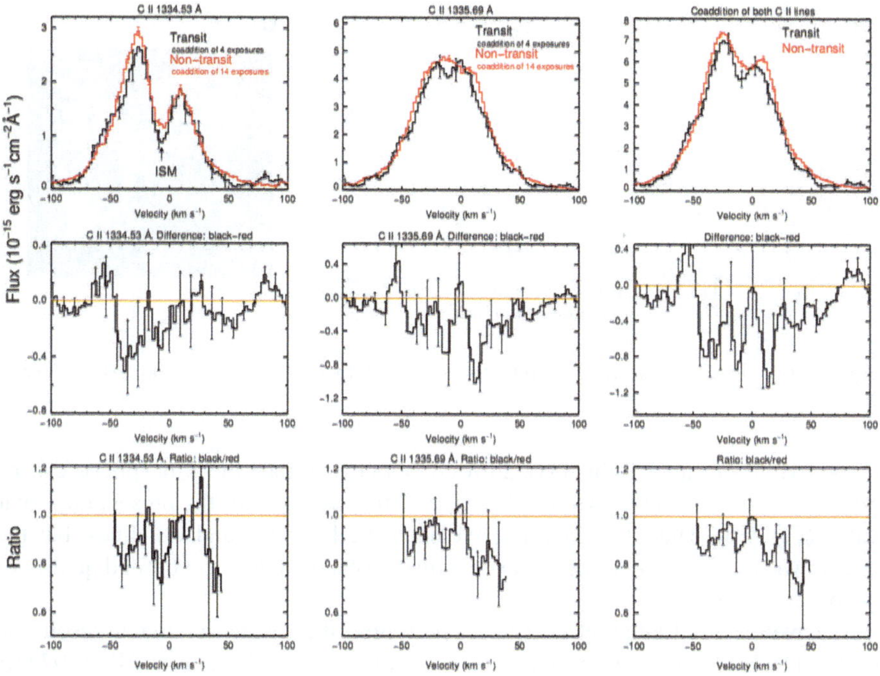

Fig. 4.4 Profiles of the C II λ 1334, 1335 Å lines of HD 209458 during transit (*black*) and out-of-transit (*red*) obtained by Linsky et al. (2010). The coaddition of the C II and the Si III λ 1206 Å lines (not shown in this figure) suggest absorption features at −10 and +15 km s^{-1} that could indicate the wind speed (After Linsky et al. 2010)

that is likely photoionised by the stellar UV and EUV emission. They computed a mass-loss rate in the range (8–40)$\times 10^{10}$ g s^{-1} and suggested that the absorption velocities may indicate the wind escape speed.

The results reported in this subsection all derive from transit light curves measured in stellar emission line flux which originates from the stellar chromosphere or transition region, see Fig. 4.5a. In the Sun we know that this emission is patchy and time-variable, as illustrated in Fig. 4.5b. In the Sun the Lyman-α flux varies by more than a factor of 10 on size-scales smaller than or comparable to Jupiter's disc (Vial et al. 2012). The transit light curve resulting from a locus across this randomly changing emission distribution is likely to be variable and unpredictable (Haswell 2010). This explains some of the disagreements amongst the FUV transit literature on HD 209458 b. The situation becomes even worse in the case of the other intensively-studied object, HD 189733, which is a K1-K2V star, and among the most active known exoplanet host stars.

Lecavelier des Etangs et al. (2010) presented three FUV transit observations of HD 189733 b obtained with the Advanced Camera for Surveys (ACS) on HST. They detected the transit signature in the Ly-α line with a transit depth of 5.05±0.75 %, which exceeds at a 3.5σ level the transit depth due to the planetary disk alone. The fit

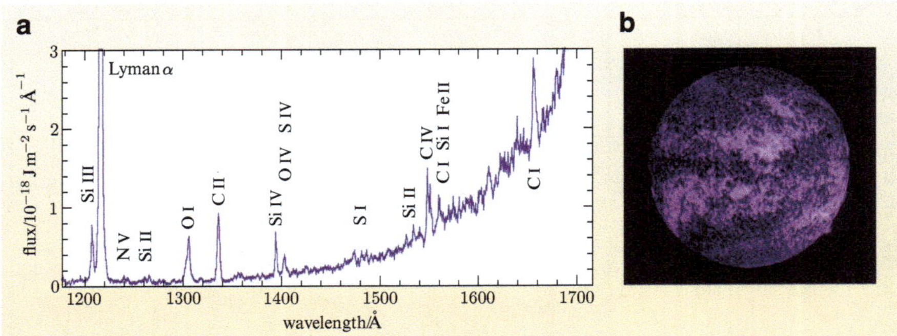

Fig. 4.5 (**a**) The FUV spectrum of HD 209458 b. (**b**) A Ly-α image of the Sun (After Haswell 2010)

to the Lyman-α transit light curve allowed them to estimate the atmospheric escape rate as 10^{10} g s^{-1} and the extreme-UV (EUV) flux as 20 times the present day solar value (Fig. 4.6). The ACS data also included the FUV O I and C II lines, but the transit was not detected at these wavelengths, probably due to the limited quality of the observations.

Lecavelier des Etangs et al. (2012) presented the results of higher resolution FUV STIS/HST observations of two primary transits of the hot Jupiter HD 189733b. The first light curve did not show evidence of a planetary transit, which came as a surprise as the bulk of the planet is indeed opaque at all wavelengths, while the second data-set showed a more regular behavior with a deep transit indicative of an extended/escaping exosphere (Fig. 4.7). This apparently erratic behavior could be attributed to stellar activity (Haswell 2010; Haswell et al. 2012). The host star is a young K-type main sequence star with a relatively strong and variable magnetic field (Fares et al. 2010) as well as surface spots (Sing et al. 2011a). These characteristics of the host star could also explain the large scatter in the unresolved ACS/HST Ly-α observations (Lecavelier des Etangs et al. 2010).

The second STIS/HST data-set of HD 189733 b (Lecavelier des Etangs et al. 2012), showed the presence of exospheric absorption at very high velocities (in the range -230 to -140 km s^{-1}) which cannot be explained by simple atmospheric escape, but requires a further acceleration mechanism such as that provided by the interaction with stellar wind protons (i.e., ENAs; Holmström et al. 2008). A further thorough analysis of the second STIS/HST data-set of HD 189733 b indicated the detection of Si III and N V in the planet exosphere (Bourrier et al. 2013b). Their absence in the first data-set was also attributed to stellar variability. Ben-Jaffel and Ballester (2013) analyzed FUV COS/HST data of the primary transit of HD 189733 b detecting an absorption of $\sim$6.4$\pm$1.8 % by neutral oxygen. Using previously published models of the planet atmosphere and the expected (using solar abundances) integrated O I column density of $\sim 8\times10^{15}$ cm^{-2}, they derived an absorption of only 3.5 %, far smaller than observed. They concluded that the observed excess absorption could be due either to an oxygen overabundance or to the presence of suprathermal broadening of the absorption lines.

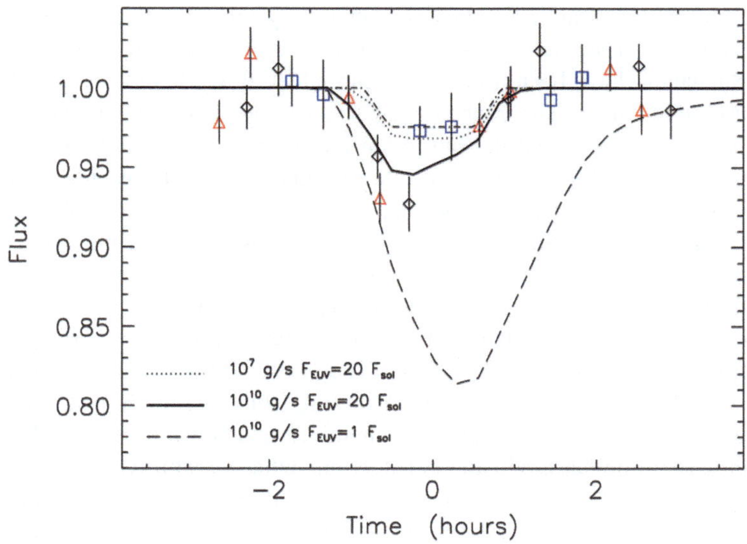

Fig. 4.6 Normalised Lyman-α line flux during the three transits (*diamond*, *triangles*, and *squares*, respectively) observed by Lecavelier des Etangs et al. (2010). The *dot-dashed line* shows the transit of an atmosphere-less planet of the same size as HD 189733b. The *dotted*, *solid*, and *dashed lines* show theoretical light curves of the total Ly-α line flux assuming different escape rates and ionising EUV fluxes, as indicated in the figure (After Lecavelier des Etangs et al. 2010)

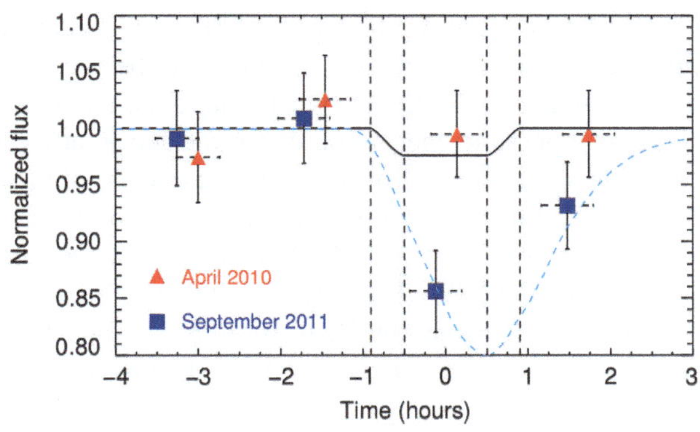

Fig. 4.7 Normalized flux of HD 189733 in the blue wing of the Ly-α line (in the range -230 to $-140\,\mathrm{km\,s^{-1}}$) as a function of the planet's orbital phase. *Red triangles* and *blue squares* show the normalized flux derived from the observations obtained in 2010 and 2011, respectively. The horizontal error bars indicate the duration of the exposure for each observation. The *thick solid line* shows the planet transit as observed in the optical. The *blue dashed line* shows the simulated transit light curve which best fits the 2011 transit observations. The synthetic light curve was computed using the same modelling scheme as in Lecavelier des Etangs et al. (2010), but assuming an ionizing EUV flux of five times the solar value, a stellar wind of protons with a temperature of $\sim 10^5$ K, velocity of $\sim 190\,\mathrm{km\,s^{-1}}$ and density of $3\times10^3\,\mathrm{cm^{-3}}$, together with an atmospheric escape rate of $10^9\,\mathrm{g\,s^{-1}}$. The *vertical dashed lines* indicate the position of the four contact points (After Lecavelier des Etangs et al. 2012)

The early K-type star 55 Cnc hosts a planetary system of at least 5 planets, one of which is the transiting hot super-Earth 55 Cnc e (Winn et al. 2011). STIS/HST and ACIS-S/Chandra have been used by Ehrenreich et al. (2012) to observe the Ly-α and X-ray light curves. The transit of 55 Cnc e was not detected, but instead a Ly-α transit signature was found in phase with the inferior conjunction of the Jupiter-like planet 55 Cnc b. The most plausible explanation is that the bulk of 55 Cnc b passes close to the stellar disc and that part of its extended atmosphere actually transits in front of the star. Nevertheless, as 55 Cnc is a rather active star, as shown by the Chandra data, further simultaneous FUV – X-ray observations are necessary to confirm this detection (Lecavelier des Etangs et al. 2010, 2012).

Kulow et al. (2014) performed Lyman-α transmission spectroscopy of the hot Neptune GJ 436 b using STIS/HST. They observed a strong variation of the Lyman-α lightcurve with a transit depth varying from 8.8 ± 4.5 % near mid-transit, to 22.9 ± 3.9 % about 2 hours after the nominal geometric egress of the planet. Exploiting the time-tag mode and considering stellar variability, they calculated a post-egress occultation of 23.7 ± 4.5 %, showing that the signature is statistically significant. The extended egress absorption is probably the signature of a comet-like tail trailing the exoplanet. They calculated the planet mass-loss rate to lie within the range of 3.7×10^6–1.1×10^9 g s^{-1}, corresponding to an atmospheric lifetime of 4×10^{11}–2×10^{14} years.

4.3.1.1 Interpretation of the Lyman-α Transit Observations

Vidal-Madjar et al. (2003), Vidal-Madjar et al. (2004), and Lecavelier des Etangs et al. (2004) interpreted the observed 15 % deep transit observed for HD 209458 b as evidence of escaping neutral hydrogen, where the radiation pressure of the host star was suggested as a possible energy source for the acceleration. However, this mechanism is not able to explain the absorption in the red part of the Lyman-α line, indicating the presence of neutral hydrogen atoms in motion towards the star. As a consequence, the interpretation of the HD 209458 b Lyman-α transit light curve given by Vidal-Madjar et al. (2003), Vidal-Madjar et al. (2004), and Lecavelier des Etangs et al. (2004) was challenged by Ben-Jaffel (2007), Ben-Jaffel (2008), and Ben-Jaffel and Sona Hosseini (2010), as well as by Holmström et al. (2008) and Ekenbäck et al. (2010).

Ben-Jaffel (2007), Ben-Jaffel (2008), and Ben-Jaffel and Sona Hosseini (2010) argued that the extra absorption compared to the transit in visible light is due to the spectral Doppler broadening effect. These studies considered a symmetrical 1D model atmosphere and obtained a symmetric absorption in both the red and the blue parts of the line. This approach led to a good fit to the data, but disregarded the asymmetry of the atmosphere of HD 209458 b and the related asymmetry in the hydrogen velocity spectrum. This asymmetry is mostly due to radiation pressure and charge-exchange, leading to the formation of a cometary-like tail of escaping atoms pointing in the antistellar direction. It should also be mentioned that the observed absorption depth defined by the broadening strongly depends on the neutral hydrogen column density. While this mechanism can explain the transit observations of HD 209458 b, it would probably fail in the case of HD 189733 b, where the column density of the neutral hydrogen is believed to be insufficient for Doppler broadening to be important (Bourrier and Lecavelier des Etangs 2013a).

In addition to radiation pressure and Doppler broadening, energetic neutral atoms (ENAs) produced by an interaction with the stellar wind were also proposed as a source for the observed extra absorption of HD 209458 b (Holmström et al. 2008). This approach was later developed further by Ekenbäck et al. (2010) as well as independently studied by Tremblin and Chiang (2013). Besides being able to explain the observed Ly-α transit light curve, ENAs allow one to indirectly study the stellar wind environment in the vicinity of an exoplanet and its magnetic field. As it is for broadening, ENAs production depends on the density of neutrals in the exosphere of a planet. However, these studies disregarded the other mechanisms mentioned above (i.e., radiation pressure and Doppler broadening), which are likely to contribute to the absorption.

The most comprehensive study so far is the one by Bourrier and Lecavelier des Etangs (2013a), who considered to some extent all the before mentioned processes. They applied their model to both HD 209458 b and HD 189733 b and estimated the mass loss rates for both exoplanets. However, they still experienced difficulties in reproducing the red part of the Ly-α absorption observed during the transit of HD 209458b.

Summarising all approaches, one can conclude that Ly-α observations can be used to determine properties of exoplanets such as neutral hydrogen column density, escape rate, and in some cases also the stellar plasma environment and the planetary magnetic field. At present, all these parameters can be estimated only indirectly.

4.3.2 Near-UV Observations

As Fig. 4.5a shows, the photospheric continuum flux of solar-type stars rises strongly with increasing wavelength through the UV spectral region. At Lyman-α there is very little continuum flux, while longwards of about 160 nm the stellar spectrum is dominated by the continuum. This is crucially important for transit studies because it means the occulted flux is more smoothly distributed across the stellar disc, hence the fraction of flux lost in transit can be reliably and straightforwardly interpreted as the fraction of the area of the stellar disc which is occulted. The near-UV (NUV; 1700–3600 Å) spectral region combines the advantages of an approximately smooth photospheric stellar light distribution with plentiful resonance lines of common atomic and ionic species; the NUV region contains more spectral features than the FUV, making the analysis slightly more complex, but more informative. Strong resonance lines of abundant species are extremely sensitive probes for the presence of diffuse absorbing gas, and have underpinned our observational characterization of exoplanet exospheres. In particular, the NUV contains resonance lines of numerous metals, including the strong Mg II h&k resonance lines at 2800 Å, which provide important information about the stellar

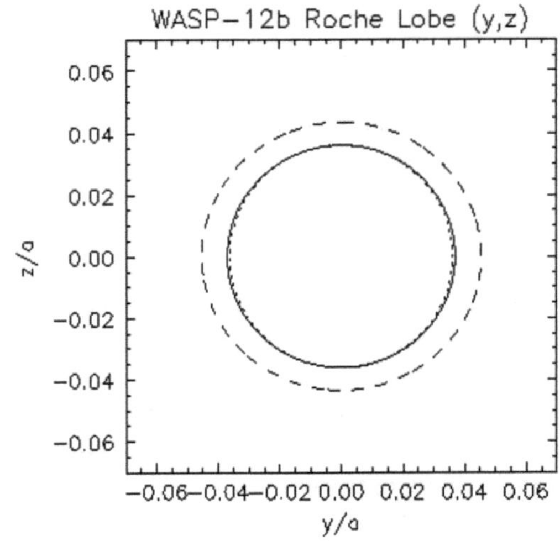

Fig. 4.8 The Roche lobe of WASP-12 b. The $y - z$ cross-section of the planet (*solid line*) fits snugly within the cross-section of the Roche lobe (*dashed line*), while the exospheric absorption indicates an effective area in transit of almost three times that of the opaque planet. The exospheric gas over-fills the Roche lobe (Adapted from Haswell et al. 2012)

photosphere and chromosphere (e.g., Hall 2008). The first NUV transit observations were made using HST/COS in 2009/2010 to observe WASP-12, a late F-type main sequence star younger than 2.65 Gyr (Fossati et al. 2010), which hosts one of the hottest, most bloated known exoplanets, orbiting extremely close to the host star (Hebb et al. 2009). The transit of WASP-12 b was observed with the NUV channel of COS twice, 6 months apart, detecting dips almost three times deeper than in the optical. In the parts of the NUV region with least stellar photospheric absorption, i.e. regions with relatively few strong absorption lines from species common in the stellar photosphere, the transit depth was only slightly enhanced over the optical transit depth. Where there is significant stellar photospheric absorption, the transit was significantly deeper. That is to say, where there are strong metal lines in the spectrum, the transit is more than 4σ deeper than the optical light curve. Figure 4.8 makes it clear that the exospheric radius indicated by these strongly absorbed regions far exceeds the $y - z$ cross-section of the planet's Roche lobe. WASP-12 b has an evaporating exosphere like those of HD 209458 b and HD 189733 b (Fossati et al. 2010; Haswell et al. 2012). Numerous atoms and ions were detected in the WASP-12 b's exosphere, including Fe II, the heaviest species yet detected during an exoplanet transit (Haswell et al. 2012). These detections confirm the hypothesis that heavy species, located in the lower thermosphere, can be dragged upwards beyond the Roche lobe by the hydrodynamically expanding thermosphere.

Vidal-Madjar et al. (2013) reported the results obtained from the analysis of three NUV transit observations of HD 209458b. They obtained a detection at the 2.1σ level ($6.2\pm2.9\,\%$) of neutral magnesium at the position of the Mg I line at $\lambda\ \sim$2853 Å in the velocity range from −62 to $-19\,\mathrm{km\,s^{-1}}$. They did not detect any atmospheric absorption at the position of the Mg II h&k resonance lines: this came as a surprise as Mg I is ionized by the FUV stellar flux. They explained the absence of Mg II absorption during transit by the presence of an efficient recombination, compensating for the photo-ionisation. This was confirmed by the required electron density being in good agreement with the most up-to-date models of the upper atmosphere of this planet (e.g., Koskinen et al. 2013). This was the first time that an already existing independent planet atmosphere model was used to interpret and understand an observational result. A comparison of these results with hydrodynamic models for the escape allowed Vidal-Madjar et al. to also conclude that the planet's magnesium abundance is roughly solar.

4.3.3 Early Ingresses

The NUV transits of WASP-12 b revealed a variable and early ingress, contrary to model expectations, but with egresses compatible with the optical ephemeris (Fossati et al. 2010; Haswell et al. 2012). Ben-Jaffel and Ballester (2013) reported the possible detection of an early-ingress in the HD 189733 b C II line at $\lambda \sim 1335$ Å transit light curve, while this feature was not detected in the O I light curve. Note, however that these same HD 189733 b data were analyzed by Haswell et al. (2012), who concluded that the transit was undetectable, due to flux variability caused by the stellar activity. See Chap. 8 by Vidotto et al. (2014) for a thorough description of the theoretical work done to explain the early ingress phenomenon.

4.3.4 X-Ray Observations of the Transit of HD 189733 b

Using the Advanced CCD Imaging Spectrometer (ACIS) instrument on Chandra, Poppenhaeger et al. (2013) detected the transit of HD 189733b in soft X-rays, deriving a transit depth of $\sim$7 %, larger than that observed in the UV (see Sect. 4.3.1) and considerably larger than the 2.41 % observed in the optical with broadband filters. This result might be compromised by possible coronal inhomogeneities, which are particularly prominent in X-rays, but the deep transit suggests the presence of a thin outer layer in the planet's exosphere that is transparent at optical and UV wavelengths, but dense enough to be opaque in the X-rays (Fig. 4.2). These observations probably reveal the presence of a layer of ionized hydrogen in the planet's exosphere.

4.4 WASP-12: An Enshrouded Planetary System

The WASP-12 b NUV spectrum revealed a remarkable anomaly in the stellar spectrum: a broad depression in place of the normally bright emission cores in the Mg II h&k resonance lines (see Fig. 4.9; Haswell et al. 2012). The line core emission is characteristic of all stars with a chromosphere, such as WASP-12, and it provides a measure of stellar activity with more active stars showing a stronger emission. WASP-12's measured flux in the line cores is zero regardless of the planet's orbital phase. The absence of Mg II core emission was completely unexpected given the

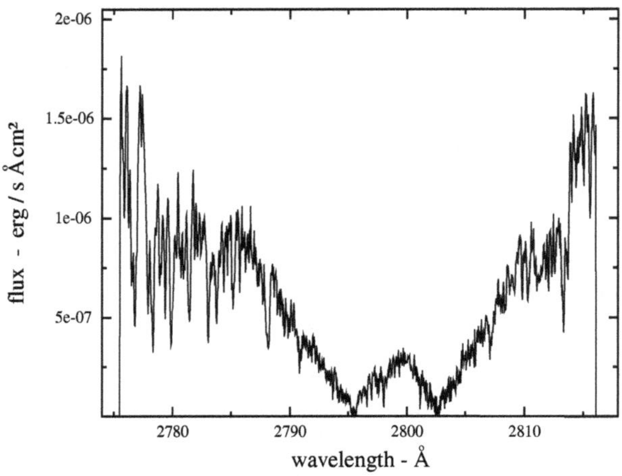

Fig. 4.9 Observed spectrum of WASP-12 obtained by averaging the five COS spectra obtained in 2010

spectral type and the age of the star. Haswell et al. (2012) and Fossati et al. (2013) investigated whether the anomalous line cores could be due to *i*) intrinsically low stellar activity; or *ii*) extrinsic absorption from either the intervening interstellar medium (ISM) or from material within the WASP-12 system itself, presumably ablated from the planet. Comparisons with other distant and inactive stars, similar to WASP-12, revealed that the Ca II H&K line cores also show broad depressions similar to the Mg II h&k line profiles. Direct radio and optical measurements of the ISM absorption along the WASP-12 line-of-sight disclosed that the ISM absorption is insufficient to produce these Ca II and Mg II depressions. Extrinsic absorption by material local to the WASP-12 system is therefore the most likely cause of the line core anomalies: gas escaping from the heavily irradiated planet could form a diffuse nebula enshrouding the entire planetary system. This gas is not noticeable at most wavelengths, but it forms a shroud which is optically thick in the cores of the very strong Mg II h&k line resonance lines, where the observed stellar flux is zero. Haswell et al. (2012) showed that planetary mass loss could plausibly produce the large column density ($\log N_{Mg\,II} = 17.30$) required to attenuate the Mg II cores.

WASP-12 also has an anomalously low stellar activity index ($\log R^{'}_{HK} = -5.50$; Knutson et al. 2010). This is a direct consequence of the extra H and K line core absorption, so similar activity index deficiencies might signal the presence of translucent circumstellar gas around other stars hosting evaporating planets. Such systems would be extreme outliers in a color $(B - V)$ vs. activity ($\log R^{'}_{HK}$) plane (Fig. 4.10). Fossati et al. (2013) identified five other systems (X0-4, CoRoT-1, WASP-13, WASP-17, WASP-18) with an activity index below $\log R^{'}_{HK} = -5.1$, a base flux level which should be a hard lower bound on the chromospheric emission of solar-like main sequence stars (Wright 2004a; Wright et al. 2004b). Any otherwise normal main sequence star, falling below this base level, must be affected by extrinsic absorption which attenuates the intrinsic core emission.

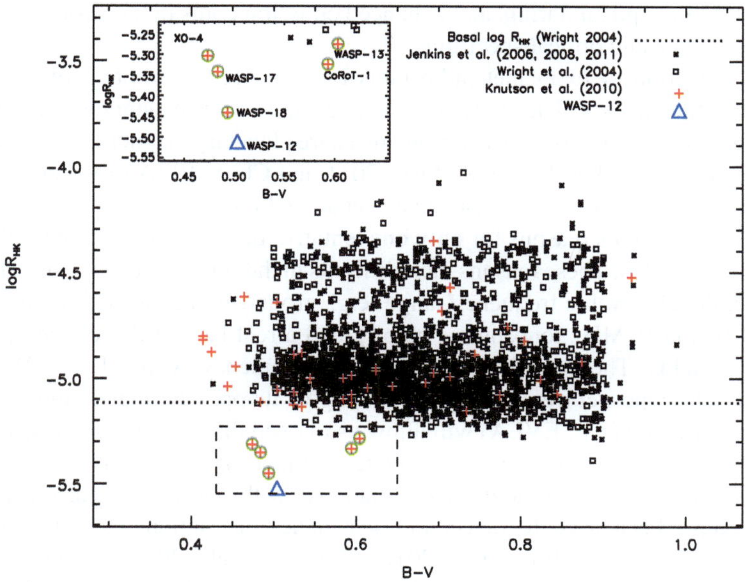

Fig. 4.10 WASP-12 (*blue triangle*) in the $B - V$ vs. $\log R'_{HK}$ plane, compared with stars with $B - V < 1.0$ as observed by Wright et al. (2004b) (*squares*), Jenkins et al. (2006), Jenkins et al. (2008), Jenkins et al. (2011) (*stars*), and Knutson et al. (2010) (*red pluses*). The *circles* indicate the positions of the planet hosting stars X0-4, CoRoT-1, WASP-13, WASP-17, and WASP-18. The *dotted line* indicates the minimum activity value within the (Wright et al. 2004b) sample, accounting for contamination by subgiants (Wright 2004a). The box in the upper left corner is an expanded view of the bottom dashed box region (After Fossati et al. 2013)

This is particularly interesting in the context of the extreme close-in rocky planets recently identified in the Kepler database. For example KIC 1255 (Rappaport et al. 2012) appears to be a low-mass rocky planet in an extreme close-in orbit undergoing catastrophic mass-loss. KIC 1255 was discovered through the transits of a putative dust-cloud surrounding the planet, which is likely to enshroud the planetary system in metal-rich gas and dust. As KIC 1255 appears to be in a short-lived evolutionary stage, close-in rocky planets, that can possibly be the remnant cores of evaporated hot Jupiters, might be plentiful in the Galaxy, and indeed other rocky bodies in orbits of less than 1 day have been identified in the Kepler data (Rappaport et al. 2013; Sanchis-Ojeda et al. 2013; Muirhead et al. 2012).

4.5 Star-Planet Interactions

By analogy with the Jupiter-Io interaction, the extreme proximity of hot Jupiter exoplanets to their host stars might well be expected to stimulate stellar activity: where the planet orbits within the stellar corona, it is hard to see how magnetic interactions cannot occur. Hallmarks of these interactions might include radio,

X-ray, UV and optical flaring, and enhanced emission cores in the strong resonance lines in the UV and optical spectrum.

The detection of exoplanet radio emission has been attempted several times, but with little success. The study of exoplanets at radio wavelengths is of extreme importance as it allows one to detect and measure planetary magnetic fields (see e.g., Grießmeier et al. 2008; Hess and Zarka 2011 and Chap. 10 (Alexeev et al. 2014), 11 (Grießmeier 2014), and 12 (Belenkaya et al. 2014)).

There are various reasons for only one tentative detection so far (Lecavelier des Etangs et al. 2013). These reasons include limits in the sensitivity of the instruments, limits in the observable frequency range (e.g., ground-based radio observations are limited by the 10 MHz ionospheric cutoff), and a real lack of detectable planetary magnetic fields. The capabilities of the Low Frequency Array (LOFAR) system of radio telescopes, currently at the beginning of its operation, and other facilities being constructed (e.g., EVLA) will allow us to overcome some of the instrumental limitations, both in terms of covered frequency range and sensitivity.

X-ray observations, sometimes simultaneous with observations at other wavelengths have been used to detect the star-planet interaction phenomenon (e.g., Kashyap et al. 2008; Miller et al. 2012). Pillitteri et al. (2010) observed the hot Jupiter HD 189733 b with the XMM-Newton telescope and observed a softening of the X-rays during secondary eclipse at the level of 3σ, as well as a very intense flare. Magnetohydrodynamic (MHD) simulations suggested that the interaction between the stellar and planetary magnetic fields might magnify the intensity of stellar flares, such as that detected during the secondary eclipse of HD 189733b. Haswell et al. (2012) showed that the NUV light curves of WASP-12 were consistent with a stellar flare occurring at ingress in the second HST visit, although direct emission in the resonance line cores was quenched by diffuse gas absorption in the line of sight to the stellar photosphere (cf. Sect. 4.4). Similar absorption in extended diffuse gas might be responsible for quenching the indicators of star-planet interactions in other systems. There has been significant work using the Ca II H&K line cores to search for evidence of close-in planets stimulating enhanced stellar activity, with inconclusive results (e.g., Canto Martins et al. 2011; Miller et al. 2012). Continued observations to build up a more extensive observational archive are needed to clarify these issues.

Conclusion

As briefly described above, relatively few years of exoplanet observations have led to a large number of ground-breaking discoveries. This success is reflected in the number of future space- and ground-based instruments for which exoplanet related observations will be one of the primary aims.

NASA's Hubble Space Telescope (HST), launched in 1990, is still producing a huge amount of crucial observations in all fields of Astronomy, and

(continued)

agency officials said in January 2013 that they plan to operate the telescope until its instruments finally give out, potentially till 2018. Other space observatories (see also Chap. 13 (Fridlund et al. 2014)) such as the James Webb Space Telescope (JWST; Clampin 2008; Belu et al. 2011), the next generation space telescope with a 6.5-m main aperture, should be launched in 2018 and it is optimized for observations at near- and infrared wavelengths. One of the primary science goals of these instruments is the atmospheric characterization of transiting exoplanets. Following the great success of its observations and the unfortunate failure of two reaction wheels, the Kepler mission has now been converted into the K2 mission with the aim of searching for planets transiting stars in the ecliptic plane (Howell et al. 2014). The K2 mission is expected to continue its current operations for the next 2–3 years. NASA has approved the Transiting Exoplanet Survey Satellite (TESS) (Ricker et al. 2009) for which the primary goal will be to identify transit events of terrestrial planets orbiting nearby stars, with a particular focus on M-dwarfs. The TESS mission is now scheduled for launch in 2017. ESA has also recently approved the PLAnetary Transits and Oscillations of stars (PLATO 2.0) (Rauer et al. 2014) satellite, a photometric planet finder mostly focussed on the search for transit events of Earth-like planets in the habitable zone of bright solar-like stars. The Characterizing ExOPlanets Satellite (CHEOPS) (Broeg et al. 2013) project was selected in October 2012 as the first S-class (small) space mission in ESA's Science Programme. The satellite's primary aim will be detecting transits of planets already discovered through radial velocity measurements, hence orbiting bright stars. The launch is scheduled for 2017. The next major UV telescope is the World Space Observatory – UltraViolet (WSO-UV) project (see Chap. 14 by Shustov et al. (2014)), a Russian-Spanish 2-m class telescope fully devoted to UV observations, both photometric and spectroscopic.

The majority of the ground-based telescopes will be instead dedicated to planet detections, rather than to atmospheric characterization (for details see Chap. 15 by Guenther (2014)).

With proper theoretical support, the observations collected by all of these instruments will lead to great improvements in our understanding of exoplanets with major advances expected in the topics of planet formation, evolution, physical and chemical properties.

Acknowledgements The authors acknowledge the support by the International Space Science Institute (ISSI) in Bern, Switzerland and the ISSI team *Characterizing stellar- and exoplanetary environments* and thank Lotfi Ben-Jaffel from the Institut Astrophysique de Paris (IAP) CNRS-UPMC, Paris and Tommi Koskinen from the Lunar and Planetary Laboratory University of Arizona, Tucson, USA for fruitful discussions.

References

Alexeev, I. I., Grygoryan, M. S., Belenkaya, E. S., Kalegaev, V. V., & Khodachenko, M. L. (2014). In H. Lammer & M. L. Khodachenko (Eds.), *Characterizing stellar and exoplanetary environments* (pp. 189). Heidelberg/New York: Springer.

Ballester, G. E., Sing, D. K., & Herbert, F. (2007). *Nature, 445*, 511.

Belu, A. R., Selsis, F., Morales, J.-C., Ribas, I., Cossou, C., & Rauer, H. (2011). *Astronomy and Astrophysics, 525*, A83.

Ben-Jaffel, L. (2007). *Astrophysical Journal, 671*, L61.

Ben-Jaffel, L. (2008). *Astrophysical Journal, 688*, 1352.

Ben-Jaffel, L., & Ballester, G. E. (2013). *Astronomy and Astrophysics, 553*, A52.

Ben-Jaffel, L., & Sona Hosseini, S. (2010). *Astrophysical Journal, 709*, 1284.

Birkby, J. L., de Kok, R. J., Brogi, M., de Mooij, E. J.W., Schwarz, H., Albrecht, S., Snellen, I. A. G. (2013). *Monthly Notices of the Royal Astronomical Society, 436*, L35.

Bourrier, V., & Lecavelier des Etangs, A. (2013a). *Astronomy and Astrophysics, 557*, A124.

Bourrier, V., Lecavelier des Etangs, A., Dupuy, H., Ehrenreich, D., Vidal-Madjar, A., Hébrard, G., Ballester, G. E., Désert, J.-M., Ferlet, R., Sing, D. K., & Wheatley, P. J. (2013b). *Astronomy and Astrophysics, 551*, A63.

Broeg, C., Fortier, A., Ehrenreich, D., Alibert, Y., Baumjohann, W., Benz, W., Deleuil, M., Gillon, M., Ivanov, A., Liseau, R., Meyer, M., Oloffson, G., Pagano, I., Piotto, G., Pollacco, D., Queloz, D., Ragazzoni, R., Renotte, E., Steller, M., & Thomas, N. (2013). *European Physical Journal Web of Conferences, 47*, 3005.

Brogi, M., Snellen, I. A. G., de Kok, R. J., Albrecht, S., Birkby, J., & de Mooij, E. J. W. (2012). *Nature, 486*, 502.

Brogi, M., Snellen, I. A. G., de Kok, R. J., Albrecht, S., Birkby, J., & de Mooij, E. J. W. (2013). *Astrophysical Journal, 767*, 27.

Brown, T. M. (2001). *Astrophysical Journal, 553*, 1006.

Canto Martins, B. L., das Chagas, M. L., Alves, S., Leao, I. C., de Souza Neto, L. P., & de Medeiros, J. R. (2011). *Astronomy and Astrophysics, 530*, A73.

Charbonneau, D., Brown, T. M., Latham, D. W., & Mayor, M. (2000). *Astrophysical Journal, 529*, L45.

Charbonneau, D., Brown, T. M., Noyes, R. W., & Gilliland, R. L. (2002). *Astrophysical Journal, 568*, 377.

Christie, D., Arras, P., & Li, Z.-Y. (2013). *Astrophysical Journal, 772*, 144.

Ciceri, S., Mancini, L., Southworth, J., Nikolov, N., Bozza, V., Bruni, I., Calchi Novati, S., D'Ago, G., & Enning, T. (2013). *Astronomy and Astrophysics, 557*, A30.

Clampin, M. (2008). *Advance Space Research, 41*, 1983.

Colón, K. D., Ford, E. B., Redfield, S., Fortney, J. J., Shabram, M., Deeg, H. J., & Mahadevan, S. (2012). *Monthly Notices of the Royal Astronomical Society, 419*, 2233.

Deming, D., Harrington, J., Seager, S., & Richardson, L. J. (2006). *Astrophysical Journal, 644*, 560.

Ehrenreich, D., Lecavelier Des Etangs, A., Hébrard, G., Désert, J.-M., Vidal-Madjar, A., McConnell, J. C., Parkinson, C. D., Ballester, G. E., & Ferlet, R. (2008). *Astronomy and Astrophysics, 483*, 933.

Ehrenreich, D., Bourrier, V., Bonfils, X., Lecavelier des Etangs, A., Hébrard, G., Sing, D. K., Wheatley, P. J., Vidal-Madjar, A., Delfosse, X., Udry, S., Forveille, T., & Moutou, C. (2012). *Astronomy and Astrophysics, 547*, A18.

Ekenbäck, A., Holmström, M., Wurz, P., Griessmeier, J.-M., Lammer, H., Selsis, F., & Penz, T. (2010). *Astrophysical Journal, 709*, 670.

Fares, R., Donati, J.-F., Moutou, C., Jardine, M. M., Grießmeier, J.-M., Zarka, P., Shkolnik, E. L., Bohlender, D., Catala, C., & Collier Cameron, A. (2010). *Monthly Notices of the Royal Astronomical Society, 406*, 409.

Fortney, J. J., Lodders, K., Marley, M. S., & Freedman, R. S. (2008). *Astrophysical Journal, 678*, 1419.
Fossati, L., Ayres, T. R., Haswell, C. A., Bohlender, D., Kochukhov, O., & Flöer, L. (2013). *Astrophysical Journal, 766*, L20.
Fossati, L., Bagnulo, S., Elmasli, A., Haswell, C. A., Holmes, S., Kochukhov, O., Shkolnik, E. L., Shulyak, D. V., Bohlender, D., Albayrak, B., Froning, C., & Hebb, L. (2010). *Astrophysical Journal, 720*, 872.
Fossati, L., Haswell, C. A., Froning, C. S., Hebb, L., Holmes, S., Kolb, U., Helling, Ch., Carter, A., Wheatley, P., Cameron, A. C., Loeillet, B., Pollacco, D., Street, R., Stempels, H. C., Simpson, E., Udry, S., Joshi, Y. C., West, R. G., Skillen, I., & Wilson, D. (2010). *Astrophysical Journal, 714*, L222.
France, K., Stocke, J. T., Yang, H., Linsky, J. L., Wolven, B. C., Froning, C. S., Green, J. C., & Osterman, S. N. (2010). *Astrophysical Journal, 712*, 1277.
Grießmeier, J.-M., Zarka, P., & Girard, J. N. (2008). *Radio Science, 46*, 1.
Guenther, E. (2014). In H. Lammer & M. L. Khodachenko (Eds.), *Characterizing stellar and exoplanetary environments* (pp. 289). Heidelberg/New York: Springer
Hall, J. C. (2008). *Living Reviews in Solar Physics, 5*, 2.
Haswell, C. A. (2010). *Transiting exoplanets*. Cambridge: Cambridge University Press
Haswell, C. A., Fossati, L., Ayres, T., France, K., Froning, C. S., Holmes, S., Kolb, U. C., Busuttil, R., Street, R. A., Hebb, L., Collier Cameron, A., Enoch, B., Burwitz, V., Rodriguez, J., West, R. G., Pollacco, D., Wheatley, P. J., & Carter, A. (2012). *Astrophysical Journal, 760*, 79.
Hebb, L., Collier-Cameron, A., Loeillet, B., Pollacco, D., Hébrard, G., Street, R. A., Bouchy, F., Stempels, H. C., Moutou, C., Simpson, E., Udry, S., Joshi, Y. C., West, R. G., Skillen, I., Wilson, D. M., McDonald, I., Gibson, N. P., Aigrain, S., Anderson, D. R., Benn, C. R., Christian, D. J., Enoch, B., Haswell, C. A., Hellier, C., Horne, K., Irwin, J., Lister, T. A., Maxted, P., Mayor, M., Norton, A. J., Parley, N., Pont, F., Queloz, D., Smalley, B., & Wheatley, P. J. (2009). *Astrophysical Journal, 693*, 1920.
Hess, S. L. G., & Zarka, P. (2011). *Astronomy and Astrophysics, 531*, A29.
Holmström, M., Ekenbäck, A., Selsis, F., Penz, T., Lammer, H., & Wurz, P. (2008). *Nature, 451*, 970.
Howell, S. B., Sobeck, C., Haas, M., Still, M., Barclay, T., Mullally, F., Troeltzsch, J., Aigrain, S., Bryson, S. T., Caldwell, D., Chaplin, W. J., Cochran, W. D., Huber, D., Marcy, G. W., Miglio, A., Najita, J. R., Smith, M., Twicken, J. D., & Fortney, J. J. (2014). *Publications of the Astronomical Society of the Pacific, 126*, 398.
Hubeny, I., Burrows, A., & Sudarsky, D. (2003). *Astrophysical Journal, 594*, 1011.
Jenkins, J. S., Jones, H. R. A., Pavlenko, Y., Pinfield, D. J., Barnes, J. R., & Lyubchik, Y. (2008). *Astronomy and Astrophysics, 485*, 571.
Jenkins, J. S., Jones, H. R. A., Tinney, C. G., Butler, R. P., McCarthy, C., Marcy, G. W., Pinfield, D. J., Carter, B. D., & Penny, A. J. (2006). *Monthly Notices of the Royal Astronomical Society, 372*, 163.
Jenkins, J. S., Murgas, F., Rojo, P., Jones, H. R. A., Day-Jones, A. C., Jones, M. I., Clarke, J. R. A., Ruiz, M. T., & Pinfield, D. J. (2011). *Astronomy and Astrophysics, 531*, A8.
Jensen, A. G., Redfield, S., Endl, M., Cochran, W. D., Koesterke, L., & Barman, T. (2012). *Astrophysical Journal, 751*, 86.
Jordán, A., Espinoza, N., Rabus, M., Eyheramendy, S., Sing, D. K., Désert, J.-M., Bakos, G. Á., Fortney, J. J., López-Morales, M., Maxted, P. F. L., Triaud, A. H. M. J., & Szentgyorgyi, A. (2013). *Astrophysical Journal, 778*, 184.
de Kok, R. J., Brogi, M., Snellen, I. A. G., Birkby, J., Albrecht, S., & de Mooij, E. J. W. (2013). *Astronomy and Astrophysics, 554*, A82.
Kashyap, V. L., Drake, J. J., & Saar, S. H. (2008). *Astrophysical Journal, 687*, 1339.
Knutson, H. A., Charbonneau, D., Allen, L. E., Fortney, J. J., Agol, E., Cowan, N. B., Showman, A. P., Cooper, C. S., & Megeath, S. T. (2007). *Nature, 447*, 183.
Knutson, H. A., Charbonneau, D., Allen, L. E., Burrows, A., & Megeath, S. T. (2008). *Astrophysical Journal, 673*, 526.

Knutson, H. A., Charbonneau, D., Burrows, A., O'Donovan, F. T., & Mandushev, G. (2009a). *Astrophysical Journal, 691*, 866.
Knutson, H. A., Charbonneau, D., Cowan, N. B., Fortney, J. J., Showman, A. P., Agol, E., & Henry, G. W. (2009b). *Astrophysical Journal, 703*, 769.
Knutson, H. A., Howard, A. W., & Isaacson, H. (2010). *Astrophysical Journal, 720*, 1569.
Knutson, H. A., Madhusudhan, N., Cowan, N. B., Christiansen, J. L., Agol, E., Deming, D., Désert, J.-M., Charbonneau, D., Henry, G. W., Homeier, D., Langton, J., Laughlin, G., & Seager, S. (2011). *Astrophysical Journal, 735*, 27.
Knutson, H. A., Lewis, N., Fortney, J. J., Burrows, A., Showman, A. P., Cowan, N. B., Agol, E., Aigrain, S., Charbonneau, D., Deming, D., Désert, J.-M., Henry, G. W., Langton, J., & Laughlin, G. (2012). *Astrophysical Journal, 754*, 22.
Koskinen, T. T., Harris, M. J., Yelle, R. V., & Lavvas, P. (2013). *Icarus, 226*, 1678.
Kulow, J. R., France, K., Linsky, J., & Loyd, R. O. P. (2014). *Astrophysical Journal, 786*, 132.
Lecavelier des Etangs, A., Vidal-Madjar, A., McConnell, J. C., & Hébrard, G. (2004). *Astronomy and Astrophysics, 418*, L1.
Lecavelier des Etangs, A., Pont, F., Vidal-Madjar, A., & Sing, D. (2008). *Astronomy and Astrophysics, 481*, L83.
Lecavelier des Etangs, A., Ehrenreich, D., Vidal-Madjar, A., Ballester, G. E., Désert, J.-M., Ferlet, R., Hébrard, G., Sing, D. K., Tchakoumegni, K.-O., & Udry, S. (2010). *Astronomy and Astrophysics, 514*, A72.
Lecavelier des Etangs, A., Bourrier, V., Wheatley, P. J., Dupuy, H., Ehrenreich, D., Vidal-Madjar, A., Hébrard, G., Ballester, G. E., Désert, J.-M., Ferlet, R., & Sing, D. K. (2012). *Astronomy and Astrophysics, 543*, L4.
Lecavelier des Etangs, A., Sirothia, S. K., Gopal-Krishna, & Zarka, P. (2013). *Astronomy and Astrophysics, 552*, A65.
Linsky, J. L., Yang, H., France, K., Froning, C. S., Green, J. C., Stocke, J. T., & Osterman, S. N. (2010). *Astrophysical Journal, 717*, 1291.
Madhusudhan, N., & Seager, S. (2013). *Astrophysical Journal, 725*, 261.
Mancini, L., Southworth, J., Ciceri, S., Fortney, J. J., Morley, C. V., Dittmann, J. A., Tregloan-Reed, J., Bruni, I., Barbieri, M., Evans, D. F., D'Ago, G., Nikolov, N., & Henning, T. (2013a). *Astronomy and Astrophysics, 551*, A11.
Mancini, L., Nikolov, N., Southworth, J., Chen, G., Fortney, J. J., Tregloan-Reed, J., Ciceri, S., van Boekel, R., & Henning, T. (2013b). *Monthly Notices of the Royal Astronomical Society, 430*, 2932.
Mancini, L., Ciceri, S., Chen, G., Tregloan-Reed, J., Fortney, J. J., Southworth, J., Tan, T. G., Burgdorf, M., Calchi Novati, S., Dominik, M., Fang, X.-S., Finet, F., Gerner, T., Hardis, S., Hinse, T. C., Jørgensen, U. G., Liebig, C., Nikolov, N., Ricci, D., Schäfer, S., Schönebeck, F., Skottfelt, J., Wertz, O., Alsubai, K. A., Bozza, V., Browne, P., Dodds, P., Gu, S.-H., Harpsøe, K., Henning, T., Hundertmark, M., Jessen-Hansen, J., Kains, N., Kerins, E., Kjeldsen, H., Lund, M. N., Lundkvist, M., Madhusudhan, N., Mathiasen, M., Penny, M. T., Prof, S., Rahvar, S., Sahu, K., Scarpetta, G., Snodgrass, C., & Surdej, J. (2013c). *Monthly Notices of the Royal Astronomical Society, 436*, 2.
Mandell, A. M., Haynes, K., Sinukoff, E., Madhusudhan, N., Burrows, A., & Deming, D. (2013). *Astrophysical Journal, 779*, 128.
Miller, B. P., Gallo, E., Wright, J. T., & Dupree, A. K. (2012). *Astrophysical Journal, 754*, 137.
Muirhead, Ph., Johnson, J., Apps, K., Carter, J., Morton, T. D., Fabrycky, D. C., Pineda, J. S., Bottom, M., Rojas-Ayala, B., Schlawin, E., Hamren, K., Covey, K. R., Crepp, J. R., Stassun, K. G., Pepper, J., Hebb, L., Kirby, E. N., Howard, A. W., Isaacson, H. T., Marcy, G. W., Levitan, D., Diaz-Santos, T., Armus, L., & Lloyd, J. P. (2012). *Astrophysical Journal, 747*, 144.
Narita, N., Suto, Y., Winn, J. N., Turner, E. L., Aoki, W., Leigh, C. J., Sato, B., Tamura, M., & Yamada, T. (2005). *Publications of the Astronomical Society of Japan, 57*, 471.
Nikolov, N., Sing, D. K., Pont, F., Burrows, A. S., Fortney, J. J., Ballester, G. E., Evans, T. M., Huitson, C. M., Wakeford, H. R., Wilson, P. A., Aigrain, S., Deming, D., Gibson, N. P., Henry,

G. W., Knutson, H., Lecavelier des Etangs, A., Showman, A. P., Vidal-Madjar, A., & Zahnle, K. (2013a). *Monthly Notices of the Royal Astronomical Society, 437*, 46.

Nikolov, N., Chen, G., Fortney, J. J., Mancini, L., Southworth, J., van Boekel, R., & Henning, T. (2013). *Astronomy and Astrophysics, 553*, A26.

Öpik, E. J. (1963). *Geophysical Journal International, 7*, 490.

Pillitteri, I., Wolk, S. J., Cohen, O., Kashyap, V., Knutson, H., Lisse, C. M., & Henry, G. W. (2010). *Astrophysical Journal, 722*, 1216.

Pont, F., Sing, D. K., Gibson, N. P., Aigrain, S., Henry, G., & Husnoo, N. (2013). *Monthly Notices of the Royal Astronomical Society, 432*, 2917.

Poppenhaeger, K., Schmitt, J. H. M. M., & Wolk, S. J. (2013). *Astrophysical Journal, 773*, 62.

Rappaport, S., Levine, A., Chiang, E., El Mellah, I., Jenkins, J., Kalomeni, B., Kite, E. S., Kotson, M., Nelson, L., Rousseau-Nepton, L., & Tran, K. (2012). *Astrophysical Journal, 752*, 1.

Rappaport, S., Sanchis-Ojeda, R., Rogers, L., Levine, A., & Winn,J. (2013). *Astrophysical Journal, 773*, 15.

Rauer, H., and the PLATO team (2014, in press). *Exp. Ast.* (arXiv: 1310.0696).

Ricker, G. R., Latham, D. W., Vanderspek, R. K., Ennico, K. A., Bakos, G., Brown, T. M., Burgasser, A. J., Charbonneau, D., Clampin, M., Deming, L. D., Doty, J. P., Dunham, E. W., Elliot, J. L., Holman, M. J., Ida, S., Jenkins, J. M., Jernigan, J. G., Kawai, N., Laughlin, G. P., Lissauer, J. J., Martel, F., Sasselov, D. D., Schingler, R. H., Seager, S., Torres, G., Udry, S., Villasenor, J. S., Winn, J. N., & Worden, S. P. (2009). *Bulletin American Astronomical Society, 41*, 729.

Sanchis-Ojeda, R., Rappaport, S., Winn,J., Levine, A., Kotson, C., & Latham, D. (2013). *Astrophysical Journal, 774*, 54.

Schneiter, E. M., Velázquez, P. F., Esquivel, A., Raga, A. C., & Blanco-Cano, X. (2007). *Astrophysical Journal, 671*, L57.

Seager, S., & Deming, D. (2010). *Annals Review Astronomy and Astrophysics 48*, 631.

Sing, D. K., Vidal-Madjar, A., Lecavelier des Etangs, A., Dèsert, J.-M., Ballester, G., & Ehrenreich, D. (2008a). *Astrophysical Journal, 686*, 667.

Sing, D. K., Vidal-Madjar, A., Désert, J.-M., Lecavelier des Etangs, A., & Ballester, G. (2008b). *Astrophysical Journal, 686*, 658.

Sing, D. K., Pont, F., Aigrain, S., Charbonneau, D., Désert, J.-M., Gibson, N., Gilliland, R., Hayek, W., Henry, G., Knutson, H., Lecavelier Des Etangs, A., Mazeh, T., & Shporer, A. (2011a). *Monthly Notices of the Royal Astronomical Society, 416*, 1443.

Sing, D. K., Désert, J.-M., Fortney, J. J., Lecavelier Des Etangs, A., Ballester, G. E., Cepa, J., Ehrenreich, D., López-Morales, M., Pont, F., Shabram, M., & Vidal-Madjar, A. (2011b). *Astronomy and Astrophysics, 527*, A73.

Sing, D. K., Huitson, C. M., Lopez-Morales, M., Pont, F., Dèsert, J.-M., Ehrenreich, D., Wilson, P. A., Ballester, G. E., Fortney, J. J., Lecavelier des Etangs, A., & Vidal-Madjar, A. (2012). *Monthly Notices of the Royal Astronomical Society, 426*, 1663.

Snellen, I. A. G., Albrecht, S., de Mooij, E. J. W., & Le Poole, R. S. (2008). *Astronomy and Astrophysics, 487*, 357.

Spiegel, D. S., Silverio, K., & Burrows, A. (2009). *Astrophysical Journal, 699*, 1487.

Tremblin, P., & Chiang, E. (2013). *Monthly Notices of the Royal Astronomical Society, 428*, 2565.

Vial, J.-C., Olivier, K., Philippon, A.A., Vourlidas, A., & Yurchyshyn, V. (2012). *Astronomy and Astrophysics, 541*, A108.

Vidal-Madjar, A., Lecavelier des Etangs, A., Désert, J.-M., Ballester, G. E., Ferlet, R., Hébrard, G., & Mayor, M. (2003). *Nature, 422*, 143.

Vidal-Madjar, A., Désert, J.-M., Lecavelier des Etangs, A., Hébrard, G., Ballester, G. E., Ehrenreich, D., Ferlet, R., McConnell, J. C., Mayor, M., & Parkinson, C. D. (2004). *Astrophysical Journal, 604*, L69.

Vidal-Madjar, A., Huitson, C. M., Bourrier, V., Dèsert, J.-M., Ballester, G., Lecavelier des Etangs, A., Sing, D. K., Ehrenreich, D., Ferlet, R., Hèbrard, G., & McConnell, J. C. (2013). *Astronomy and Astrophysics, 560*, A54.

Vidotto, A. A., Bisikalo, D. V., Fossati, L., & Llama, J. (2014). In H. Lammer & M. L. Khodachenko (Eds.), *Characterizing stellar and exoplanetary environments* (pp. 153). Heidelberg/New York: Springer

Winn, J. N., Matthews, J. M., Dawson, R. I., Fabrycky, D., Holman, M. J., Kallinger, T., Kuschnig, R., Sasselov, D., Dragomir, D., Guenther, D. B., Moffat, A. F. J., Rowe, J. F., Rucinski, S., & Weiss, W. W. (2011). *Astrophysical Journal, 737*, L18.

Wright, J. T. (2004a). *Astronomical Journal, 128*, 1273.

Wright, J. T., Marcy, G. W., Butler, R. P., & Vogt, S. S. (2004b). *Astrophysical Journal Supplement, 152*, 261.

Zahnle, K., Marley, M. S., Freedman, R. S., Lodders, K., & Fortney, J. J. (2009). *Astrophysical Journal, 701*, L20.

Zhou, G., & Bayliss, D. D. R. (2012). *Monthly Notices of the Royal Astronomical Society, 426*, 2483.

Chapter 5
Types of Hot Jupiter Atmospheres

Dmitry V. Bisikalo, Pavel V. Kaygorodov, Dmitry E. Ionov, and Valery I. Shematovich

Abstract Hot Jupiters, i.e. exoplanet gas giants, having masses comparable to the mass of Jupiter and semimajor axes shorter than 0.1 AU, are a unique class of objects. Since they are so close to the host stars, their atmospheres form and evolve under the action of very active gas dynamical processes caused by the gravitational field and irradiation of the host star. As a matter of fact, the atmospheres of several of these planets fill their Roche lobes, which results in a powerful outflow of material from the planet towards the host star. The energy budget of this process is so important that it almost solely governs the evolution of hot Jupiters gaseous envelopes. Based on the years of experience in the simulations of gas dynamics in mass-exchanging close binary stars, we have investigated specific features of hot Jupiters atmospheres. The analytical estimates and results of 3D numerical simulations, discussed in this Chapter, show that the gaseous envelopes around hot Jupiters may be significantly non-spherical and, at the same time, stationary and long-lived. These results are of fundamental importance for the interpretation of observational data.

5.1 Introduction: Exoplanet Gaseous Envelopes

Hot Jupiters, have a number of outstanding features, caused mostly by their proximity to the host star: e.g., gas outflowing from the planet's atmosphere to the star, as it happens in close binary stars. In addition, the short distance between the planet and the star leads to a large planet's orbital velocity: if this exceeds the local sound speed a bow-shock forms ahead of the planet. These effects substantially change the kind of interaction between the planet's gaseous envelope (atmosphere) and the stellar wind.

Observations of hot Jupiters, conducted with the Hubble Space Telescope (HST), prove the existence of complex physical processes in the gaseous envelopes of these planets (see Chap. 4 (Fossati et al. 2014) and Chap. 8 (Vidotto et al. 2014)).

D.V. Bisikalo (✉) • P.V. Kaygorodov • D.E. Ionov • V.I. Shematovich
Institute of Astronomy of the Russian Academy of Sciences, Moscow, Russian Federation
e-mail: bisikalo@inasan.ru; pasha@inasan.ru; ionovd@inasan.ru; shematov@inasan.ru

H. Lammer, M. Khodachenko (eds.), *Characterizing Stellar and Exoplanetary Environments*, Astrophysics and Space Science Library 411,
DOI 10.1007/978-3-319-09749-7_5

For example, the observations of HD 209458b (Vidal-Madjar et al. 2003, 2008; Ben-Jaffel 2007) reveal a transit absorption depth at the position of the Lyman-α line of 9–15 % (see also Chap. 4 (Fossati et al. 2014)), though the bulk of the planet produces a primary transit depth of only 1.8 %. This reveals that the planet is embedded into an extended gaseous envelope. This conclusion was confirmed by primary transit observations of this planet at far-UV spectral lines of carbon, oxygen and silicon (Vidal-Madjar et al. 2004; Ben-Jaffel and Sona Hosseini 2010; Linsky et al. 2010), where the transit depth reached 8–9 %. The existence of extended gaseous envelopes, surrounding hot Jupiters, was also confirmed with observations of the hot Jupiters HD 189733b (Lecavelier Des Etangs et al. 2010) and WASP-12b (Fossati et al. 2010a,b; Haswell et al. 2012). In addition, the near-UV lightcurve of the primary transit of WASP-12b shows a variability in the shape and timing of the ingress.

To correctly explain existing and upcoming observational data we need to understand what physical effects are most significant and what observable features they might lead to. In this Chapter we introduce the reader to some of the major physical processes determining the properties of the gaseous envelopes of hot Jupiters. This Chapter is based on the results, obtained in the past 2 years on the study of the atmospheres of hot Jupiters (Ionov et al. 2012; Bisikalo et al. 2013a,b,c). However, to generalize the presentation of the results we add general information on close binary stars which provide several similarities to a star-planet system.

5.2 Outflow of the Hot Jupiter Atmosphere Caused by the Gravity of the Host Star

A system consisting of a star and a hot Jupiter may be regarded as a close binary system with an extremely low mass ratio. This approach is very fruitful, since it allows one to use the experience gained in the past on such systems. As a matter of fact, the general assumptions used to study binary stars can be adapted to describe a star–hot Jupiter system. Let us consider the force field in a binary system composed by a star and a hot Jupiter with masses M_* and M_{pl}, respectively (see, e.g. Boyarchuk et al. (2002); Bisikalo et al. (2013d)). Let us also reasonably assume that the orbits of the components are circular and their proper rotations are synchronized with the orbital motion $\Omega_* = \Omega_{pl} = \Omega = 2\pi/P_{orb}$, where P_{orb} is the orbital period. In addition, we consider the Roche approximation, in which the internal density of both the components sharply grows towards the center. This allows one to assume the star and planet as point masses and therefore to describe their gravitational potentials in the frame of classical Newtonian mechanics.

We further introduce a Cartesian coordinate system (x, y, z), rotating counter-clockwise along with the binary system and having the origin in the center of the star. The x axis is directed along the line, connecting the centers of the components, the z axis is perpendicular to the orbital plane and parallel to $\boldsymbol{\Omega}$, while the y axis completes the right-handed coordinate system.

In the considered case of point masses, a potential Φ, describing the force field in the system, is known as the Roche potential. Since the components move in accordance with Kepler's third law

$$G(M_* + M_{pl}) = A^3\Omega^2,$$

where G is the gravitational constant, A is the orbital separation, and the center of mass of the system is

$$R_{cm} = \frac{M_{pl}}{M_* + M_{pl}}A,$$

one can write the potential up to a constant as

$$\Phi = -\frac{GM_*}{\sqrt{x^2 + y^2 + z^2}} - \frac{GM_{pl}}{\sqrt{(x - A)^2 + y^2 + z^2}}$$
$$-\frac{1}{2}\Omega^2\left(\left(x - A\frac{M_{pl}}{M_* + M_{pl}}\right)^2 + y^2\right). \quad (5.1)$$

In the left panel of Fig. 5.1 we display equipotentials in the equatorial plane xy ($z = 0$) of the system having a mass ratio $q = M_{pl}/M_* = 1$. In the right panel of Fig. 5.1 we display equipotentials in the frontal plane xz ($y = 0$) of the system. Fig. 5.1 shows also the adopted coordinate system (x, y, z). As shown in Fig. 5.1, the equipotentials close to the centers of the components (marked by dashed lines) are almost spherical. When going farther away from the center of a component, the gravitational influence of the secondary grows, and the equipotentials become ellipsoids, stretched along the x axis. Within this setup, rotation results in the compression of the equipotentials along the z axis.

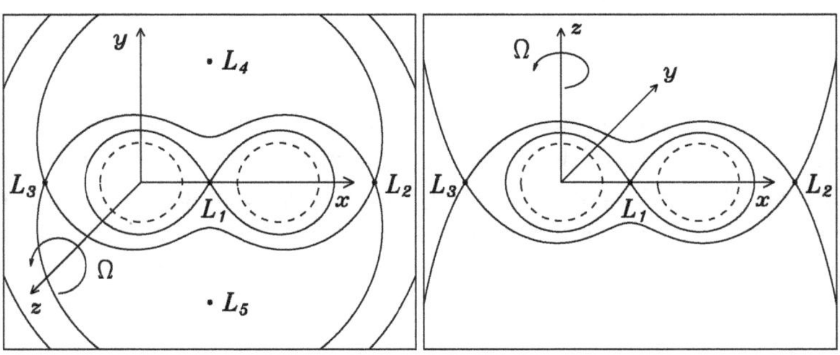

Fig. 5.1 Roche equipotentials in the equatorial (xy, *left panel*) and frontal (xz, *right panel*) planes for a binary system with the mass ratio $q = M_{pl}/M_* = 1$. The *dashed line* denotes the equipotential, containing the point $(0.3A, 0, 0)$. The positions of the Lagrangian points $L_1, \ldots, L_5$ are marked as well as the adopted coordinate system

The Roche potential has five libration points, called Lagrangian points (see Fig. 5.1). The position of these points can be determined from the following condition:

$$\nabla \Phi = 0. \tag{5.2}$$

All five Lagrangian points are located in the equatorial plane; three of them (L_1, L_2 and L_3) lie on the x axis and are the inflection points of Φ, while the L_4 and L_5 points are the maxima of Φ.

The equipotential, containing the inner Lagrangian point L_1, encloses two contiguous volumes, known as critical surfaces or Roche lobes. The term Roche lobe is of particular importance in astronomy: for an object (star or planet's atmosphere) whose boundary surface is inside the Roche lobe, one finds a stationary configuration where the gradient of the Roche potential is balanced by the gradient of gas pressure. Once the boundary of a component has reached the critical surface, because of the pressure gradient, mass outflow occurs at the inner Lagrangian point, where the total force (gravity of each component plus centrifugal force) is equal to zero.

Let us now derive the parameters of the outflow forming in the vicinity of the L_1 point. We define the density and the temperature of the planet's atmosphere at the L_1 point as ρ_{L_1} and T_{L_1}, respectively. We can estimate the rate of mass loss through the vicinity of the L_1 point as

$$\dot{M}_{pl} = S\rho_{L_1} v_{L_1}, \tag{5.3}$$

where S is the effective cross-section of the stream, ρ_{L_1} is the density averaged over the cross-section, and v_{L_1} is the gas velocity. The outflow through the vicinity of the inner Lagrangian point is similar to the free gas expansion into vacuum from a vessel with a pinhole. This means that the velocity of the flow through the L_1 point is approximately equal to the sound speed in the planet's atmosphere ($v_{L_1} \simeq c_s$). As a consequence relation Eq. (5.3) can be written as

$$\dot{M}_{pl} = S\rho_{L_1} c_s. \tag{5.4}$$

To determine the size of the stream (cross-section), flowing through the L_1 point one should estimate the size of the area where the kinetic energy of a particle, having the velocity c_s (or specific kinetic energy $\sim c_s^2$), is large enough to escape from the Roche lobe of the planet. Equating the difference between the potential energy in the plane $x = x_{L_1}$ and the specific kinetic energy, one can derive an equation that describes the stream's shape in the vicinity of the L_1 point, in analogy to Savonije (1979)

$$\Delta\Phi = c_s^2. \tag{5.5}$$

Expanding this relation into a Tailor series over y and z, and taking into account that $\nabla\Phi|_{(x_{L_1},0,0)} = 0$, one obtains an equation which describes the elliptical shape of the stream. Following simple mathematical transformations one can calculate (see, e.g., Boyarchuk et al. (2002); Bisikalo et al. (2013d)) the area of the stream's cross-section in the vicinity of the L_1 point and, hence, the mass loss rate of the planet's atmosphere.

However, astronomers use more commonly a more simple relation to determine the rate at which an astrophysical object, filling its Roche lobe, loses mass. For this purpose one has to introduce the effective (or volume) radius of the Roche lobe. This radius R_{L_1} is the radius of a sphere having the same volume as the Roche lobe. Besides, one needs to also define the degree of Roche lobe overfilling $\Delta R = R - R_{L_1}$, where R, in this case, is the radius of the planet's atmosphere. According to Pringle and Wade (1985), the mass loss rate for a star with an adiabatic atmosphere depends on the degree of overfilling as follows:

$$\dot{M}/M_{pl} = (\Delta R/R_{L_1})^3 \sqrt{\frac{GM_{pl}}{R_{L_1}^3}} \,. \tag{5.6}$$

The atmospheres of hot Jupiters should be fully convective, so that one can consider them adiabatic and use therefore Eq. (5.6) to estimate the mass loss rate of an exoplanet. To make these estimates one needs to define a surface that can be regarded as the boundary of the planet's atmosphere. Simple physical considerations prompt us to use for this purpose the exobase, i.e. the level at which the mean free path of atmospheric particles becomes equal to the scale height. Equation (5.6) shows that even a small overfilling results in a considerable outflow. For instance, for a Jupiter-like planet orbiting at a distance $A = 0.045$ AU, the degree of overfilling $\Delta R/R_{L_1} = 0.1$ results in a mass loss rate $\dot{M} \approx 4 \times 10^{-8}\, M_{Jup}/\mathrm{s}$. This means that if $\dot{M}$ is constant with time the planet will completely lose its mass in about 300 days.

The possibility for planetary material to outflow through the L_1 point was first considered in Vidal-Madjar et al. (2003), Lai et al. (2010), and Li et al. (2010). For the typical exoplanet WASP-12b, where the degree of overfilling is $\Delta R/R_{L_1} \approx 0.16$, Eq. (5.6) gives a very high mass loss rate and therefore a very short planet's lifetime. On the other hand, the fact that we observe many hot Jupiters with an atmosphere overfilling their Roche lobe is in strong contrast with the estimates of the lifetimes of planets' atmospheres given above. Obviously, the use of a simplified expression (5.6) gives only an estimate of the lifetime of a planet, because it does not take into account the time-dependence of the mass loss rate. Nevertheless, further considerations show that these estimates are quite correct. Indeed, the decrease of the mass of a gas giant planet leads to an increase of the Roche lobe overfilling, and, consequently, an increase of the mass loss rate. Thus, the value, derived from Eq. (5.6) can be considered as a maximum estimate of the lifetime of a gas giant planet. On the other hand, for a planet with a rocky core the mass loss rate will decrease with time. However, in this case the mass of the atmosphere is not large enough and one can use Eq. (5.6) to obtain a reasonable estimate of the time period in which the planet actively loses mass.

Finally, assuming that the estimates of hot Jupiters' lifetimes given above are correct, one needs to find physical processes which would prevent the gas outflow from the planet's atmosphere. Both the stellar wind and the dynamical pressure, caused by the fast orbital motion of the planet inside the stellar wind material, may prevent the gas to outflow. Let us now consider these processes in detail.

5.3 Interaction of Hot Jupiter Atmospheres with Stellar Winds

Let us consider the motion of a hot Jupiter immersed in the gas of the stellar wind. To estimate how much the environment influences the planet's atmosphere in the gas dynamical sense we can take three important physical values: gas density of the stellar wind, velocity of the stellar wind, and proper (orbital) velocity of the planet inside the stellar wind. Here we consider only gas dynamical processes, supposing that the influence of the radiative pressure of the host star, its magnetic field and the magnetic field of the planet is much weaker.

Parameters of the stellar winds of stars other than the Sun are poorly known (see also Chap. 2 (Wood et al. 2014)). Thus, hereafter we consider a solar-like wind. Without any loss of generality, let us also assume a solar-twin star and a Jupiter-twin planet, situated close to the star. In gas dynamics the fundamental question concerns the ratio between the velocity of the gas incoming on the planet's atmosphere (incoming flow), and the sound speed. In the adopted coordinate system, the interaction between the planet and the stellar wind may be considered as an incidence of two flows on the planet, one is exactly the stellar wind, the other is the flow caused by the orbital motion of the planet. An analysis of the radial profiles of the wind

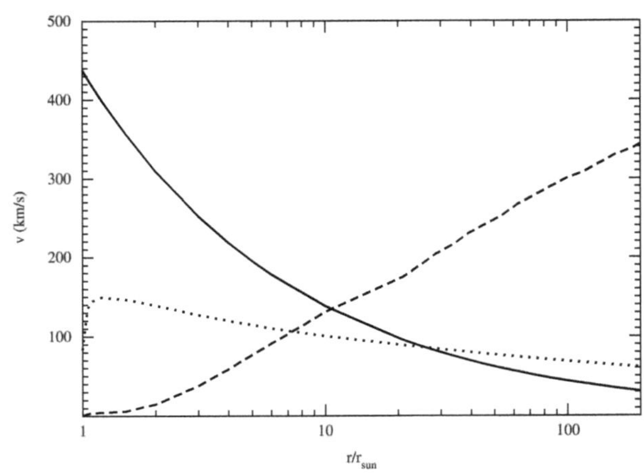

Fig. 5.2 Radial profiles of the stellar wind velocity (*dashed line*) and local sound speed (*dotted line*) determined for the Sun (Withbroe 1988). The *solid line* shows the orbital velocity of the planet in the Sun-Jupiter system as a function of the distance to the star

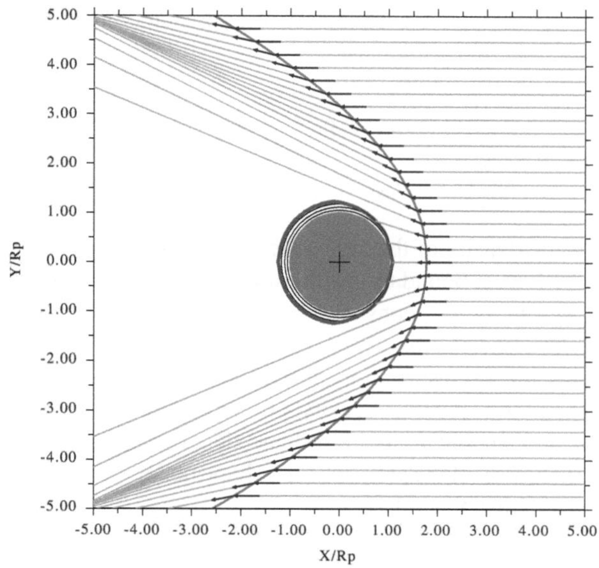

Fig. 5.3 Location of the shock wave (*solid line*) and contact discontinuity (*bold line* around the planet), forming near a hot Jupiter as a result of supersonic flow past the planet by the gas of the stellar wind. *Stream lines* and velocity vectors before and after passing the shock front are also shown. The *cross* denotes the planet's center of mass, while the *shaded circle* indicates the radius of the atmosphere

velocity (dashed line), planet's orbital velocity (solid line), and local sound speed (dotted line) presented in Fig. 5.2 shows that the planet's orbital motion is always supersonic: either because of the short orbital separation ($<10R_*$) and therefore large orbital velocity or because of the proper velocity of the stellar wind at larger distances. Supersonic incoming flow on a planet surrounded by a gaseous envelope must result in the formation of a bow-shock in the gas of the stellar wind, followed by a contact discontinuity, a boundary between the gas of the stellar wind and the gas of the planet's atmosphere. Behind the planet, there is a region of reduced pressure, called a rarefaction wave. The structure of the forming flow pattern is schematically shown in Fig. 5.3, where one can see the location and shape of the shock wave (solid line), contact discontinuity (bold line around the planet), stream lines, and velocity vectors of the wind before and after passing the shock front.

The shape and position of the contact discontinuity can be calculated analytically by using the momentum conservation law that in the purely gas dynamical case is (Landau and Lifshitz 1966)

$$\rho_1 v_1^2 + p_1 = \rho_2 v_2^2 + p_2 \,, \tag{5.7}$$

where ρ_1, ρ_2 are the densities, v_1, v_2 the velocities, and p_1, p_2 the pressures on both sides of the discontinuity. In the simplest case, neglecting the non-spherical shape of the gaseous envelope, one can calculate densities and pressures in the upper atmosphere, on the basis of the condition of hydrostatic equilibrium of a perfect gas in the gravity field of a point mass

$$\begin{aligned} \rho_{atm}(r) &= \rho_0 \cdot \exp\left\{-\frac{G\,M_{pl}}{R_{gas}T_{atm}}\left(\frac{1}{r_0}-\frac{1}{r}\right)\right\}, \\ p_{atm}(r) &= \rho_{atm}\,R_{gas}\,T_{atm}, \end{aligned} \tag{5.8}$$

where $\rho_{atm}(r)$ is the density of the atmosphere at a radius r, ρ_0 is the density of the atmosphere at the lowest boundary r_0 (as a rule, ρ_0 is set at the photometric radius of the planet), R_{gas} is the gas constant, T_{atm} is the temperature of the atmosphere, and p_{atm} is the atmospheric pressure.

By substituting the values of atmospheric density and pressure to the left-hand side of Eq. (5.7) and the values of density, pressure and velocity of the stellar wind, incoming on the atmosphere, to the right-hand side, one can derive an equation that determines the distortion of the atmosphere's shape under the action of the incoming flow (Baranov and Krasnobaev 1977)

$$p_{atm}(r) = \rho_w {\mathbf{v_w}}^2 \cos^2(\mathbf{n}, \mathbf{v_w}) + p_w, \tag{5.9}$$

where ρ_w is the wind density, $\mathbf{v_w}$ is the wind velocity, and $\mathbf{n}$ is the unit vector normal to the surface of the atmosphere. Equation (5.9) allows one to determine the shape of the windward portion of the atmosphere, directly interacting with the stellar wind. The head-on collision point, where $\cos(\mathbf{n}, \mathbf{v_w}) = 1$, is situated at the shortest distance from the planet's center.

The shock wave front lies at some distance from the contact discontinuity where the shortest distance is achieved at the head-on collision point, otherwise called shock standoff distance Δ. In a recent work by Verigin et al. (2003) a semi-empirical formula allowing the determination of this value was derived

$$\Delta = 1.1 r_{cd} \frac{(\gamma - 1)M^2 + 2}{(\gamma + 1)M^2}, \tag{5.10}$$

where γ is the adiabatic index, M is the Mach number, and r_{cd} is the distance between the planet's center and the contact discontinuity. The shape of the shock wave is determined by the following equation:

$$y^2(x) = 2R_s(r_s - x) + b_s(r_s - x)^2, \tag{5.11}$$

where R_s is the radius of the curvature of the shock, $r_s = r_{cd} + \Delta$ is the distance between the planet's center and the shock at the head-on collision point, and b_s is the bluntness of the shock wave. The values R_s and b_s are also found from semi-empirical relations:

$$R_s = \Delta \cdot \left(1 + \sqrt{\frac{8}{3}\epsilon}\right) / \epsilon \,, \tag{5.12}$$

$$b_s = \frac{1}{M^2 - 1} \,, \tag{5.13}$$

where ϵ is the compression ratio at the shock wave, calculated from:

$$\epsilon = \frac{(\gamma - 1)M^2 + 2}{M^2(\gamma + 1)} \,. \tag{5.14}$$

It is important to note that Eqs. (5.10) and (5.11), describing the shape and position of the bow-shock, are derived in the assumption that the shape of the windward portion of the contact discontinuity is approximated by an equation similar to Eq. (5.11)

$$y^2(x) = 2R_{cd}(r_{cd} - x) + b_{cd}(r_{cd} - x)^2 \,, \tag{5.15}$$

where b_{cd} is the bluntness of the contact discontinuity surface and R_{cd} is the radius of its curvature. Equations (5.11) and (5.15) are given in the form $y^2(x)$, because the atmosphere and bow shock are symmetrical with respect to the x axis.

As we noted above, the shape of the contact discontinuity may be derived from Eq. (5.9). When approximating it with Eq. (5.15) one can derive r_{cd}, R_{cd} and b_{cd}. Then, substituting them into Eq. (5.10), one can find Δ (the standoff distance of the shock wave), and obtain the shape of the wave using Eq. (5.11). Note that Eqs. (5.9) and (5.15) determine only the shape of the windward portion of the atmosphere. The shape of the leeward portion depends on the pressure distribution in the rarefaction wave propagating behind the planet. However, for the sake of simplicity one can presume that the leeward part of the atmosphere is governed by the equilibrium pressure with the environment and has a spherical shape (as shown in Fig. 5.3).

The above considerations apply in the case when the atmosphere is influenced only by the planet's gravity and by the pressure forces caused by the stellar wind incoming on the planet. For the class of exoplanets considered here one actually cannot neglect additional forces such as the gravity of the host star and the centrifugal force. Moreover, as we have shown above, the planet fills its Roche lobe, which inevitably leads to an outflow of gas from the planet onto the star and results in gas dynamic processes whose correct consideration may be conducted only by solving the complete system of gas dynamic equations. Nonetheless, the equations given above allow one to find important estimates and explain a number of physical processes taking place in the system. In particular, they allow one to determine all the possible types of gaseous envelopes around hot Jupiters and, hence, establish a classification.

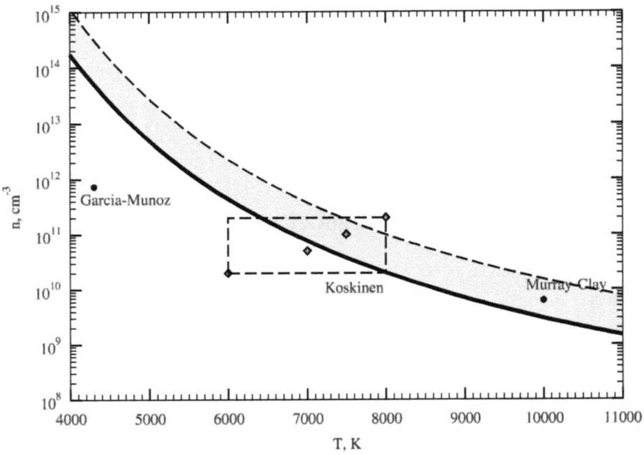

Fig. 5.4 Parameters of the atmosphere (temperature and gas number density at the photometric radius of the planet) of the HD 209458b hot Jupiter whose values allow the existence of one of the three determined types of planets' envelopes. The parameters corresponding to a closed atmosphere lie below the *solid line*. In the *shaded region* of the considered parameter space a quasi-closed atmosphere may exist, where the outflow through Lagrangian points is prevented by the dynamical pressure of the stellar wind. The region of outflowing (open) atmospheres is situated above the *dashed line*. The *points* denote the values of the parameters obtained for this planet in García Muñoz (2007) and Murray-Clay et al. (2009). The *rectangular* region encloses the range of parameters estimated in Koskinen et al. (2013) for the upper atmosphere of HD 209458b. The *diamonds* are the values for which we conducted gas dynamic simulations

5.4 Classification of Hot Jupiter Envelopes

Let us now use the equations and estimates obtained in the previous sections and consider possible configurations of hot Jupiters' gaseous envelopes. Equation (5.9) allows one to find the location of the Head-on Collision Point (HCP) with respect to the planet's center. If the HCP is inside the Roche lobe of the planet then no mass loss takes place. This atmosphere may therefore be considered as totally closed. If the HCP is outside the Roche lobe then the planet atmosphere outflows from L_1 and L_2.[1] In Fig. 5.4 the solid line separates two regions of the parameter space for a totally closed atmosphere (below) and an outflowing atmosphere (above), calculated for the HD 209458b hot Jupiter under the assumption that the stellar wind parameters correspond to those of the Sun (Withbroe 1988).

The analysis, performed in Bisikalo et al. (2013a) and Bisikalo et al. (2013b), shows that, for a certain set of the parameters, the outflows through the L_1 and L_2 points may be stopped by the dynamic pressure of the stellar wind. One can theoretically estimate a distance, at which the stream from L_1 decelerates down

[1]The atmosphere, when expanding, approaches the L_1 and L_2 points (say, uncork these points) at almost the same time. Thus, hereafter we consider a criterion of corking/uncorking only for the L_1 point, keeping in mind that the outflow starts/end through both points.

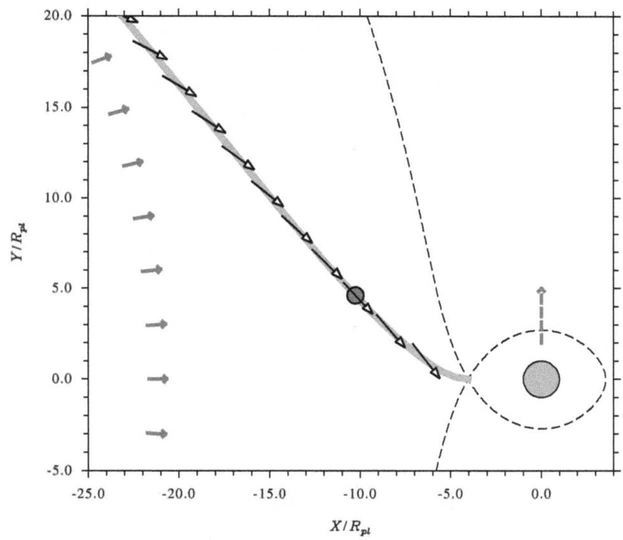

Fig. 5.5 Ballistic trajectory of the stream from the L_1 point (*solid grey line*). The *dashed line* denotes the isosurface of the Roche potential, passing through L_1. The planet's center is at the point with coordinates (0,0). The *black arrows*, crossing the stream's trajectory, denote the direction of the stellar wind in the coordinate system rotating along with the star/planet system. The *solid grey arrows* denote the radial direction of the stellar wind (the star is on the left), while the *dashed grey arrow* shows the direction of the planet's orbital motion. The *circle* on the trajectory corresponds to a point at which the stream and stellar wind are collinear

to zero, and find a criterion of its complete stopping by solving Eq. (5.7) with the stream's parameters (density ρ_{str}, velocity $\mathbf{v}_{str}$, and pressure[2] $p_{str} = \rho_{str} \cdot R_{gas} \cdot T_{atm}$) on one hand and the parameters of the stellar wind on the other hand. We are interested in finding a criterion for the stream stopping but not for its deflection, so we consider only the points on the trajectory of the stream that may potentially become the head-on collision point, i.e. locations where the velocity vectors of the stream and incoming stellar wind are collinear. In Fig. 5.5 we schematically show the ballistic trajectory of the stream from the L_1 point. The black arrows, crossing the stream's trajectory, show the directions of the stellar wind at the corresponding points of the stream in a coordinate system rotating along with the star/planet system. In this figure one can see that at a certain point on the trajectory (in Fig. 5.5 this point is denoted by the circle; let us call it collinearity point) the stream direction becomes collinear to the stellar wind, which allows solving Eq. (5.7).

It is easy to show that this point must exist for any set of parameters of the stellar wind. Indeed, at the very beginning of the outflow, when leaving the L_1 point, the

[2]Our assumption that the temperature of the stream is constant all along its trajectory is justified by the fact that the stream and the upper planet's atmosphere are heated by the radiation of the star in the same way.

stream moves directly towards the star; its radial velocity is equal to the sound speed and the tangential velocity component is zero. Afterwards, the stream is deflected by the Coriolis force in the direction of the planet's orbital motion. At the minimal distance from the star (in the periastron of the stream's trajectory) the radial velocity of the stream becomes zero for a non-zero tangential component. This means that in the part of the trajectory between L_1 and the periastron the velocity of the stream changes its direction from purely radial to purely tangential. On the other hand, in a rotating coordinate system, the wind velocity is also a composition of radial v_r (for simplicity we assume it to be constant) and tangential v_t components, the latter depending on the distance from the star as

$$v_t = \Omega \cdot r\,, \tag{5.16}$$

where r is the distance from the center of mass of the system that almost coincides with the center of the star. If the radial velocity is infinitely large we can neglect the tangential component and therefore find the point of collinearity at the L_1 point where the stream moves in the radial direction. If the radial velocity of the wind is zero the point of collinearity is situated at the periastron of the stream. Correspondingly, for any value of the wind's radial velocity v_r one can find a point of collinearity on the stream's trajectory from L_1 to the periastron. For the assumed properties of the stellar wind and with a fixed set of system parameters, the location of the point of collinearity depends only on the wind's radial velocity. In Fig. 5.6 we show the wind's radial velocity v_r, needed to find the point of collinearity at a certain distance from the planet, projected onto the limb of the star. The estimates are obtained for the hot Jupiter WASP-12b. If we assume that the stellar wind has the parameters of the solar wind (dashed line) the point of collinearity for WASP-

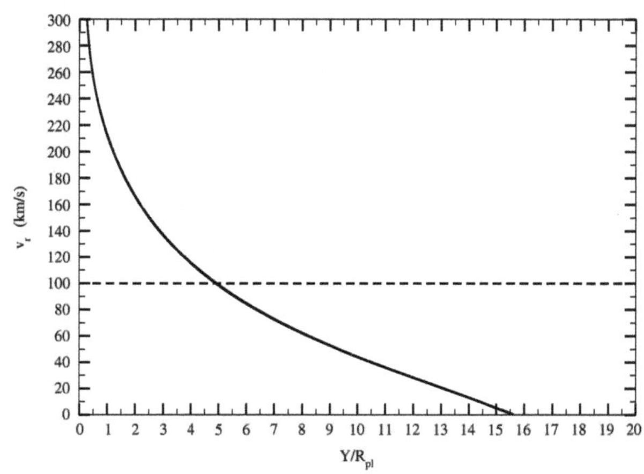

Fig. 5.6 Radial velocity of the stellar wind needed to have the point of collinearity at a certain distance as projected onto the limb of the star, for the WASP-12 system. The *dashed line* denotes the radial velocity of the stellar/solar wind at the distance of the planet

12b is located at a distance of ~5 planetary radii. For the hot Jupiter HD 209458b this point is located at ~4.6 R_{pl} ahead of the planet as projected on the stellar limb. By ignoring gas dynamical processes, the point of collinearity corresponds to the location of the head-on collision point of the bow shock.

If at the point of collinearity the dynamic pressures of the stream and wind are balanced (Eq. (5.7)) the stream stops and the flow pattern becomes stationary. If the sum of the gas and dynamic pressures of the stream overwhelms that of the wind, the stream will continuously move towards the star, because there is no second collinearity point on the trajectory from L_1 to the periastron. After the periastron passage the gas of the stream joins the envelope of the star (and/or accretion disk), without returning back onto the planet's atmosphere. Finally, if at the collinearity point the dynamic pressure of the stellar wind dominates, the wind keeps deflecting the stream until equilibrium pressure given in Eq. (5.7) sets in; at this location the head-on collision point settles.

Let us now determine the parameters for which the wind's dynamic pressure is not enough to stop the stream, i.e. the atmosphere is completely open and may be lost by the planet in a short time. To find these parameters one needs to determine a critical value of the atmospheric density $\rho_0^*(T_{atm})$ separating solutions with quasi-closed atmospheres ($\rho_0 < \rho_0^*$), where the outflow is present but stopped by the stellar wind, from those of open atmospheres ($\rho_0 > \rho_0^*$). In this procedure, several important physical effects must be taken into account. Since the stream accelerates under the effect of the stellar gravity, its density decreases while moving towards the star. Using the known stream acceleration law for a binary system (Lubow and Shu 1975) and the Bernoulli equation, one can determine the density decrease rate along the stream's trajectory. We also assume here that the radial velocity of the stellar wind is constant. Thus, from the flux conservation law, the density of the stellar wind should decrease proportionally to the square distance from the star:

$$\rho_w = \rho_{w*} \cdot \left(\frac{R_*}{r}\right)^2 ,$$

where ρ_{w*} is the wind density at the radius of the star. For the given parameters of the stellar wind, one finds the gas density at the point where the stream stops on the basis of the condition of equilibrium of dynamic pressures (5.7) at this point. One can further determine the gas density at L_1 and, assuming that the distribution of the atmospheric parameters is in equilibrium up to L_1, calculate the atmospheric density at the photometric radius ρ_0. The values of the critical density depend on the presumed temperature of the atmosphere (see Eq. (5.8)). The critical density is therefore a function of the atmospheric temperature $\rho_0^*(T_{atm})$.

The estimates, obtained for HD 209458b are shown in Fig. 5.4. The critical values of density, separating quasi-closed and open atmospheres, are denoted by the dashed line. For the parameters above the line, an atmosphere is completely open and is therefore subject to an enormous mass loss rate. For the parameters in the shaded region between two curves, we expect to have a quasi-closed non-spherical atmosphere. It is interesting to note that almost all the estimates of the atmospheric parameters of HD 209458b obtained by Koskinen et al. (2013) are in the region of closed or quasi-closed atmospheres.

The fundamental question is whether the introduced classification and criteria, separating different types of atmospheres, apply in general to all hot Jupiters. The possible problem is that the performed analysis, from which we derived the critical parameters of density and temperature (see Fig. 5.4), is approximate, because it disregards a number of gas dynamical effects.

- Our analysis neglects possible deviations of the stream's trajectory from a ballistic one. The approximation used here applies only for streams much denser than the surrounding environment. If this is not the case, the HCP may shift back and forth around the L_1, requiring to re-estimate ρ_0^*. However, the comparison of our analytical estimates with the results of the gas dynamic simulations shows that, due to the supersonic velocity of the stream, gas dynamical effects lead to a negligible deviation of the stream's trajectory requiring rather small corrections of the obtained criteria, only within a few percent.
- The other significant assumption made here is that the temperature and cross-section of the stream remain constant all along the trajectory. Considering the irradiation from the star as the main source of heating, the temperature of the stream may remain constant. However, the stream is bounded by the hot shock wave, its temperature may grow from the L_1 point toward HCP. The cross-section of the stream may also vary due to the compression of the stream by the stellar wind and to temperature variations. It is very difficult to take into account such effects, though conducting numerical simulations with the correct description of radiative transfer is not easier. The effects of radiative transfer included into the simulations may, indeed, significantly change the basic parameters of the atmospheres. However the criteria, separating the types of the atmospheres, in general, should remain qualitatively the same.
- Besides, we used here a simplified model of the stellar wind with a zero tangential velocity, constant radial velocity, and a density decreasing proportionally to the square distance from the star. These approximations work well in a system where the star rotates synchronously (or have no strong magnetic field that is capable to twist the wind) and the stream stops quickly allowing one to neglect the gradient of the radial velocity. This seems to be the case for most planet-hosting stars.
- Finally, the solution may significantly change if the gas escaping from the atmosphere forms a torus-like envelope or disk. In this case the problem becomes non-linear, i.e. the gas of the atmosphere mixes with the gas of the wind and changes its parameters from orbit to orbit. In this approach the problem cannot be solved analytically, though the main consequence of this effect is a variation of the parameters of the gaseous environment around the planet. Also in this case, the given criteria, separating different types of atmospheres, change only quantitatively while the general classification remains valid.

Summarizing the considerations discussed in this section, one can conclude that, depending on the location of the head-on collision point, hot Jupiters atmospheres may be classified in two classes: (1) if the HCP is inside the planet's Roche lobe, the envelope is almost spherical as the atmosphere is only slightly distorted by the tidal action of the star and the interaction with the stellar wind; (2) if the HCP is outside the planet's Roche lobe, outflow through the L_1 and L_2 points occurs and the envelope becomes substantially asymmetric. The latter type may also be split into two sub-classes. If the dynamic pressure of the wind is sufficient to suppress the more powerful outflow through the inner Lagrangian point L_1 then, as shown in Bisikalo et al. (2013a), a stationary, quasi-closed envelope forms in the system. If the wind is not capable to stop the stream from L_1 a non-spherical open envelope forms. To finally examine whether these conclusions on the types of hot Jupiter atmospheres are physically relevant we now consider the results of 3D gas dynamic simulations of typical hot Jupiters.

5.5 Shapes of Hot Jupiter Atmospheres as Obtained from 3D Numerical Simulations

This section presents the results of 3D numerical gas dynamic simulations of the envelope of the hot Jupiter HD 209458b (Bisikalo et al. 2013b). To conduct the simulations we adopted the model described in Bisikalo et al. (2013a) and Bisikalo et al. (2013b). We consider a system that consists of a star with $M_* = 1.1 M_\odot$ and $R_* = 1.1 R_\odot$, and a planet with $M_{pl} = 0.64 M_{Jup}$ and $R_{pl} = 1.32 R_{Jup}$. We also assume that the components of this binary system, having the separation $A = 0.045\,\text{AU}$, move in circular orbits with a period $P_{orb} = 3.5^d$. The linear velocity of the planet in this system is 141 km s^{-1}. Given the parameters of the stellar wind, assumed to be that of the Sun at the distance to the planet ($T_w = 7.3 \cdot 10^5$ K, $n_w = 1.4 \cdot 10^4\,\text{cm}^{-3}$, $v_w = 100\,\text{km/s}$ (Withbroe 1988)), the flow of the stellar wind is slightly subsonic, with a Mach number $M = 0.99$. However, taking into account the supersonic orbital motion of the planet with $M = 1.4$, the resulting velocity of the planet relative to the stellar wind appears to be significantly supersonic with a Mach number $M = 1.75$.

The flow is described by the 3D system of equations of gravitational gas dynamics, closed by the equation of state of perfect monatomic gas. In this model we neglect non-adiabatic processes of heating and cooling. To solve the system we use a TVD Roe-Osher scheme with the Einfeldt correction. This method is described in detail in Boyarchuk et al. (2002), Bisikalo et al. (2004), and Bisikalo et al. (2013d).

Table 5.1 Parameters of the atmosphere used to model the hot Jupiter HD 209458b. The temperature and gas number density are that at the photometric radius.

Model No.	T_{atm}, K	n_0, 10^{10} cm^{-3}
1	6,000	2
2	7,000	5
3	7,500	10
4	8,000	20

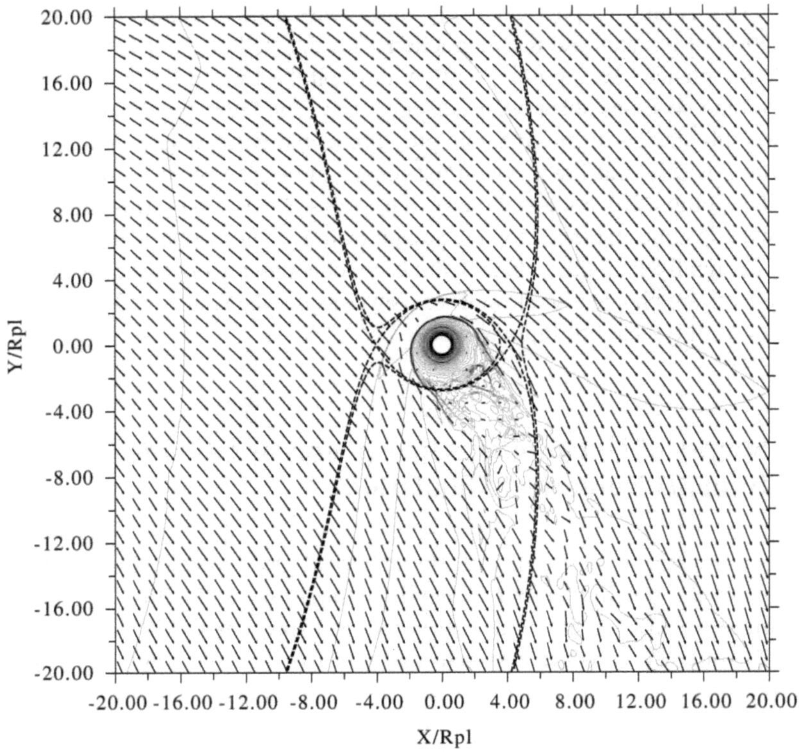

Fig. 5.7 Model 1: flow pattern in the vicinity of HD 209458b. The planet, with the center of mass at the point (0,0), is denoted by the *white circle*. The plot shows isolines of density and velocity vectors in the equatorial plane of the system. The spatial scale unit is R_{pl}. The *dashed lines* denote the Roche equipotentials containing the L_1 and L_2 points

The parameters of the planet's atmosphere ρ_0 and T_{atm} are set in accordance to the most recent (at the moment of writing this text) estimates, obtained for this planet (Koskinen et al. 2013). The parameters taken into account here are denoted by diamonds in Fig. 5.4 and are listed in Table 5.1. Note that for the given range of parameters, all three types of atmospheres discussed in the previous section may exist: Models 1 and 2 correspond to a closed atmosphere, while Models 3 and 4 result in a quasi-closed and open atmosphere, respectively. The results of the numerical simulations for the four sets of parameters are shown in Figs. 5.7–5.10.

Figures 5.7–5.10 show that the flow patterns strongly differ from case to case. As expected from the analytical consideration described above, in Model 1 (Fig. 5.7) we have obtained a closed atmosphere flown around by the stellar wind. Here one can see the formation of a symmetrical bow shock that is almost spherical near the HCP and tends to the Mach cone far from this point. The contact discontinuity, enclosing the planet's atmosphere, is totally within the Roche lobe of the planet. As a whole, the planet's atmosphere very little differs from a sphere. The mass loss rate

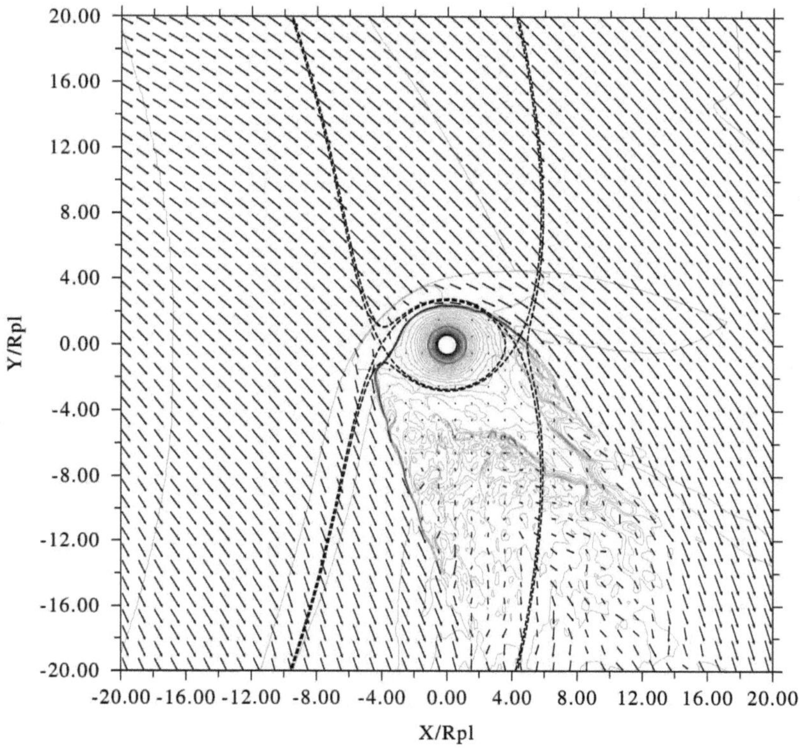

Fig. 5.8 Model 2: as in Fig. 5.7, but for model 2

from the atmosphere $\dot{M}$ in this Model is less than $\sim 1 \times 10^9$ g/s (Cherenkov et al. 2014).

In Model 2 (Fig. 5.8) the shape of the atmosphere is significantly non-spherical. In this case the HCP is shifted farther away from the planet in comparison to Model 1, but it is still located within the Roche lobe of the planet. In Fig. 5.8 one can clearly see two ledges, directed towards the L_1 and L_2 points, which results in important modifications of the shapes of the shock wave and contact discontinuity. In addition, the trail behind the planet (a region edged by the bow shock) is much broader than in Model 1. It is interesting to note that in this case we see no outflow towards the star from L_1 while we obtain a weak outflow through the L_2 point. The total mass loss rate from the atmosphere in this Model is $\dot{M} \approx 1.2 \times 10^9$ g/s (Cherenkov et al. 2014). Thus, the calculated envelope is partly open in spite of the fact that the analytical considerations lead to a closed atmosphere for this set of the parameters (see Fig. 5.4). From this we can deduce that the precision of the analytical estimates is tens of percents, which can be explained by the fact that, when obtaining these estimates, we disregarded gas dynamic effects.

In Fig. 5.9, 5.10 we plot the results for Models 3 and 4, respectively. The results for Model 4 (Fig. 5.10) and Model 3 are shown when the end of the stream in both

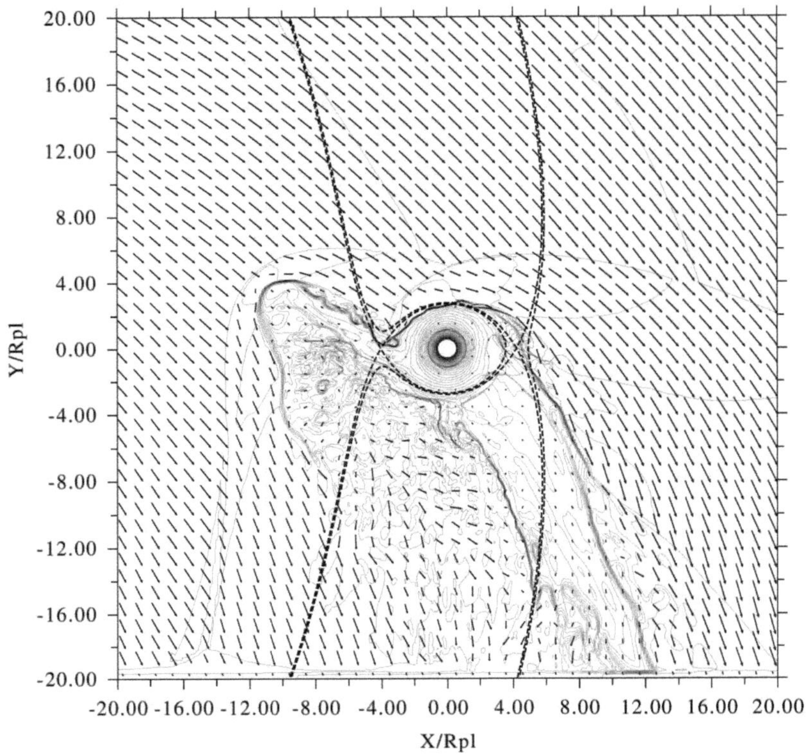

Fig. 5.9 Model 3: as in Fig. 5.7, but for model 3

the solutions is located at approximately the same distance from the planet. The flow structure in the system has qualitatively changed in comparison to Models 1 and 2. One can clearly see two streams, powerful from the L_1 point, directed to the star, and less powerful though noticeable, from L_2. According to the angular momentum conservation law these two streams are deflected in opposite directions, along the orbital motion (stream from L_1) and against it (stream from L_2). Unlike in solutions, typical for outflows in close binary stars (see, e.g., Bisikalo et al. (2004)), the regions of formation of these streams are rather extended. The stream from the L_1 point originates in a broad region between the Lagrangian point and the upper edge of the Roche lobe. The region of the stream's origin near the L_2 point is approximately of the same size. However, farther away along the streams, the flow structures are significantly different. While the stream from the L_1 point retains its shape almost unchanged and even gradually narrows, the stream from L_2, in contrast, greatly expands. The density isolines show that the stream from L_1 is much denser than that from L_2 at the same distance from the planet.

The solutions for Models 3 and 4 differ fundamentally from each other. In Model 4 the stream from the L_1 does not stop and keeps moving towards the star, i.e. the planet's atmosphere is open. This is well shown in Fig. 5.11,

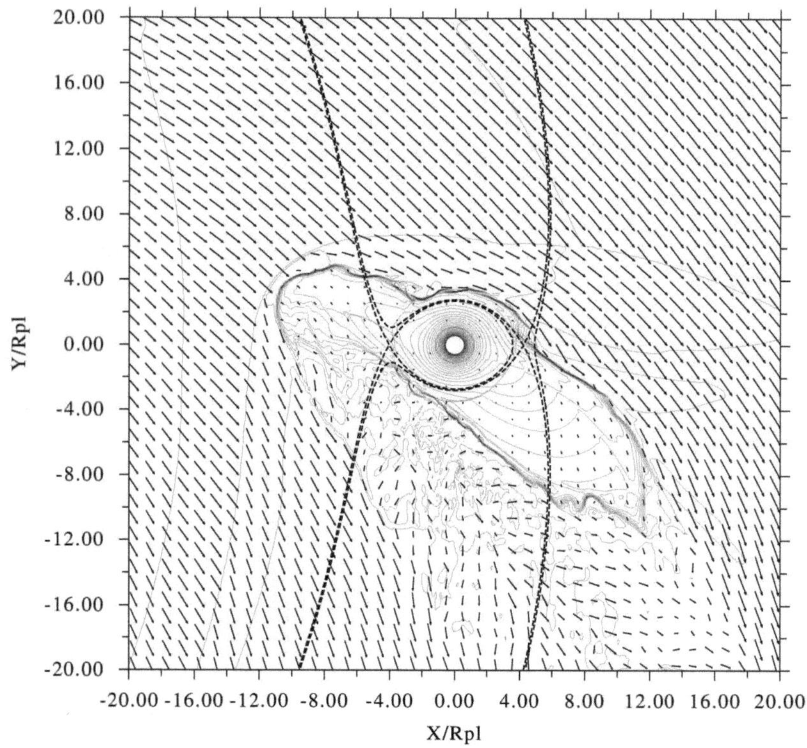

Fig. 5.10 Model 4: as in Fig. 5.7, but for model 4. This plot shows the solution at the time in which the end of the stream is at approximately the same distance from the center of the planet as in Model 3

where we plot the solution for Model 4 when the stream approaches the boundary of the computational domain. In the solution for Model 4 we have obtained a large mass loss rate of $\dot{M} \approx 3 \times 10^{10}$ g/s (Cherenkov et al. 2014). It is possible that, by analogy with close binary stars, in such systems acretion disks or tori of dense material may form. In Model 3 the atmosphere is quasi-closed, i.e. the streams are stopped by the stellar wind at certain distances. The weak outflow is observed along the discontinuity with a total mass loss rate of $\dot{M} \approx 3 \times 10^{9}$ g/s (Cherenkov et al. 2014). The shock wave and contact discontinuity in this solution have complex shapes. The asymmetric shape of the planet's envelope (consisting of atmosphere and ledge-like flows from L_1 and L_2) results in the formation of a pronounced double-humped structure of the wave, or, possibly, two shock waves, one around the atmosphere, the other around the stream from the L_1 point. The head-on collision point is on the tip of the stream from the L_1 point. However, when the wind approaches the planet, the atmosphere starts to influence the shock, bends it, pushes it out of the planet, and, finally, tends to form the second hump on the shock (or, even, the second shock). It is noticeable that the matter flow, moving

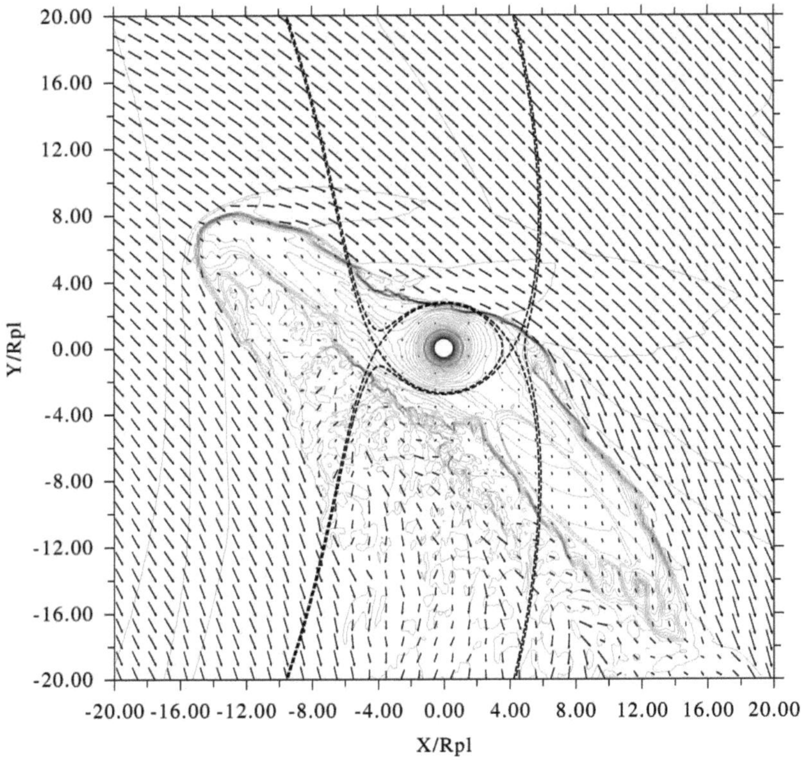

Fig. 5.11 Model 4: as in Fig. 5.7, but for model 4. This plot shows the solution at the time in which the stream has already passed the point of collinearity and is approaching the boundary of the computational domain

along the shock from the head-on collision point, gets substantially disturbed when approaching the dip between the humps. In particular, the vortices forming in this dip result in the erosion of the contact discontinuity and mixing of the stellar wind and atmosphere gas. In the solution, one also can see a number of short and weakly pronounced shock waves. Besides, the planet leaves a very broad trail because in the considered case the shock wave bounds not only the atmosphere but both the streams from the L_1 and L_2 points.

Assuming similar properties for the hot Jupiter discovered so far, one could expect that the obtained solutions (asymmetric, quasi-closed, long-lived envelope) are typical. As a further test we consider the flow structure near the hot Jupiter WASP-12b (Bisikalo et al. 2013a). WASP-12 is a late F-type main sequence star with $M_* = 1.35 M_\odot$ and $R_* = 1.6 R_\odot$ (Fossati et al. 2010b). The star hosts a transiting hot Jupiter, WASP-12b, with a mass of $M_{pl} = 1.41 \pm 0.1 M_{Jup}$ and radius of $R_{pl} = 1.74 \pm 0.09 R_{Jup}$ (Chan et al. 2011). WASP-12b revolves in a rather circular orbit (Campo et al. 2011) with the period of $\sim$1.09 days (Hebb et al. 2009) at a distance of 0.0229 AU ($\sim$3 stellar radii) from its host star.

As in the case of HD 209458b, there is reason to believe that the planet's atmosphere fills the Roche lobe. Indeed, the distance between the center of the planet and the L_1 point is only $1.85R_{pl}$. Taking into account the rather high temperature of the planet's upper atmosphere up to 10^4 K (Lai et al. 2010) we expect a very high degree of Roche lobe overfilling ($\Delta R/R \approx 0.16$). Besides, in analogy to the case of HD 209458b, the proper velocity of the stellar wind is subsonic ($M = 0.85$). However, the fast orbital motion of the planet ($M = 1.97$) leads to a significantly supersonic total planet's velocity relative to the stellar wind ($M = 2.14$).

The results of 3D gas dynamic simulations of the interaction between WASP-12b and its host star Bisikalo et al. (2013a) show that the overfilling of the planet's Roche lobe causes an important outflow of the atmosphere from the L_1 and L_2 points and the formation of a non-axisymmetric envelope near the planet. The supersonic motion of the planet and its envelope in the gas of the stellar wind results in the formation of a bow shock with a complex shape. The dynamic pressure of the stellar wind suppresses the active mass loss through the vicinities of the L_1 and L_2 points and leads to a steady-state and long-lived flow structure. This shows that the configuration with a quasi-closed asymmetric envelope might be typical for hot Jupiters.

Conclusion

As we see in Sects. 5.1–5.4 the classification of hot Jupiter envelopes is based on the degree of overfilling of the planet's Roche lobe. If the atmosphere of a planet overfills its Roche lobe powerful gas dynamic motions, as outflows from the Lagrangian points L_1 and L_2, occur in the system. The energy budget of these flows is rather large. As the stream from the L_1 point would constantly accelerate in the gravitational field of the host star, gas dynamic processes must dominate over the flow structure in the system. The radiative pressure of the star and/or magnetic fields of the star and planet influence the flow structure, but only marginally. Let us consider for example the influence of a planetary magnetic field. Despite the fact that no magnetic field has yet been detected on an exoplanet, there is a reason to believe that they should exist and can therefore influence flow structures (see, e.g., Vidotto et al. (2013)). Assuming that the magnetic axis of the planet is coaxial to the rotation axis, then the magnetic field acts isotropically in the equatorial plane of the system. For the discussed properties of hot Jupiters the solution in the equatorial plane is of fundamental importance, because this plane contains the stream from the L_1 point. As a result the isotropic influence of an even strong magnetic field in the equatorial plane (note that the magnetic field acts as extra-pressure) does not qualitatively change the solution with an asymmetric envelope but changes the spatial scale of the flow structure. The radiative pressure and magnetic

(continued)

field of the star act in an anisotropic manner and they may, in principle, distort the shape of the envelope, but for solar-type stars these phenomena are too weak to produce noticeable effects.

The degree of overfilling of the planet's Roche lobe depends on the temperature and density of the upper atmosphere. Hot Jupiters orbit close to their host stars and therefore their upper atmospheres are exposed to intensive plasma flows and irradiation from the host star. This results in the formation of dense and extended thermospheres and ionospheres. The upper atmosphere adsorbs mostly soft X-rays and hard UV photons, which noticeably increases its temperature (Yelle 2004; García Muñoz 2007; Koskinen et al. 2010, 2013). Besides, the ion composition of the upper atmosphere changes as $H_2 \rightarrow H \rightarrow H^+$. This is an additional factor to take into account in the formation of an extended envelope. Indeed, because of the dissociation of molecules in the atmosphere there is no efficient mechanisms of radiative cooling leading to a growth of the specific scale height. Thus, given the high temperatures of hot Jupiter upper atmospheres (several thousand Kelvin) the probability for such a planet to overfill its Roche lobe is extremely high.

The results of analytical considerations and 3D numerical simulations point towards the possibility that three types of gaseous envelopes around hot Jupiters may exist: spherical closed, asymmetric quasi-closed, and asymmetric open. Note that the range of atmospheric parameters obtained, for instance, for HD 209458b allows all the three types of envelopes to occur. However, it is highly probable (see Fig. 5.4) that the atmospheres of HD 209458b and similar exoplanets are asymmetric quasi-closed. In this case, the flow structure and the shape of the envelope are governed by outflows from hot Jupiter atmospheres.

The very first attempts to use an asymmetric envelope for the interpretation of observational data allowed us to explain various abnormal phenomena observed in hot Jupiters such as WASP-12b and HD 209458b. The asymmetric envelope and bow shock (see Bisikalo et al. (2013a)) are responsible for the early ingress observed in the UV and for the abnormally deep transit observed at wavelength of specific resonance lines of WASP-12b (see Fossati et al. (2010a) and Chap. 8 (Vidotto et al. 2014)) The existence of two flows, moving behind the bow shock in opposite directions from the head-on collision point (see Ionov et al. (2012) and Fig. 5.3) allowed us to explain the double-humped profiles of the C II and Si III absorption lines in the spectrum of HD 209458b (Linsky et al. 2010). As a matter of fact, the presence of asymmetric envelopes is in agreement with the observations.

We would like to stress that the existence of asymmetric envelopes around hot Jupiters and the presence of bow shocks of complex shapes drastically changes our understanding of matter distribution around this kind

(continued)

of planets. Consequently, this improves our capability to correctly interpret the observational results, but it opens new horizons in the field.

Acknowledgements The authors acknowledge the support by the International Space Science Institute (ISSI) in Bern, Switzerland and the ISSI team *Characterizing stellar- and exoplanetary environments* and thank L. Fossati from the Argelander-Institut für Astronomie der Universität Bonn, Germany, Lotfi Ben-Jaffel from the Institut Astrophysique de Paris (IAP) CNRS-UPMC, Paris and Tommi Koskinen from the Lunar and Planetary Laboratory University of Arizona, Tucson, USA for fruitful discussions. The authors also acknowledge the support by the RFBR projects 12-02-00047 and 14-02-00215.

References

Baranov, V. B., & Krasnobaev, K. V. (1977). *Hydrodynamic theory of a cosmic plasma*. Moscow: Izdatel Nauka.

Ben-Jaffel, L. (2007). *Astrophysical Journal Letters, 671*, L61.

Ben-Jaffel, L., & Sona Hosseini, S. (2010). *Astrophysical Journal, 709*, 1284.

Bisikalo, D. V., Boyarchuk, A. A., Kaigorodov, P. V., Kuznetsov, O. A., & Matsuda, T. (2004). *Astronomy Reports, 48*, 449.

Bisikalo, D. V., Kaygorodov, P. V., Ionov, D. E., Shematovich, V. I., Lammer, H., & Fossati, L. (2013a). *Astrophysical Journal, 764*, 19.

Bisikalo, D. V., Kaigorodov, P. V., Ionov, D. E., & Shematovich, V. I. (2013b). *Astronomy Reports, 57*, 715.

Bisikalo, D. V., Kaygorodov, P. V., & Ionov, D. E. (2013c). In N. V. Pogorelov, E. Audit, G. P. Zank (Eds.), *Numerical modeling of space plasma flows* (Astronomical Society of the Pacific Conference Series, vol 474, pp. 41). San Francisco: Astronomical Society of the Pacific.

Bisikalo, D. V., Zhilkin A. G., & Boyarchuk A. A. (2013d). *Gas dynamic close binary stars (in Russian)*. Moscow: Physmatlit.

Boyarchuk, A. A., Bisikalo, D. V., Kuznetsov, O. A., & Chechetkin, V. M. (2002). Mass Transfer in Close Binary Stars (Advances in astronomy and astrophysics, Vol. 6, pp 1–365). London and New York: Francis & Taylor.

Campo, C. J., Harrington, J., Hardy, R. A., Stevenson, K. B., Nymeyer, S., Ragozzine, D., Lust, N. B., Anderson, D. R., Collier-Cameron, A., Blecic, J., Britt, C. B. T., Bowman, W. C., Wheatley, P. J., Loredo, T. J., Deming, D., Hebb, L., Hellier, C., Maxted, P. F. L., Pollaco, D., & West, R. G. (2011). *Astrophysical Journal, 727*, 125.

Chan, T., Ingemyr, M., Winn, J. N., Holman, M. N., Sanchis-Ojeda, R., Esquerdo, G., & Everett, M. (2011). *Astrophysical Journal, 141*, 179.

Cherenkov, A. A., Bisikalo, D. V., & Kaigorodov, P. V. (2014). *Astronomy Reports, 58*, 679.

Fossati, L., Haswell, C. A., Froning, C. S., Hebb, L., Holmes, S., Kolb, U., Helling, C., Carter, A., Wheatley, P., Collier Cameron, A., Loeillet, B., Pollacco, D., Street, R., Stempels, H. C., Simpson, E., Udry, S., Joshi, Y. C., West, R. G., Skillen, I., & Wilson, D. (2010a). *Astrophysical Journal Letters, 714*, L222.

Fossati, L., Bagnulo, S., Elmasli, A., Haswell, C. A., Holmes, S., Kochukhov, O., Shkolnik, E. L., Shulyak, D. V., Bohlender, D., Albayrak, B., Froning, C., & Hebb, L. (2010b). *Astrophysical Journal, 720*, 872.

Fossati, L., Haswell, C. A., Linsky, J. L., & Kislyakova, K. G. (2014). In H. Lammer & M. L. Khodachenko (Eds.), *Characterizing stellar and exoplanetary environments* (pp. 59). Heidelberg/New York: Springer.
García Muñoz, A. (2007). *Planetary and Space Science, 55*, 1426.
Haswell, C. A., Fossati, L., Ayres, T., France, K., Froning, C. S., Holmes, S., Kolb, U. C., Busuttil, R., Street, R. A., Hebb, L., Collier Cameron, A., Enoch, B., Burwitz, V., Rodriguez, J., West, R. G., Pollacco, D., Wheatley, P. J., & Carter, A. (2012). *Astrophysical Journal, 760*, 79.
Hebb, L., Collier-Cameron, A., Loeillet, B., Pollacco, D., Hébrard, G., Street, R. A., Bouchy, F., Stempels, H. C., Moutou, C., Simpson, E., Udry, S., Joshi, Y. C., West, R. G., Skillen, I., Wilson, D. M., McDonald, I., Gibson, N. P., Aigrain, S., Anderson, D. R., Benn, C. R., Christian, D. J., Enoch, B., Haswell, C. A., Hellier, C., Horne, K., Irwin, J., Lister, T. A., Maxted, P., Mayor, M., Norton, A. J., Parley, N., Pont, F., Queloz, D., Smalley, B., & Wheatley, P. J. (2009). *Astrophysical Journal, 693*, 1920.
Ionov, D. E., Bisikalo, D. V., Kaygorodov, P. V., Shematovich, V. i. (2012). In M. T. Richards & I. Hubeny (Eds.), *From interacting binaries to exoplanets: Essential modelling tools* (IAU symposium, Vol. 282, pp. 545). Cambridge: Cambridge University Press
Koskinen, T. T., Yelle, R. V., Lavvas, P., & Lewis, N. K. (2010). *Astrophysical Journal, 723*, 116.
Koskinen, T. T., Harris, M. J., Yelle, R. V., & Lavvas, P. (2013). *Icarus 226*, 1678.
Lai, D., Helling, C., & van den Heuvel, E. P. J. (2010). *Astrophysical Journal, 721*, 923.
Landau, L. D., & Lifshitz, E. M. (1966). *Hydrodynamik, Lehrbuch der theoretischen Physik*. Berlin: Akademie-Verlag.
Lecavelier Des Etangs, A., Ehrenreich, D., Vidal-Madjar, A., Ballester, G. E., Désert, J. M., Ferlet, R., Hébrard, G., Sing, D. K., Tchakoumegni, K. O., & Udry, S. (2010). *Astronomy and Astrophysics, 514*, A72.
Li, S. L., Miller, N., Lin, D. N. C., & Fortney, J. J. (2010). *Nature, 463*, 1054.
Linsky, J. L., Yang, H., France, K., Froning, C. S., Green, J. C., Stocke, J. T., & Osterman, S. N. (2010). *Astrophysical Journal, 717*, 1291.
Lubow, S. H., & Shu, F. H. (1975). *Astrophysical Journal, 198*, 383.
Murray-Clay, R. A., Chiang, E. I., & Murray, N. (2009). *Astrophysical Journal, 693*, 23.
Pringle, J. E., & Wade, R. A. (1985). *Interacting binary stars* (Cambridge astrophysics series). Cambridge: Cambridge University Press.
Savonije, G. J. (1979). *Astronomy and Astrophysics, 71*, 352.
Verigin, M., Slavin, J., Szabo, A., Gombosi, T., Kotova, G., Plochova, O., Szegö, K., Tátrallyay, M., Kabin, K., & Shugaev, F. (2003). *Journal of Geophysical Research, 108*, 1323.
Vidal-Madjar, A. , Lecavelier des Etangs, A., Désert, J.M., Ballester, G. E., Ferlet, R., Hébrard, G., & Mayor, M. (2003). *Nature, 422*, 143.
Vidal-Madjar, A., Désert, J. M., Lecavelier des Etangs, A., Hébrard, G., Ballester, G. E., Ehrenreich, D., Ferlet, R., McConnell, J. C., Mayor, M., & Parkinson, C. D. (2004). *Astrophysical Journal Letters, 604*, L69.
Vidal-Madjar, A., Lecavelier des Etangs, A., Désert, J. M., Ballester, G. E., Ferlet, R., Hébrard, G., & Mayor, M. (2008). *Astrophysical Journal Letters, 676*, L57.
Vidotto, A. A., Jardine, M., Morin, J., Donati, J. F., Lang, P., & Russell, A. J. P. (2013). *Astronomy and Astrophysics, 557*, A67.
Vidotto, A. A., Bisikalo, D. V., Fossati, L., & Llama, J. (2014). In H. Lammer & M. L. Khodachenko (Eds.), *Characterizing stellar and exoplanetary environments* (pp. 153). Heidelberg/New York: Springer.
Withbroe, G. L. (1988). *Astrophysical Journal, 325*, 442.
Wood, B. E., Linsky, J. L., & Güdel, M. (2014). In H. Lammer & M. L. Khodachenko (Eds.), *Characterizing stellar and exoplanetary environments* (pp. 19). Heidelberg/New York: Springer.
Yelle, R. V. (2004). *Icarus, 170*, 167.

Chapter 6
Suprathermal Particles in XUV-Heated and Extended Exoplanetary Upper Atmospheres

Valery I. Shematovich, Dmitry V. Bisikalo, and Dmitry E. Ionov

Abstract The photolysis of hydrogen-rich atmosphere of a close-in exoplanet by the extreme ultraviolet radiation of the parent star leads to the formation of the suprathermal particles (i.e., particles with an excess of kinetic energy), primary photoelectrons in the $H_2/H/He$ ionization and hydrogen atoms in the H_2 dissociation and dissociative ionization processes. These particles with excess kinetic energies are an important source of thermal energy in the upper atmosphere of the hydrogen-rich exoplanets. In the contemporary aeronomical models the kinetics and transfer of hot hydrogen atoms and fresh photoelectrons were not calculated in detail, because they require solving of the Boltzmann equation for a non-thermal population of these particles. This chapter estimates the effect of the XUV radiation of the parent star on the production of the suprathermals in the $H_2 \rightarrow H$ transition region in the upper atmosphere of a hydrogen-rich exoplanet. Partial deposition rates of the stellar XUV radiation due to the photolytic processes in the $H_2 \rightarrow H$ transition region in the upper atmosphere of HD 209458b were calculated. The Monte Carlo model developed by authors was used to calculate the collisional kinetics and the transport of photoelectrons in the atmosphere of HD209458b. Using this model the partial deposition rates of the stellar XUV radiation due to the electron impact processes in the $H_2 \rightarrow H$ transition region in the upper atmosphere of HD209458b were calculated. This allowed us to estimate the heating rate of the atmospheric gas by photoelectrons in the upper atmosphere of exoplanet. For the first time the heating efficiency η with and without taking into account the photoelectron impact processes in the $H_2 \rightarrow H$ transition region in the hydrogen-rich atmosphere of exoplanet was calculated. Using the numerical stochastic model for a hot planetary corona the kinetics and transfer of suprathermal hydrogen atoms in the upper atmosphere and the emergent flux of atoms evaporating from the atmosphere were investigated. The latter is estimated as $5.8 \times 10^{12}\,\mathrm{cm}^{-2}\mathrm{s}^{-1}$ for a moderate stellar activity level of UV radiation, which leads to a planetary atmosphere evaporation rate of 5.8×10^9 g/s due to the process of the dissociation of H_2. This estimate shows that suprathermal hydrogen atoms provide a significant contribution to the observational estimate of $\sim 10^{10}$ g/s for the atmospheric loss rate of HD 209458b.

V.I. Shematovich (✉) • D.V. Bisikalo • D.E. Ionov
Russian Academy of Sciences, Institute of Astronomy, Moscow, Russian Federation
e-mail: shematov@inasan.ru; bisikalo@inasan.ru; ionovd@inasan.ru

H. Lammer, M. Khodachenko (eds.), *Characterizing Stellar and Exoplanetary Environments*, Astrophysics and Space Science Library 411,
DOI 10.1007/978-3-319-09749-7_6

6.1 Introduction: Short-Wavelength Radiation Effects in Upper Atmospheres

More than 1,000 exoplanets are known today (http://exoplanets.eu). The detection of hydrogen- and volatile-rich exoplanets at orbital distances <1 AU opens questions regarding their upper atmosphere structure and the stability against escape of atmospheric gases. Since $\geq$40 % of all discovered exoplanets are orbiting their host stars at distances closer than the orbit of Mercury, the atmospheres of these bodies evolve in much more extreme environments than what is known from the planets in our Solar System. More intense stellar X-ray, soft X-ray, extreme ultraviolet (EUV, the radiation between 1 and 100 nm contains soft X-rays and EUV and is considered as XUV) radiation and particle fluxes at such close orbital distances will change to a great extent the upper atmosphere structure of these objects.

The photolysis of hydrogen-rich atmosphere of a close-in exoplanet by the extreme ultraviolet radiation of the parent star leads to the formation of the suprathermal particles (i.e., particles with an excess of kinetic energy), primary photoelectrons in the $H_2/H/He$ ionization and hydrogen atoms in the H_2 dissociation and dissociative ionization processes. These particles with excess kinetic energies are an important source of thermal energy in the upper atmosphere of the hydrogen-rich exoplanet. This Chapter estimates and summarizes the effect of the XUV radiation of the host star on the upper atmosphere and the production of suprathermal atoms in the $H_2 \rightarrow H$ transition region in the upper atmosphere of a hydrogen-rich exoplanet.

6.2 Aeronomy of Suprathermal Atoms in Planetary Upper Atmospheres

Suprathermal atoms and molecules are generally considered to be particles with kinetic energies above 5–10k_BT, where T is the temperature of the ambient atmospheric gas. Suprathermal (or hot) particles are produced in various physical and chemical processes whose products have an excess kinetic energy owing to exothermic chemistry induced by photons and electrons and atmospheric sputtering by the solar wind, magnetospheric or pick-up ions. Dissociative recombination, dissociation by ultraviolet photons and electrons, and exothermic chemical reactions are accompanied by the release of energy on the order of several eV; part of this energy can be stored as the internal excitation of the products (Wayne 1991; Marov et al. 1996; Johnson et al. 2008). Charge exchange and atmospheric sputtering induced by energetic plasma ions can result in much larger energy transfers, producing hot particles with energies up to several hundred eVs (Johnson et al. 2008). If the production rate of these particles, which are typically suprathermal, is faster than thermalization, then a steady-state fraction of them is formed.

Interest in studies of the role of suprathermal particles in the physics and chemistry of planetary and satellite upper atmospheres has increased significantly in the last decades (Wayne 1991; Shizgal and Arkos 1996; Marov et al. 1996; Johnson et al. 2008). In particular, hot particles produced in upper atmospheric layers have been shown to play an important role in the chemistry and energetics of the upper atmosphere. Specifically, they

- Lead to local changes in the chemical composition, because the non-equilibrium rate coefficients of the chemical reactions (particularly with high activation energies) between suprathermal particles and the ambient atmospheric gas are much larger than those at thermal energies (Shematovich et al. 1994, 1999);
- Produce nonthermal atmospheric emission features (Hubert et al. 1999, 2001);
- Form hot planetary coronae (Nagy et al. 1990; Shematovich et al. 1994, 1999, 2005) and enhance nonthermal atmospheric losses (Shizgal and Arkos 1996; Johnson et al. 2008).

The stochastic simulation method had been developed to investigate the formation, kinetics, and transport of suprathermal particles for the hot planetary and satellite coronas by (Shematovich et al. 1994). This approach was first used to study the formation of the hot oxygen geocorona (Shematovich et al. 1994, 1999, 2005), taking into account the exothermic chemistry (Gerard et al. 1995) and the precipitation of magnetospheric protons and high-energy O^+ ions from the ring current (Bisikalo et al. 1995). A stochastic modeling approach was also applied to the study of the hot hydrogen corona at Jupiter (Bisikalo et al. 1996), formed by electron precipitation and the induced exothermic chemistry.

6.2.1 Hot Planetary Coronae

The uppermost layer of a planetary atmosphere, where the density of neutral particles is vanishingly low, is commonly called the exosphere or the planetary corona. Since the atmosphere is not completely bound to the planet by the planetary gravitational field, light atoms, such as hydrogen and helium, with sufficiently large velocities can escape from the upper atmosphere into interplanetary space. This process is commonly called Jeans escape, and depends on the temperature of the ambient atmospheric gas at an altitude where the atmospheric gas is virtually collisionless (Chamberlain and Hunten 1987), i.e., at the exobase. The heavier carbon, nitrogen, and oxygen atoms can escape from the atmospheres only through nonthermal processes such as photo- and electron impact dissociation, charge exchange, atmospheric sputtering, and ion pick-up (Johnson et al. 2008). Current theories of planetary coronae are based mainly on ground-based and space observations of exospheric emission features such as the 121.6 nm Ly-α and 102.6 nm Ly-β hydrogen lines, the 58.4 nm helium line, and the 130.4 and 135.6 nm atomic oxygen lines. The Mariner observations indicated the presence of hot hydrogen, and the Pioneer Venus UV spectrometer data established the presence of hot oxygen and carbon atoms at Venus (Johnson et al. 2008). These observations, together with in situ mass-spectrometer measurements, allow the density and temperature height profiles of the exospheric components to be constructed. The measurements reveal that planetary coronas contain both a fraction of thermal neutral particles with a mean kinetic energy corresponding to the exospheric temperature and a fraction of hot neutral particles with mean kinetic energy much higher than the exospheric temperature (Marov et al. 1996; Johnson et al. 2008).

These observations and models required a different description of the uppermost atmospheric layers than that provided by the thermal, collisionless models of planetary exospheres (Chamberlain and Hunten 1987). The existence of a hot atom component is a manifestation of the importance of non-thermal processes in planetary and satellite atmospheres (Marov et al. 1996; Johnson et al. 2008).

6.2.2 Suprathermal Neutral Particles

Suprathermal atoms and molecules are produced in various physical and chemical processes whose products have an excess kinetic energy. The main sources of suprathermal particles in the rarefied gas of planetary atmospheres are the following processes

- The charge-exchange between high-energy magnetospheric ions and neutral atmospheric gas components;
- The dissociative recombination of molecular ions with ionospheric electrons;
- The dissociation and dissociative ionization by ultraviolet solar radiation and magnetospheric plasma;
- The exothermic ion–molecule and neutral chemical reactions;
- The sputtering (or knock-on) of the atmospheric gas by magnetospheric plasma;
- The nonthermal desorption from the surfaces of the aerosol and dust fractions.

The dissociative recombination, the dissociation by solar ultraviolet photons and energetic electrons, and the exothermic chemical reactions which can be written as follows

$$\begin{cases} AB^{+} + e \rightarrow A^{*}_{hot} + B^{*}_{hot} \\ AB + h\nu(e) \rightarrow A^{*}_{hot} + B^{*}_{hot} + (e) \\ C + D \rightarrow A^{*}_{hot} + B^{*}_{hot} \end{cases} \tag{6.1}$$

are accompanied by the release of energy on the order of several eVs; part of this energy can be stored as the internal excitation of the products (Wayne 1991).

$$\begin{cases} A_{th} + B^{+}_{hot}(E) \rightarrow A^{+}_{th} + B_{hot}(E) \\ A_{th} + B^{+}_{hot}(E) \rightarrow A^{+}_{th} + B_{hot}(E' \leq E) \\ AC_{th} + B^{+}_{hot}(E) \rightarrow A^{+}_{th} + C_{hot} + B_{hot}(E' \leq E) \end{cases} \tag{6.2}$$

The charge exchange of high-energy ions and the sputtering of the atmospheric gas by magnetospheric plasma can result in much higher energy, producing hot particles with energies up to several hundred eVs (Johnson et al. 2008).

The fresh suprathermal particles lose their excess kinetic energy in elastic and inelastic collisions with the ambient atmospheric gas

$$A_{th} + B_{hot}(E) \rightarrow A_{hot}(E' < E) + B_{hot}(E'' \leq E - E'). \quad (6.3)$$

These processes are generally considered in the linear approximation, where it is assumed that the ambient gas is in thermal equilibrium and that the perturbation of its state by suprathermal particles is small. If, however, the production rate of hot particles is high, then the nonlinear kinetic approximation should be used, because the secondary particles are also produced with suprathermal energies. Accordingly, their subsequent collisions with the ambient gas lead to the cascade formation of new hot particles and, as a result, to significant perturbations of the thermal state of the atmospheric gas. Therefore, the kinetics of hot particles can be rigorously described only at the microscopic level using the Boltzmann kinetic equation.

6.2.3 *Kinetic Description of Suprathermal Particles*

Let the rarefied gas of a planetary atmosphere be composed of $\alpha_i, i = 1, \ldots, S$ atoms and molecules in a physical volume V. Each particle of the component α_i (an atom, a molecule, and/or their ion) has its own mass m_i, position $\mathbf{r}_i \in V$, velocity $\mathbf{c}_i$, and set of quantum numbers $\mathbf{z}_i$ for each of the possible internal excitation levels. These chemically distinct components collisionally interact through $m = 1, \ldots, M > 1$ chemical reactions in accordance with the dynamic schemes

$$m : \alpha_i(\mathbf{c}_i, \mathbf{z}_i) + \alpha_j(\mathbf{c}_j, \mathbf{z}_j) \rightarrow \alpha_k(\mathbf{c}'_k, \mathbf{z}_k) + \alpha_l(\mathbf{c}'_l, \mathbf{z}_l). \quad (6.4)$$

For the generality of describing this chemically reactive system, we will consider reactions shown in Eq. (6.4) as a collisional process that includes both elastic ($\alpha_i = \alpha_k$ and $\alpha_j = \alpha_l$), and inelastic ($\alpha_i = \alpha_k$, $\alpha_j = \alpha_l$, but $\mathbf{z}_i \neq \mathbf{z}_k$ and/or $\mathbf{z}_j \neq \mathbf{z}_l$ internal excitation levels differ) and chemically reactive ($\alpha_i \neq \alpha_k$ and/or $\alpha_j \neq \alpha_l$) collisions. The probabilities of reactions Eq. (6.4) are specified by the scattering functions $g_{ij} d\sigma_m = |c_i - c_j| d\sigma_m(|c_i - c_j|, \Omega) d\Omega$, where $d\sigma_m$ are the differential scattering cross section for reactions Eq. (6.4) $g_{ij} = |c_i - c_j|$ is the relative velocity, and Ω is the solid angle of the scattering. Each channel of the collisional process Eq. (6.4) has the corresponding (elastic $\sigma_m^{(el)}$, inelastic $\sigma_m^{(in)}$, and chemically reactive $\sigma_m^{(r)}$) scattering cross sections, with $\sigma_m = \sigma_m^{(el)} + \sigma_m^{(in)} + \sigma_m^{(r)}$. The velocities c'_k, c'_j of the product particles from reaction m can be calculated from the laws of conservation of mass, momentum, and total energy of the interacting molecules and their direction in laboratory frame is determined by the probability density $d\sigma_m/\sigma_m$.

The evolution of a chemically reactive system at the microscopic level of description can be determined by solving the system of Boltzmann kinetic equations

$$\frac{\partial F_{\alpha_i}}{\partial t} + \mathbf{c}\frac{\partial F_{\alpha_i}}{\partial \mathbf{r}} + \frac{\mathbf{Y}}{m_\alpha}\frac{\partial F_{\alpha_i}}{\partial \mathbf{c}} = Q_{\alpha_i} + \sum_m J_m^{\alpha_i}(F_{\alpha_i}, F_{\alpha_j}), i, j = 1, \ldots, S, \quad (6.5)$$

together with initial and boundary conditions for the atmospheric gas in volume V subjected to the external force fields $\mathbf{Y}$ of the planet, and under such physical assumptions as gas rarefaction and finite or rapidly decreasing particle interaction radii in collisions (Marov et al. 1996). Here, we microscopically describe the state of the gas by using the velocity and internal excitation state distributions functions for the gas particles $F_{\alpha_i}(t, \mathbf{r}, \mathbf{c}) = n_{\alpha_i}(t, \mathbf{r}, \mathbf{z}) f_{\alpha_i}(t, \mathbf{r}, \mathbf{c})$, where $n_{\alpha_i}(t, \mathbf{r}, \mathbf{z})$ is the number density of the particles in state z, and $f_{\alpha_i}(t, \mathbf{r}, \mathbf{c})$ is the single-particle velocity distribution function normalized to unity. The source functions Q_{α_i} specify the suprathermal-particle production rates in the collisional processes Eqs. (6.1) and (6.2). The collision integrals on the right-hand sides of the kinetic equations describe the change in gas state due to any collisions and are written in the standard form.

The chemical kinetics of a rarefied atmospheric gas at the microscopic level of description is completely determined by means of the dynamical and probabilistic characteristics of molecular collisions – the scattering functions and the distributions of the colliding particles in translational and internal degrees of freedom. The chemical evolution of the atmospheric gas with the production of suprathermal particles has a complex structure of the kinetic rates of translational and internal energy exchange. One can discriminate between the following characteristic cases

- Linear case: chemistry of suprathermal (hot) particles – a situation where suprathermal particles (hot subsystem) are a small admixture that weakly perturbs the thermal state of the ambient atmospheric gas (thermal subsystem), i.e., the source functions in Eq. (6.5) are much smaller than collision integrals. Thus, the thermal and hot subsystems are described by using the balance gasdynamic equations for the thermal components and the kinetic equations Eq. (6.5) for the suprathermal components, which contain the partially averaged collision integrals between particles of the different subsystems.
- Nonlinear case: microscopic non-equilibrium kinetics – a situation where the characteristic microscopic and macroscopic timescales of the change in parameters for all gas components are comparable. In this case, the state of the gas is determined by the solution of the basic system Eq. (6.5) of nonlinear kinetic equations, and, accordingly, the distribution functions depend explicitly on time.

The flow of gas in a planetary atmosphere is most accurately described either by a mixed kinetic system, where the perturbations of the thermal state of the ambient atmospheric gas by suprathermal particles are small, or by a completely kinetic system Eq. (6.5) of Boltzmann equations, where these perturbations are significant. A number of methods have been developed in the kinetic theory of rarefied gases to investigate the gas dynamics and kinetics in states close to the local thermal equilibrium (Bird 1994). Highly non-equilibrium systems are difficult to analyze because of the mathematical complexity of the Boltzmann kinetic equations (nonlinearity and high multiplicity of the collision integrals), which requires using new and sophisticated approaches developed in the field of rarefied gas dynamics.

6.2.4 *The Stochastic Kinetic Equation for Suprathermal Particles*

A very promising approach is the development of discrete mathematical models that use the probabilistic interpretation of collisions in an ensemble of model particles. The Direct Simulation Monte Carlo (DSMC) method (Bird 1994) and its modification for studying non-equilibrium processes in the planetary atmospheres (Shematovich et al. 1994; Marov et al. 1996; Shamatovich 2004) belong to this class of approaches. A stochastic discrete model to investigate the formation, kinetics, and transport of suprathermal particles in a planetary atmosphere should take into account the following peculiarities of the flow of atmospheric gas:

- The local mean free time and path for suprathermal particles gas should be taken as the characteristic time and space scales at the molecular level of describing the state of the gas in the planetary corona;
- The parameters of the atmospheric gas change strongly in a planetary atmosphere from the collision-dominated regime of gas flow in the dense thermosphere to the virtually collisionless (free-molecule) regime of flow in the exosphere;
- Significant differences between the densities of the suprathermal particles produced in the chemical and magnetospheric plasma sputtering processes and the ambient atmospheric gas are commonly observed.

Therefore the following approaches must be used in constructing a numerical model of hot planetary coronae:

- The splitting of the solution of the basic kinetic system Eq. (6.5) in physical processes into the simulation steps for the suprathermal particle sources, the collisional thermalization of these particles, and the free molecular transport of suprathermal particles in the planetary corona on a discrete time scale;
- The stochastic simulation of the formation of suprathermal particles and their local kinetics by using analog Monte Carlo algorithms with statistical weights;
- The calculation of the trajectories of suprathermal particles in the planetary corona by using finite- difference algorithms.

Basing on the theory of random processes, the formation, kinetics and transport of the suprathermal particles in the atmospheric gas can be described by the following stochastic kinetic equation (Shematovich et al. 1994; Marov et al. 1996; Shamatovich 2004)

$$\frac{\partial}{\partial t}\varphi(\mathbf{X},t) = V^{-1}\sum_{m}\sum_{i,j}\int g_{ij}d\sigma_m[\varphi(\mathbf{X}_{ij}^{\mathbf{m}},t) - \varphi(\mathbf{X},t)]. \tag{6.6}$$

This equation is linear with respect to the probability density distribution $\varphi(\mathbf{X},t)$ for state $\mathbf{X} = [\ldots,\mathbf{c}_i,\ldots,\mathbf{c}_j,\ldots]$ of the gas at time t and is called the stochastic (or master) kinetic equation for the chemical kinetics of a rarefied gas. The Eq. (6.6) describes the evolution of a homogeneous jump-like Markovian process (Shematovich et al. 1994; Marov et al. 1996; Shamatovich 2004; Shematovich 2007, 2008).

6.2.5 The Analogue Monte Carlo Method of Solving the Stochastic Kinetic Equation

The direct methods of solving the stochastic kinetic equation consist in setting up and solving a system of equations for the probabilities of all possible paths of the state of a chemically reactive rarefied gas. Unfortunately, this direct procedure can be performed only for a few very simple chemical systems (van Kampen 1984) and involves enormous computational difficulties for real systems of chemical reactions. The Monte Carlo method, which consists in generating a sample of paths for the state of a chemically reactive gas, is an efficient tool for studying complex chemical systems in the stochastic approximation. The path generation procedure is much simpler – a sequence of transitions between the states of a chemically reactive gas and transition-separating times should be drawn based on the proper probability distributions. Such procedure is an analogue Monte Carlo algorithm for solving the stochastic kinetic equation (Marov et al. 1996; Shamatovich 2004). In the numerical realizations of the stochastic model the following recent developments in the theory and practice of DSMC method was used:

- An effective approximation of the majorant frequency (Ivanov and Rogazinskij 1988; Shamatovich 2004), where the collision probability for the chosen pair is estimated from the maximum possible frequencies, is used in choosing the next transition;
- The multichannel nature of the selected reaction is taken into account for the transition to be realized; i.e., this transition is treated as the simultaneous drawing of all possible (elastic, inelastic, and chemically reactive) channels. For each of them the corresponding weight is transferred proportional to the ratio of the partial cross section for a given channel to the total cross section of the collisional process;
- Since the algorithmic steps of throwing in suprathermal particles, in accordance with the source functions, and drawing the collisional transitions are accom-

panied by the formation of new model particles, it is necessary to control the total number of model particles in the numerical model. An efficient method for this control is the so-called clustering of model particles (Rjasanow et al. 1998; Shamatovich 2004), where groups of model particles with similar parameters are combined into a single particle with weighted parameters. This procedure allows the total number of model particles to be controlled.

6.2.6 Current Progress on Hot Atom Corona Modeling

The creation and loss to space of hot hydrogen, carbon, nitrogen and oxygen atoms at terrestrial planets are due to:

Exothermic photochemistry: include both direct photo- and electron impact dissociation of atmospheric molecules, exothermic ion-molecular chemistry, and, in particular, dissociative recombination of molecular ions.

The photo- and electron impact dissociation of molecules, as well as the dissociative recombination of the molecular ions result in substantial densities of suprathermal H, C, N, and O atoms in the upper atmospheres of the terrestrial planets. These hot atoms, in turn, react with the ambient atmospheric gas triggering hot atom chemistry. Subsequently, the transport of suprathermal atoms to exospheric heights leads to the formation of hot atomic coronae around Venus, Earth, and Mars. It has been well established by both observations and theoretical calculations (see reviews Nagy et al. 1990; Shizgal and Arkos 1996, Johnson et al. 2008 and references within) that hot atoms is an important constituent in the transition region between upper thermosphere and exosphere at terrestrial planets.

A number of theoretical model calculations of hot hydrogen, carbon, nitrogen, and oxygen populations have appeared to compare with direct observations of hot oxygen (Johnson et al. 2008). The presence of an extended neutral corona plays an important role in mass loading and slowing down the solar wind at Venus and Mars. The numerical analysis of the processes of formation, collisional kinetics, and transport of suprathermal atoms in the transition region of the upper atmosphere of the terrestrial planet is based on the solution of the Boltzmann kinetic equation (Shematovich et al. 1994, 1999). In such models it is important to use the differential cross sections, because these molecular data are critical parameters in the calculations of the thermalization rate of suprathermal atoms in collisions with the ambient atmospheric gas. It is known, that the calculated differential cross sections for elastic collisions of hot atoms with the main atmospheric constituents – O, N_2, O_2, are characterized by a strong peak at small scattering angles for energies below 5 eV. Consequently, it was found by Krestyanikova and Shematovich (2005), Krestyanikova and Shematovich (2006) and Gröller et al. (2010) that such scattering angle distributions resulted in a lower rate of energy loss by the suprathermal oxygen atoms, and consequently, in higher escape rates as compared with the models utilizing an isotropic distribution of scattering angle in hard sphere model of elastic collisions (Fox and Hać 1997, 2009).

Ion sputtering: a fraction of the ions produced in the corona and ionosphere re-impact the neutral atmosphere with enough energy to eject atmospheric neutrals.

Such precipitating ions can cause massive sputtering of the atmosphere (Johnson et al. 2008). These ions of exospheric origin are accelerated by the solar wind and the interplanetary magnetic field. Pick-up ions follow helical trajectories along interplanetary magnetic field lines draped across Venus and Mars and can either be swept away or re-impact the atmosphere with significant amounts of energy (up to 1 keV). Through momentum transfer collisions they can excite other atmospheric atoms and molecules, so enhancing escape and populating the hot corona. There are several sources of uncertainty in present estimates of the sputtering effect on the Venusian and Martian atmospheres such as the energy spectra of ionospheric pick-up ions, and the energy dependencies of differential cross sections for collisions of high-energy ions with atmospheric species. Furthermore, there is a feedback which determines escape, i.e., an inflated corona can, in principle, lead to greater loss rates and higher pick-up ion formation, but this is mitigated by the fact that the solar wind plasma will be deflected at larger distances from the Martian exobase, reducing the coronal heating and expansion. Describing this feedback process will be critical for determining the loss of atmosphere in earlier epochs.

Ion escape and ionospheric outflow: ions produced by photoionization, electron impact, and charge-exchange in the planetary corona are dragged away along solar magnetic field lines wrapping the planet leading to their loss. An ionospheric outflow is realized when ionospheric matter is accelerated by the convective electric field induced by the solar wind.

Without shielding by the planetary magnetic field, the solar wind plasma transfers momentum to atoms and ions on high ballistic trajectories and they can be swept away from the planet by the solar wind. There are two main sources for the pickup of new ions by the solar wind. The first is ionization of neutral particles inside the corona by UV photons, electron impact or charge exchange. Such a flux is composed mainly of O^+, H^+ and C^+. The second is the ionospheric planetary wind; that is, the outflow of ions produced above the photochemical equilibrium region and below the ionopause. Coupled ionosphere and thermosphere models by Fox et al. (2008) place constraints on atmospheric loss through ion outflow, because the relative rates of ion loss are determined by ion-neutral chemistry. The inclusion of exothermic chemistry results in the production of hot atoms through processes that involve ions, such as the dissociative recombination of O_2^+, N_2^+, CO^+, and NO^+, which yields fragments of suprathermal energies.

6.3 Suprathermals in the Extended Atmosphere of the Hot Jupiter HD 209458b

Many of the recently discovered exoplanets have relatively large masses and show similarities to Jupiter and Saturn in our Solar System. However usually they orbit their host stars at distances closer 0.1 a.u., therefore such giant planets are called hot Jupiters. Direct observations of giant exoplanets are scarce because a planetary signal is difficult to isolate from the much stronger signal from the star. Nonetheless, the orbits of some exoplanets lie on a line of sight with the Earth, which allows for the study of the absorption spectrum of the atmosphere of the planet transiting in front of the star (see Chap. 4, Fossati et al. (2014)). The most interesting results were obtained in the UV range (Vidal-Madjar et al. 2003, 2004; Ben-Jaffel 2007; Ben-Jaffel and Sona Hosseini 2010; Ehrenreich et al. 2008; Linsky et al. 2010). The sufficiently strong weakening of the stellar emission of HD 209458 in the HI Lyman-α line and in the lines of atomic oxygen at 130.5 nm and the ionized carbon at 133.5 nm was observed (see Chap. 4, Fossati et al. (2014) for details). The observed absorption in the UV resonance lines was significantly deeper than one caused by the the planetary disc alone indicating the presence of an extended atmosphere around HD 209458b. The planet itself shields only 1.5 % of stellar radiation in the visible range (Ballester et al. 2007), which means that the planet is surrounded by a neutral hydrogen cloud as large as about 3.3 planetary radii (R_p). The Hill sphere radius for this exoplanet, which determines the area where the gravitational attraction of the planet dominates over the gravitational attraction of the central star, equals to 4.08 R_p. The observed size of the extended hydrogen atmosphere is comparable to this value. Atoms and molecules reaching the boundary of the Hill sphere (or Roche lobe) can leave the atmosphere; therefore, strong outflows can occur (see for details Chap. 5, Bisikalo et al. (2014)). Ballester et al. (2007) reported the results of observations of HD 209458b with the STIS imaging spectrograph onboard the Hubble Space Telescope (HST) in a broad UV-to-optical spectral range. They detected a new feature in the planetary atmosphere absorption spectrum at 356–390 nm that they interpreted as Balmer-continuum absorption by hot (suprathermal) neutral hydrogen atoms in the upper atmosphere of the planet.

To interpret the observational results cited above, aeronomical models for the physical and chemical processes in the upper atmosphere of hot Jupiters were developed (Yelle 2004, 2006; García Muñoz 2007; Penz et al. 2008; Koskinen et al. 2010, 2013). These aeronomic models agreed with the hypothesis of Lammer et al. (2003, 2009) that close-in hydrogen-rich exoplanets experience dynamic atmospheric expansion and outflow up to their Roche lobes with mass-loss rates in the order of $\sim(1-5)\times10^{10}\,\mathrm{g\,s^{-1}}$. Based on the results of the observational and aeronomical models, the following different interpretations were proposed for the physical and evolutionary state of the upper atmosphere of HD 209458b.

Evaporating upper atmosphere: The close proximity of the planet to its parent star (0.045 AU) leads to the strong insolation of its upper atmosphere by UV stellar radiation. Lammer et al. (2003) were the first to provide a model of exoplanetary upper atmospheres, which is based on an approximate solution of the heat balance equation in planetary thermosphere and found that hydrogen-rich upper atmospheres of Jupiter-type gas giants in close orbital distances will be heated to a few 10^4 K, so that hydrostatic conditions cannot be valid anymore, because the planet's upper atmosphere will expand dynamically upwards. The expanding atmosphere produces strong outflows, i.e., the atmosphere starts to evaporate (Vidal-Madjar et al. 2003, 2008; Lecavelier des Etangs et al. 2004; Lecavelier Des Etangs et al. 2010; Lecavelier des Etangs et al. 2012; Bourrier and Lecavelier des Etangs 2013). The corresponding large spatial size of the extended atmosphere leads to its large optical depth in the Ly-α line. The possible atmospheric mass-loss rate is estimated at about 10^{10} g s^{-1} (Ehrenreich et al. 2008; Linsky et al. 2010).

Charge-exchange reactions between the stellar wind and neutral planetary corona: Under conditions of the strong insolation of planetary upper atmosphere by UV stellar radiation the exobase can move to locations which are above a possible magnetopause or even at the Roche lobe distance (e.g., Lammer et al. 2003, 2009; Lecavelier des Etangs et al. 2004; Erkaev et al. 2007), so that the upward-flowing neutral gas can interact with the dense stellar plasma flow (Holmström et al. 2008; Ekenbäck et al. 2010; Lammer et al. 2011). The observed Ly-α profile may also be explained by the interaction of stellar wind protons with the neutral atoms of the planetary atmosphere. Then, the protons of the solar wind become the source of the observed high-velocity neutral hydrogen atoms (Holmström et al. 2008; Ekenbäck et al. 2010).

Gas-dynamic interaction between the stellar wind and neutral planetary corona: Usually the orbital velocities of hot Jupiters are supersonic relative to the stellar wind, resulting in the formation of a bow shock. Gas-dynamical modeling by Bisikalo et al. (2013a,b) (see also Chap. 5, Bisikalo et al. (2014)) shows that the gaseous envelopes around hot Jupiters can belong to two classes, depending on the position of the collision point. If the collision point is inside the Roche lobe of the planet, the envelopes have almost spherical shapes of classical atmospheres, slightly distorted by the influence of the star and interactions with the stellar-wind gas; if the collision point is located outside the Roche lobe, outflows from the vicinity of the Lagrangian points L_1 and L_2 arise, and the envelope becomes substantially asymmetrical. The latter class of objects can also be divided into two types. If the dynamical pressure of the stellar-wind gas is high enough to stop the most powerful outflow from the vicinity of the inner Lagrangian point L_1, a closed non-spherical envelope forms in the system. If the wind is unable to stop the outflow from L_1, an open non-spherical envelope forms. Using the typical hot Jupiter HD 209458b as an example, it was shown by Bisikalo et al. (2013a,b) that all three types of atmospheres could exist within the range of estimated parameters of this planet. Since different types of envelopes have different observational manifestations, determining the type of the envelope of HD 209458b could apply additional constrains on the parameters of this exoplanet. For most probable case of non-spherical close atmosphere the loss rate was estimated as 4.0×10^9 g/s (see Chap. 5, Bisikalo et al. (2014))

Currently, all of these explanations together with several other possible hypotheses regarding the formation and evolutionary status of the extended upper atmosphere of HD 209458b are being actively discussed, but full clarity has not yet been achieved. Further observations of the transiting planet are needed.

6.4 Heating Efficiency in Hydrogen-Dominated Upper Exoplanet Atmospheres

The flux of stellar XUV emission photons incident upon a planetary atmosphere of hydrogen-dominated composition photoionizes the gas producing a flux of high energy photoelectrons, which again deposit their energy into the gas. In a partially neutral medium, electrons ionize, excite, and dissociate atomic and molecular species, as well as heat the gas through Coulomb collisions. In determining these energy deposition events, we must account for all the possible degradation histories of the energetic electrons. When the stopping medium is only partially neutral, electron-electron interactions contribute to the electron energy degradation, and a significant portion of the fast electron energy is deposited into the stopping medium as heat. As the fractional ionization rises, more and more of the electron energy heats the gas, and the excitation and ionization yields decrease.

Most of the estimates of the extrasolar hot Jupiters escape rates are based on a total conversion of the absorbed stellar XUV energy into powering escape. Therefore, it is important to assign the fraction of stellar XUV radiation that goes into the heating of upper atmospheres of extrasolar giant planets, and to evaluate how deep in an atmosphere stellar photons, e.g. in the EUV spectral range, keep releasing part of their energy as heat. To estimate the effect of XUV emission from solar-type stars of different ages an accurate description of radiative transfer and photoelectron energy deposition is required. In the studies by Cecchi-Pestellini et al. (2006, 2009) it was shown that X-rays give an important contribution to the heating of hydrogen-dominated planetary atmospheres.

6.4.1 Photolytic and Electron-Impact Processes in the Upper Atmosphere

The incoming stellar XUV flux decreases due to absorption in the thermosphere, which results in dissociation and ionization and, hence, in heating of the upper atmosphere. The extreme UV radiation of the star is absorbed by atmospheric gas and leads to the excitation, dissociation, and ionization of different components of the atmosphere. In a case of $H_2/H/He$ atmosphere the following photolytic processes were taken into account

$$H_2 + h\nu, (e_p) \rightarrow \begin{cases} H(1s) + H(1s, 2s, 2p) + (e_p) \\ H_2^+ + e + (e_p) \\ H(1s) + H^+ + e + (e_p) \end{cases} \tag{6.7}$$

$$H, He + h\nu, (e_p) \rightarrow H^+, He^+ + e + (e_p). \tag{6.8}$$

In the photoionization processes Eqs. (6.7) and (6.8) photoelectrons with energies sufficient for the subsequent ionization and excitation of atomic and molecular hydrogen are produced. The energy of the ionizing quanta by definition exceeds the ionization potential, and its excess produces electrons with an excess of kinetic energy and ions in excited states. The differential photoelectron production rate $q_e(E,z)$ at a given altitude z in the upper atmosphere is defined by the following expression:

$$\begin{aligned} q_e(E,z) &= \textstyle\sum_k q_e^{(k)}(E,z) \\ q_e^{(k)} &= \textstyle\sum_l n_k(z) \int_0^{\lambda_i} d\lambda I_\infty(\lambda) exp(-\tau(\lambda,z))\sigma_k^i p_k(\lambda, E_{k,l}). \end{aligned} \tag{6.9}$$

where the optical thickness τ is given by

$$\tau(\lambda,z) = \sum_k \sigma_k^a(\lambda) \int_z^\infty n_k(z')dz',$$

and n_k is the neutral component k number density; and $k, \sigma_k^i(\lambda)$ and $\sigma_k^a(\lambda)$ are the ionization and absorption cross sections, respectively, dependent on the wavelength λ. In expression Eq. (6.9) , we use the relative yields $p_k(\lambda, E_{k,l})$ and the potential of ionization $E_{k,l}$ for the electronically excited states of the ion. The energy of the forming photoelectron is $E = E_\lambda - E_{k,l}$, where E_λ is the energy of the photon and λ_k is the wavelength corresponding to the ionization potential of the k_{th} neutral component. $I_\infty(\lambda)$ is the incident stellar radiation flux at the wavelength λ. In formula Eq. (6.9) $q_e^{(k)}$ is a partial (by neutral species) differential production rate of photoelectrons in the photoionization processes. Spectra of the stellar XUV fluxes currently are not well known (Lammer et al. 2012), therefore in the calculations below, we used a flux of solar radiation in the wavelength range of 1–115 nm for the moderate-activity solar spectrum model from Huebner et al. (1992) scaled for the distance of 0.047 a. u. equal to the semi-major axis of the HD 209458b exoplanet. The relative yields for excited ionic states, absorption, and ionization cross sections are taken from the same work for the main atmospheric components H_2, H, and He. The input parameters taken from paper Huebner et al. (1992) and used in the numerical model are shown in Fig. 6.1. These newly formed electrons are transported in the thermosphere where they lose their kinetic energy in elastic, inelastic and ionization collisions with the ambient atmospheric gas:

$$e(E) + X \rightarrow \begin{cases} e(E') + X \\ e(E') + X^* \\ e(E') + X^+ + e(E_s) \end{cases} \tag{6.10}$$

where E and $E'(< E)$ are the kinetic energies of the primary electron before and after a collision; $X = H_2, H, He$; X^* and X^+ are atmospheric species in excited

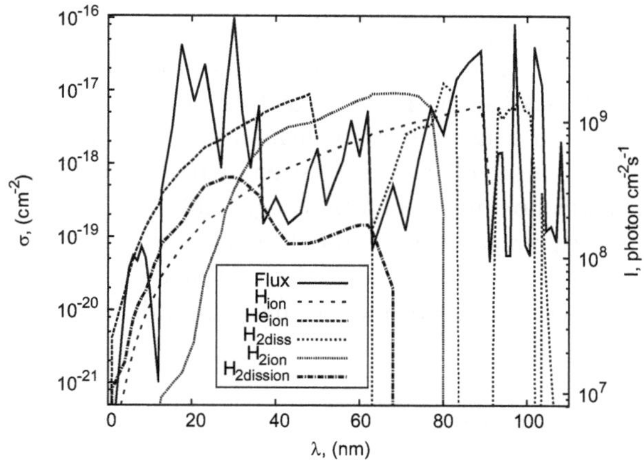

Fig. 6.1 The input parameters – the moderate-activity solar spectrum model of the solar XUV radiation scaled for the HD 209458b orbital distance (*solid line*), and ionization and dissociation cross sections for the main atmospheric components H_2, H, and He have been taken from the study of Huebner et al. (1992) and used in the numerical model

and ionized states; E_s is the energy of the secondary electron formed in the ionizing collision. We consider the following neutral excited states for the main atmospheric species:

- Excitation and dissociative excitation of $H_2^* = H_2($ rot, vib, electronic states A^3, $B^3, C^3, B^1, C^1, E^1, B'^1, D^1, D'^1, B''^1, \Pi_s, Ly-\alpha)$;
- Direct ionization of $H_2 \rightarrow H_2^+$;
- Dissociative ionization of $H_2 \rightarrow H^+ + H$;
- Excitation of $He^* = He$ (21 electronic states at energies between 20.61 and 23.91 eV);
- Direct ionization of $He \rightarrow He^+$;
- Excitation of $H^* = H$ (9 states 1s2p–1s10p);
- Direct ionization of $H \rightarrow H^+$.

If the collision produces ionization, a secondary electron is created and is randomly assigned an isotropically distributed pitch angle and an energy, using an integral form of the formula of Green and Sawada (1972) and Jackman et al. (1977) based on the laboratory results of Opal et al. (1971)

$$\int_0^{E_s} \sigma_{i,j}(E_p, E')dE' = A(E_p)\Gamma(E_p)\left[\tan^{-1}\left(\frac{E_s - T_0(E_p)}{\Gamma(E_p)}\right) + c\right],$$

where $\sigma_{i,j}(E_p, E')$ is the state-specific cross section for species i and state j at primary electron energy E_p and secondary electron energy E_s, $A(E_p)$, $\Gamma(E_p)$, and $T_0(E_p)$ are fitting functions defined by the tabulated parameters of Jackman et al.

(1977), and $c = \tan^{-1}\left[\frac{T_0(E_p)}{\Gamma(E_p)}\right]$. Energy E_s of the secondary electron produced by an ionization collision is calculated according to the procedure described by Garvey and Green (1976), Jackman et al. (1977) and Garvey et al. (1977). For inelastic collisions, the forward scattering approximation was used: it is assumed that the phase function from these collisions is so strongly peaked in the forward direction that angular redistribution by this process is negligible. Below 100 eV there can be considerable backscatter, particularly from forbidden excitation transitions, but the flux becomes so isotropic and the relative size of the elastic cross sections becomes so large that this has little effect on the final pitch angle distribution.

6.4.2 Kinetic Equation

The fresh electrons lose their excess kinetic energy in collisions with the ambient atmospheric particles. Their kinetics and transport is described by the kinetic Boltzmann equation (Shematovich 2008)

$$\frac{\partial}{\partial \mathbf{r}} f_e + \frac{\mathbf{Y}}{m_e}\frac{\partial}{\partial \mathbf{v}} f_e = Q_{e,photo}(v) + Q_{e,secondary}(v) + \sum_{M=H,He,H_2} J(f_e, f_M), \qquad (6.11)$$

where $f_e(\mathbf{r}, \mathbf{v})$, and $f_M(\mathbf{r}, \mathbf{v})$ are the velocity distribution functions for electrons, and for the species of the ambient gas, respectively. The left side of the kinetic equation describes the transport of electrons in the planetary gravitational field $\mathbf{Y}$. In the right-hand side of the kinetic equation the $Q_{e,photo}$, photo term describes the formation rate of primary electrons due to photoionization, while the $Q_{e,secondary}$, secondary term describes the rate of formation of the secondary electrons. The elastic and inelastic scattering terms J for electron collisions with ambient atmospheric species are written in a standard form. It is assumed that the ambient atmospheric gas is characterized by the local Maxwellian velocity distribution functions.

6.4.3 Numerical Model

The Direct Simulation Monte Carlo (DSMC) method is an efficient tool to solve atmospheric kinetic systems in the stochastic approximation (Shematovich et al. 1994; Bisikalo et al. 1995; Gerard et al. 2000). The details of the algorithmic realization of the numerical model were given earlier in Shematovich et al. (1994) and Bisikalo et al. (1995). In the numerical simulations, the evolution of the system of modelling particles due to collisional processes and particle transport is calculated from the initial to the steady state. In order to minimize boundary effects, the lower boundary is set at altitudes where atmosphere is collision-dominated and the upper boundary is fixed at altitudes where the atmospheric gas flow is practically collisionless. The relative importance of the collisional processes is governed by their cross sections. In this particular realization of the model, we used experimental and calculated data for the cross sections and distributions of the scattering angles in the elastic, inelastic, and ionization collisions of electrons with H_2, He, and H taken from the following sources: (a) for electron collisions with H_2 we used the AMDIS database (https://dbshino.nfs.ac.jp) and the study by Shyn and Sharp (1981); and (b) for electron collisions with He and atomic hydrogen, we used the H_2 NIST database (http://physics.nist.gov/PhysRef.Data/Ionization/) and the works of Jackman et al. (1977) and Dalgarno et al. (1999).

6.4.4 Energy Deposition of the Stellar Soft X-Ray and EUV Radiation

The XUV radiation of the host star is absorbed by atmospheric gas and leads to the excitation, dissociation, and ionization of different components and to the thermal heating of the atmosphere. In a case of $H_2/H/He$ atmosphere the partial deposition rates of the stellar XUV radiation due to the photolytic processes Eqs. (6.7) and (6.8) in the $H_2 \rightarrow H$ transition region in the upper atmosphere of HD209458b can be calculated in accordance with formula Eq. (6.9) as follows

$$
\begin{aligned}
&W_{h\nu}(z) = \textstyle\sum_k W_{h\nu}^{(k)}(z)\\
&W_{h\nu}^{(k)} = \textstyle\sum_l n_k(z) \int_0^{\lambda_i} d\lambda E_\lambda I_\infty(\lambda) exp(-\tau(\lambda, z)) \sigma_k^a p_k(\lambda, E_{k,l}).
\end{aligned}
$$

where $W_{h\nu}(z)$ and $W_{h\nu}^{(k)}(z)$ are the local total and partial deposition rates of stellar XUV radiation in the upper atmosphere. The rate $W_{pe}(z)$ of kinetic energy storage in the primary or fresh photoelectrons is equal to

$$
\begin{aligned}
&W_{pe}(z) = \textstyle\sum_k W_{pe}^{(k)}(z)\\
&W_{pe}^{(k)} = \textstyle\sum_l n_k(z) \int_0^{\lambda_i} d\lambda (E_\lambda - E_{k,l}) I_\infty(\lambda) exp(-\tau(\lambda, z)) \sigma_k^i p_k(\lambda, E_{k,l}).
\end{aligned}
$$

Using the Monte Carlo model of Shematovich (2010) the partial deposition rates of the energy of the accompanying flux of the primary photoelectrons due to the electron impact processes Eq. (6.10) in the $H_2 \rightarrow H$ transition region in the planetary upper atmosphere can be calculated. This finally allows us to estimate the heating rate W_T by photoelectrons in the planetary upper atmosphere and to calculate the heating efficiency coefficient η which is a critical parameter in aeronomical models (Yelle et al. 2008) (see also the Chap. 7, Kislyakova et al. 2014). Heating efficiency η usually is defined as the ratio of the absorbed energy accumulated as the gas heat to the deposited energy of the stellar radiation. We calculated the heating efficiency in accordance with this strict definition, namely,

$$\eta_{h\nu}(z) = \frac{W_T(z)}{W_{h\nu}(z)}.$$

Also the simplified definition is sometimes used as η is an approximate ratio between kinetic energy stored by the fresh (primary) photoelectrons and the deposited energy of the stellar radiation. Then

$$\eta_{pe}(z) \approx \frac{W_{pe}(z)}{W_{h\nu}(z)}.$$

6.4.5 Calculations of Heating Efficiency Height Distribution

Calculations of the energy deposition of the stellar XUV radiation were made in the $H_2 \rightarrow H$ transition region ($1.04R_p < R < 1.2R_p$) in the upper atmosphere of HD209458b. Height profiles of the main neutral constituents H_2, H, and He were adopted from the aeronomical model by Yelle (2004). The rate of the transition of the stellar XUV radiation and photoelectron energy into the internal energy of the atmospheric gas in each of the photolytic and electron-impact reactions is calculated in the model. Separately the energy of the suprathermal photoelectrons which turns into the heat is also calculated. Thus, the results of the simulation allow us to determine the total efficiency of heating and efficiency of heating by photoelectrons and to understand which processes to the greatest extent affect the heating of the atmosphere.

In Fig. 6.2 the component-dependent deposition rates of the stellar XUV radiation due to the photolytic processes Eqs. (6.7) and (6.8) in the $H_2 \rightarrow H$ transition

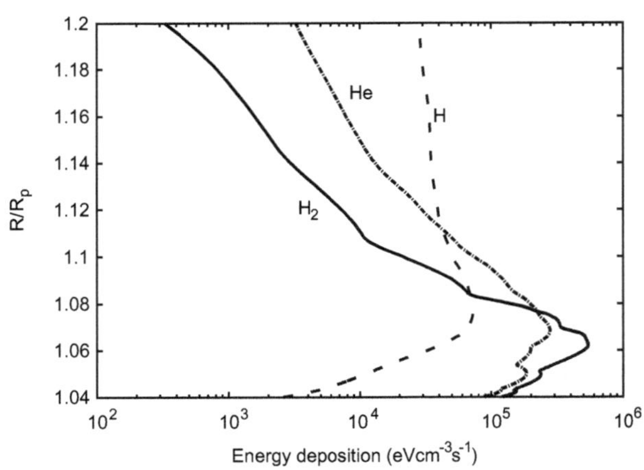

Fig. 6.2 Component-dependent deposition rates of the stellar XUV radiation due to the photolytic processes Eqs. (6.7) and (6.8) in the $H_2 \rightarrow H$ transition region in the upper atmosphere of HD209458b. Neutral $H_2/H/He$ atmosphere was adopted from the aeronomical model by Yelle (2004)

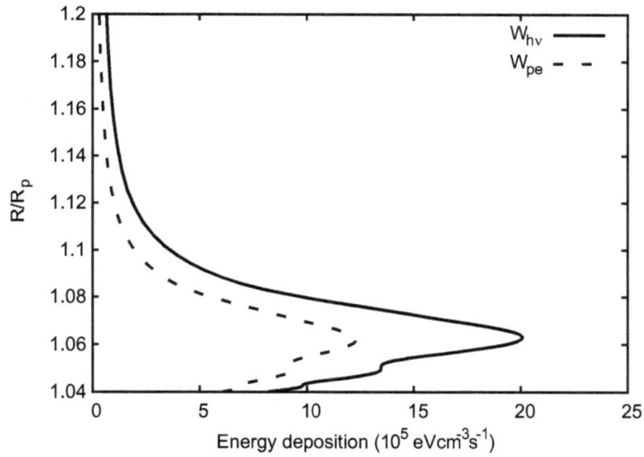

Fig. 6.3 Total deposition rate $W_{h\nu}$ (*solid curve*) of the stellar XUV radiation and the rate W_{pe} (*dashed curve*) of energy accumulation by the fresh photoelectrons due to the photolytic processes Eqs. (6.7) and (6.8) in the $H_2 \rightarrow H$ transition region in the upper atmosphere of HD209458b

region in the upper atmosphere of HD209458b are shown. It can be seen, that H_2 photo-dissociation and H_2 and He photo-ionization are the main channels of the total stellar XUV radiation energy deposition near the lower boundary of the transition region.

In Fig. 6.3 the total deposition rate $W_{h\nu}$ (solid curve) of the stellar XUV radiation and the rate W_{pe} (dashed curve) of energy accumulation by the fresh photoelectrons due to the photolytic processes Eqs. (6.7) and (6.8) in the $H_2 \rightarrow H$ transition region in the upper atmosphere of HD209458b are given.

The partial deposition rates of the stellar XUV radiation due to the photoelectron impact processes Eq. (6.10) in the $H_2 \rightarrow H$ transition region in the upper atmosphere of HD209458b given in Fig. 6.4. It can be seen that both excitation of internal states and ionization of molecular and atomic hydrogen are the dominant channels of the photoelectron energy deposition.

Finally, in Fig. 6.5 the height profiles of the rate W_{pe} (solid curve) of energy accumulation in the photo-ionization processes Eqs. (6.7) and (6.8) and the heating rate by photoelectrons W_T (dashed curve) are shown. These values allow us to calculate the heating efficiency η defined as a ratio of the absorbed energy accumulated as a gas heat to the deposited energy of the stellar XUV radiation. Therefore, in Fig. 6.6 the heating efficiency $\eta_{h\nu}$ with (solid curve) and η_{pe} without (dashed curve) taking into account the photoelectron impact processes Eq. (6.10) in the $H_2 \rightarrow H$ transition region in the upper atmosphere of HD209458b are presented. It is seen that the total heating efficiency is height-dependent with peak value approaching 0.3.

From the presented results it can be seen how is realized the heating of the upper atmosphere of hot Jupiter by the stellar XUV radiation, when the effect

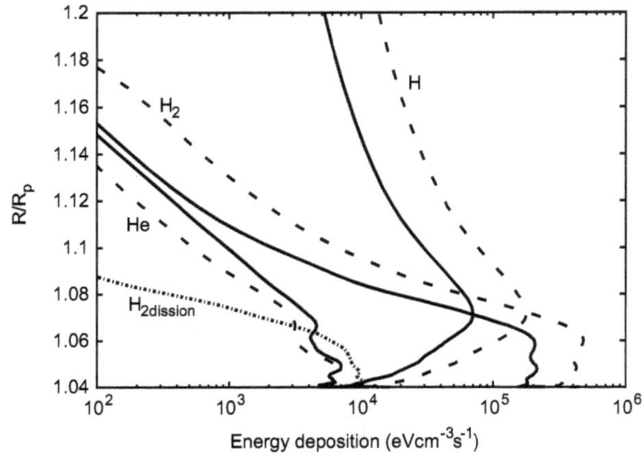

Fig. 6.4 Partial deposition rates of the stellar XUV radiation due to the electron impact processes Eq. (6.10) in the $H_2 \rightarrow H$ transition region in the upper atmosphere of HD209458b. *Solid lines* – ionization, *dashed lines* – excitation of the internal states, *dash-dotted line* – dissociative ionization of H_2

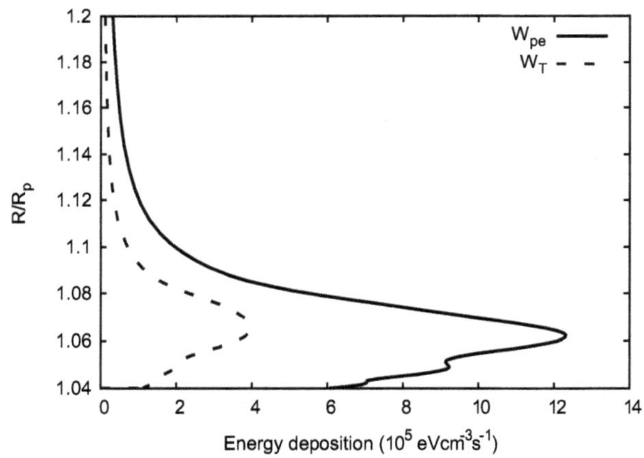

Fig. 6.5 The rate W_{pe} of energy accumulation in the photoionization processes and heating rate W_T by photoelectrons in the $H_2 \rightarrow H$ transition region in the upper atmosphere of HD209458b

of the electron-impact processes Eq. (6.10) with participation of the suprathermal photoelectrons is taken into account. The height profiles of heating efficiency by the stellar XUV radiation were calculated in self-consistent way. It was shown that this value does not exceed 0.2 almost everywhere in the $H_2 \rightarrow H$ transition region of the hydrogen-rich thermosphere of the exoplanet. The correct account of photoelectrons reduces the efficiency of heating by 3–4 times.

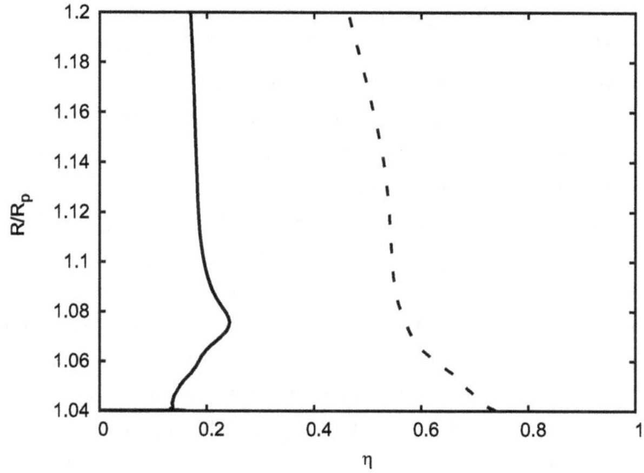

Fig. 6.6 The heating efficiency $\eta_{h\nu}$ with (*solid curve*) and η_{pe} without (*dashed curve*) taking into account the photoelectron impact processes Eq. (6.10) in the $H_2 \rightarrow H$ transition region in the upper atmosphere of HD209458b

6.5 Suprathermal Fraction of Atomic Hydrogen

One of the important sources of suprathermal hydrogen atoms in the upper atmosphere of HD 209458b are the processes of the dissociation and ionization of molecular hydrogen by the stellar XUV radiation. No detailed studies of suprathermal hydrogen atom kinetics and transfer were performed at the molecular level (Yelle et al. 2008), because this requires to solve numerically the kinetic Boltzmann equation. In this study, we consider the effect of molecular hydrogen dissociation by UV radiation and the subsequent flux of photoelectrons on the suprathermal atomic hydrogen production in the $H_2 \rightarrow H$ transition region and the formation of the corresponding outflow in the upper atmosphere of HD 209458b. For this, we calculate the production rate and the energy spectrum of hydrogen atoms which are produced with an excess of kinetic energy during the dissociation of H_2. Then, we use the stochastic hot planetary corona model by Shamatovich (2004) to study the kinetics and transfer of the suprathermal hydrogen atoms in the expanded upper atmosphere and to estimate the atmospheric mass-loss rate.

6.5.1 Molecular Hydrogen Dissociation in the Upper Atmosphere of HD 209458b

The thermal regime and escape rate in the atmospheres of hot Jupiters significantly depend on the chemical composition of the atmosphere. In contrast to the planets of the Solar System that have atmospheric compositions stable over geological timescales, the composition of giant exoplanets may change rapidly as a consequence of the influence of the intense radiation of the parent star. Moreover, the thermal regime and the composition of the atmosphere are tightly connected through heating and cooling (Yelle 2004; García Muñoz 2007). Aeronomical models such as that of Yelle (2004) generally assume that reaching temperatures of several thousand Kelvin at particle densities on the order of $\sim 10^{10}\,\mathrm{cm}^{-3}$ leads to the thermal dissociation of molecular hydrogen, $H_2 + M \rightarrow H + H + M$, and to the formation of the $H_2 \rightarrow H$ transition region in the inner thermosphere of the exoplanet. In the upper thermospheric layers, the photoionization of atomic hydrogen starts to play the dominant role. Correspondingly, the composition of the upper atmosphere changes with altitude as H_2, H, H^+, which is an additional factor forming the extended atmosphere, since such a change in the composition is accompanied by the increase of the characteristic scale height. The model of the upper neutral atmosphere consists primarily of H_2, H, and He and was taken from model by Yelle (2004); we use it to trace the kinetic properties of suprathermal hydrogen atoms. In the work of García Muñoz (2007) it was shown that the concentrations of atomic oxygen and ionized carbon observed in the vicinity of the exoplanet HD 209458b (Vidal-Madjar et al. 2004) are more than an order of magnitude lower than those in the upper atmosphere.

The dissociation processes such as photodissociation, collisional dissociation, dissociative ionization, etc., are the primary sources of thermal and suprathermal fragments of molecules in electronically excited states in the upper planetary atmospheres (Wayne 1991). Although a hydrogen molecule is simple, the UV and/or electron collisional dissociation of H_2 may occur in several ways:

$$H_2 + h\nu(e_p) \rightarrow H_2^* \rightarrow H(1s) + H(1s, 2s, 2p, \ldots) + (e_p) + \Delta E_{dis} \qquad (6.12)$$

If as the result of the absorption of radiation, electrons are excited to unbound or antibound orbits and their excitation energy exceeds the binding energy of the molecule, the molecule may dissociate. This mechanism allows for the photodissociation of molecules excited either to a continuum bound state or immediately to an unbound (repulsive) state. The photodissociation cross sections for these processes are usually smooth functions dependent on the wavelength; therefore, low (0.05–0.1 nm) spectral resolution data on the incident flux and cross sections are sufficient to estimate the dissociation rate (Fox et al. 2008). Another important mechanism is pre-dissociation, when an absorbed photon excites the molecule into a bound state of electron excitation, from which the subsequent non-radiative transition into an unbound state is possible. The dissociation rate depends on the absorption rates at the wavelengths of the selected line transitions and the pre-dissociation probabilities (Fox et al. 2008).

The H_2 dissociation energy is 4.48 eV (which corresponds to 276.9 nm), but the photoabsorption cross sections at wavelengths longer than 111.6 nm are negligibly small. In planetary atmospheres, H_2 photodissociation by UV photons in the spectral range of 84.5–111.6 nm occurs mainly through dipole transitions from the ground state $X^1\sum_g^+(\nu)$ into excited bound $B^1\sum_u^+(\nu')$, $C^1\Pi_u(\nu')$, $B'^1\sum_u^+(\nu')$ and $D^1\Pi_u(\nu')$ states. From these states, H_2 molecules may either transfer to the discrete levels of the ground state, radiating away the excess energy, or to the ground state continuum with the subsequent dissociation into two ground state hydrogen atoms. The pre-dissociation probabilities for these levels are in the range of 0.1–0.15 (Abgrall et al. 1997). At shorter wavelengths of 84.5 nm, direct absorption into electron excitation continua dominate the photodissociation processes. These processes have sufficiently high cross sections and as a result of the dissociation lead to the formation of a hydrogen atom in the ground state H(1s) and a hydrogen atom in the excited H(2s,2p) state (Glass-Maujean 1986).

The excess kinetic energy ΔE_{dis} of the atoms produced through the photodissociation of the hydrogen molecule Eq. (6.12) was calculated as the difference between the energy of an absorbed UV photon, the energy of electronically excited state, and the dissociation energy. In the case of the dissociation of H_2 by photoelectrons, the excess energy was found through the distributions calculated by Ajello et al. (1985). The distributions calculated for several electron beam energies show the populations of relatively slow thermal energy (0–1 eV) and fast high-energy (1–10 eV, with the main peak at about 4 eV) atoms. For dissociative ionization by photoelectrons,

$$H_2 + e_p \rightarrow H^+ + e + H(1s, 2s, 2p, \ldots) + (e_p) + \Delta E_{disi}, \tag{6.13}$$

excess kinetic energy was obtained through the distributions calculated by van Zyl and Stephen 1994. Suprathermal hydrogen atoms created through processes Eqs. (6.12) and (6.13) lose their energy in elastic collisions with the basic neutral components of the surrounding atmospheric gas:

$$H_h(E) + H_{th}, He_{th}, H_{2th} \rightarrow H_h(E' < E) + H_{th}, He_{th}, H_{2th} \tag{6.14}$$

It should be noted that for suprathermal energies of hydrogen atoms, the energy transfer efficiency between hot and thermal atoms through elastic scattering is largely determined by the phase functions – distributions by the scattering angle. Experimental and numerical simulation data (Hodges and Breig 1991; Krstić and Schultz 1999a,b) show that these distributions peak at low scattering angles, although the total cross sections are relatively high. The effectiveness of the energy transfer is thus strongly dependent on the collision energy. These properties of elastic collisions with thermal H_2, He, and H to a high degree determine of the parameters of the fraction of suprathermal hydrogen atoms in the upper photosphere of HD 209458b.

6.5.2 *Kinetics of Suprathermal Hydrogen Atoms*

As hydrogen atoms appear in the process of dissociation with an excess of kinetic energy, their distribution in the $H_2 \rightarrow H$ transition region of the hydrogen-rich atmosphere of exoplanet is determined by the Boltzmann equation with a photochemical source function

$$\frac{\partial F_H}{\partial t} + \mathbf{c}\frac{\partial F_H}{\partial \mathbf{r}} + \frac{\mathbf{Y}}{m_H}\frac{\partial F_H}{\partial \mathbf{c}} = \sum_s Q_s^H + \sum_m J_m^H(F_H, F_m), \tag{6.15}$$

together with the initial and boundary conditions for the atmospheric gas in the volume V influenced by the gravitation field Y of the planet. Like in Eq. (6.5) we use a microscopic description of the suprathermal hydrogen atom population by the distribution function $F_H(t, \mathbf{r}, \mathbf{c}) = n_H(t, \mathbf{r}) f_H(t, \mathbf{r}, \mathbf{c})$, where $n_H(t, \mathbf{r})$ is the number density of suprathermal particles, and $f_H(t, \mathbf{r}, \mathbf{c})$ is a normalized one-particle velocity distribution function. Source functions $Q_s^H(t, \mathbf{r}, \mathbf{c})$ set the formation rates for the suprathermal atoms in photochemical reactions Eqs. (6.12) and (6.13) and are usually written as follows:

$$Q_s^H(t, \mathbf{r}, \mathbf{c}) = q_s^H(E) f_s^H(t, \mathbf{r}, \mathbf{c}). \tag{6.16}$$

Here, $Q_s^H(E) =< |c_i - c_j| \sigma_s(E) >$ is the differential suprathermal hydrogen atom production rate by a certain photochemical source s at the particle collision energy E; the function $f_H(t, \mathbf{r}, \mathbf{c})$ gives the normalized velocity distribution for the fresh particles formed with an excess of kinetic energy. The collision integrals at the right-hand sides of the kinetic equations describe the gas-state changes due to chemical reactions and equal

$$J_m^H(F_H, F_m) = \int g_{ij} d\sigma_m dc_j [F_H(c_i') F_m(c_l') - F_H(c_i) F_m(c_j)], \tag{6.17}$$

where g_{ij} is the relative velocity and $d\sigma_m$ is the cross-section of the elastic scattering of suprathermal hydrogen atoms elastically colliding with hydrogen and helium. The scattering cross sections were taken from the works by Hodges and Breig (1991) and Krstić and Schultz (1999a,b). For thermal components, we used Maxwellian distributions with the local temperature and density values calculated by the aeronomical model.

To estimate the source function Eq. (6.16) for the suprathermal hydrogen atoms, it is necessary to calculate the dissociation and ionization rates of atmospheric gas caused by UV stellar radiation, the photoelectron production rate, and the molecular hydrogen dissociation and dissociative ionization rates caused by electron collisions. Therefore the suprathermal hydrogen atom production rates $q_s^H(E, z)$ through the dissociation and dissociative ionization of H_2 by extreme UV radiation were calculated using formula Eq. (6.9) with photodissociation and dissociative photoionization cross sections, respectively.

6.5.3 Calculation Results

The calculations were conducted in the $H_2 \rightarrow H$ transition region in the upper atmosphere of HD 209458b at an altitude interval of $(1.0 - 1.1) \times R_p$ km. At the lower boundary of the studied region the mean free path of the hydrogen atoms is much smaller than the density scale height, and, correspondingly, the suprathermal hydrogen atoms are locally thermalized in the elastic collisions with the surrounding atmospheric gas. At the upper boundary, the mean free path is close to the local density scale height and, correspondingly, suprathermal atoms may escape if their kinetic energies are higher than the local escape energy. The calculation domain is divided into cells of a size of about the local mean free path length for suprathermal hydrogen atoms. The production rates and energy spectra of suprathermal hydrogen atoms produced by processes Eqs. (6.12) and (6.13) due to the impact of stellar UV radiation and the accompanying photoelectron flux are shown in Fig. 6.7. The upper panel shows the formation rates of suprathermal hydrogen atoms in the processes of dissociation and dissociative ionization of molecular hydrogen by

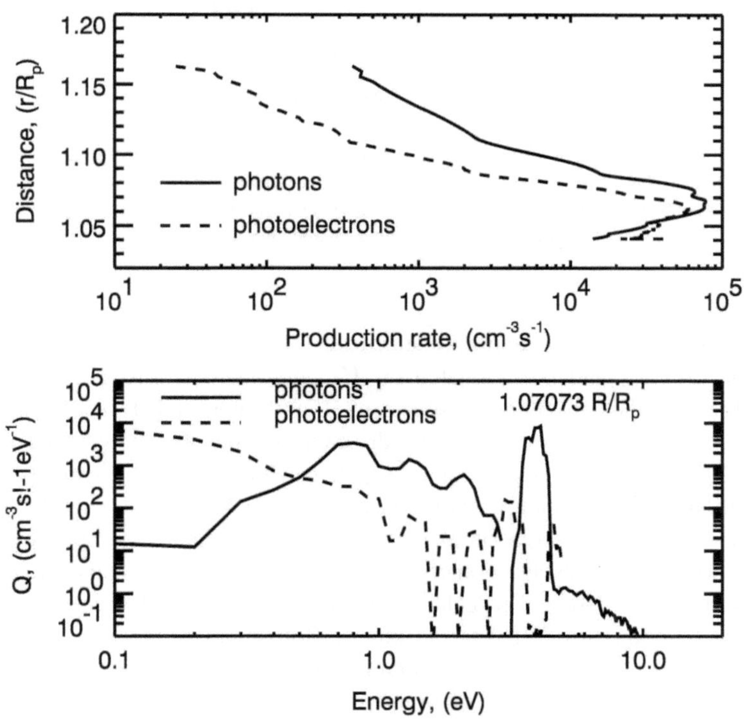

Fig. 6.7 *Upper panel*: suprathermal hydrogen atom production rate via dissociation by UV photons (*solid line*) and photoelectrons (*dashed line*). *Lower panel*: energy spectrum at altitude of 1.07 R_p for the suprathermal atoms formed by the photo- (*solid line*) and electron impact (*dashed line*) H dissociation processes

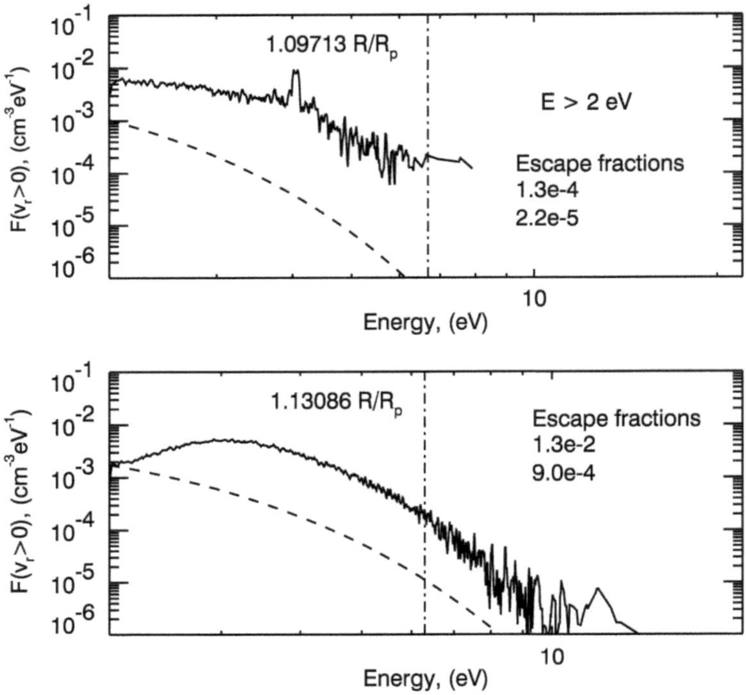

Fig. 6.8 One-particle distribution function for upward moving suprathermal hydrogen atoms at altitudes of $1.1R_p$ (*upper panel*) and $1.13R_p$ (*lower panel*). The *dashed curves* show the local equilibrium distributions of atomic hydrogen corresponding to the parameters of the model by Yelle (2004). *Vertical dot-dashed lines* mark the escape energies of hydrogen atoms at the given altitudes

UV photons (solid curve) and photoelectrons (dashed curve). The formation rate peaks at 1.07 R_p and is dominated by H_2 photodissociation. The lower panel shows the hydrogen atom energy spectrum at at altitude of 1.07 R_p for hydrogen atoms formed in the processes of the dissociation and dissociative ionization of H_2 by XUV-photons and photoelectrons. It can be seen that the photoelectron dissociation produces mainly hydrogen atoms with kinetic energies lower than 1 eV (slow fraction dissociation channel). Nonetheless, there is also a significant contribution from the fast fraction channel producing hydrogen atoms with energies in the range of 1–10 eV. The calculated formation rates and spectra were used as source functions Eq. (6.16) in the kinetic Boltzmann equation Eq. (6.15). The solution to the kinetic equation was obtained numerically with the stochastic model of Shamatovich (2004). The calculations were carried out for steady-state conditions in the daytime upper atmosphere in the substellar point. In Fig. 6.8, one-particle distribution functions are shown for suprathermal hydrogen atoms moving up at altitudes of $1.1R_p$ (upper panel) and $1.13R_p$ (lower panel). The dashed curves in Fig. 6.8 correspond to the local equilibrium of the one-particle distributions of

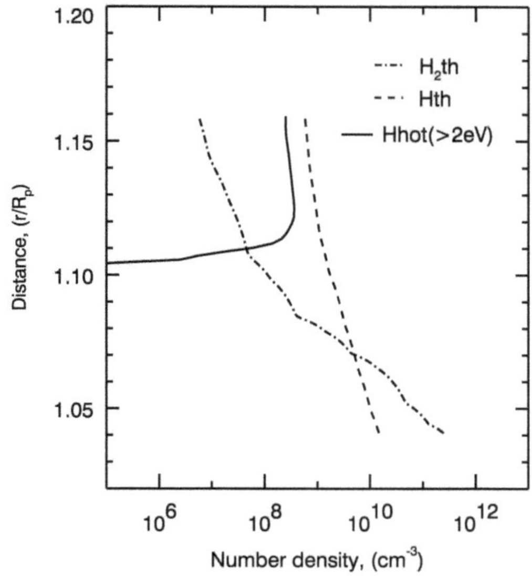

Fig. 6.9 Thermal and suprathermal hydrogen density distribution with altitude for the $H_2 \rightarrow H$ transition region of the upper atmosphere of HD 209458b

atomic hydrogen calculated based on the temperature profile from the model of Yelle (2004). The vertical dot-dashed lines show the escape energies of hydrogen atoms at the given altitudes. As we are interested in the estimate of the escape rates of hydrogen atoms due to the process of the dissociation of H_2, in Fig. 6.8 the distribution function is presented only in the suprathermal energy range above 2 eV. Simulations show that the distribution functions significantly deviate from the local equilibrium distributions. At a height of $1.07 R_p$, close to the region of the maximal production of hydrogen atoms as a result of the dissociation of H_2, a significant fraction of hydrogen atoms with energies high enough to escape the gravitational field (about 6.9 eV) is produced, in contrast to the local equilibrium distribution. The distribution functions shown in Fig. 6.8 allow us to estimate the number density of suprathermal hydrogen atoms produced in the dissociation of H_2, as well as to compare it with the height distribution of the thermal neutral components of the atmosphere. The altitude profile of suprathermal (with energies above 2 eV) hydrogen atoms is shown in Fig. 6.9. For comparison, we also show the densities of thermal atomic and molecular hydrogen corresponding to the aeronomical model of Yelle (2004). The calculations show that a steady-state fraction of suprathermal hydrogen atoms with energies above 2 eV is formed only in the outermost parts of the transition region, where occur most of the collisions of suprathermal hydrogen with neutral hydrogen atoms. As follows from the analysis of the distribution functions presented in Fig. 6.8, the concentration of suprathermal hydrogen atoms, due to the process of the dissociation of H_2, is here several times higher than the concentration of atmospheric hydrogen with energies above 2 eV. The calculations of the distribution functions presented in Fig. 6.8 show that dissociation processes in Eqs. (6.12) and (6.13) of molecular hydrogen are accompanied by the formation in the $H_2 \rightarrow H$ transition region and transfer to the uppermost atmospheric layers of

Fig. 6.10 Energy spectrum of the suprathermal H atoms escaping the upper atmosphere of HD 209458b formed by dissociation processes

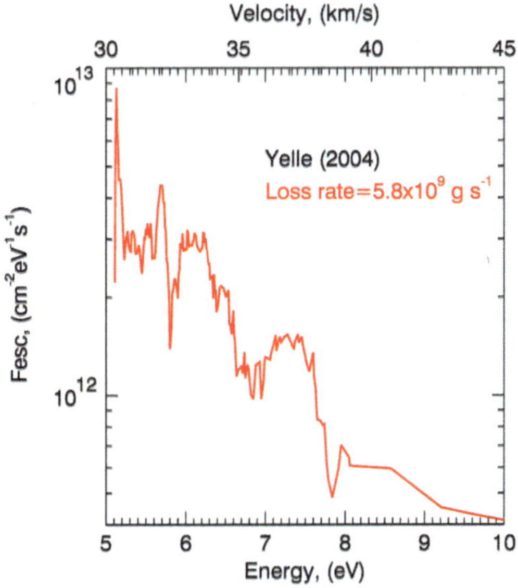

the moving upwards suprathermal hydrogen atoms with energies exceeding the local escape energy. In Fig. 6.10, we show the energy spectrum of the stream of hydrogen atoms leaving the atmosphere of HD 209458b through the upper boundary of the transfer region at an altitude of $\sim 1.2R_p$ as a result of the dissociation processes. For moderate stellar activity in the considered range of UV radiation, the flux of escaping particles is $5.8 \times 10^{12}\,\mathrm{cm^{-2}\,s^{-1}}$. Averaged over the upper atmosphere, this flux corresponds to an atmospheric mass-loss rate of about 5.8×10^9 g/s due to the dissociation of H_2, which is somewhat lower than the observational estimate of about $10^{10}\,\mathrm{g\,s^{-1}}$ (Ehrenreich et al. 2008).

Our atmospheric mass-loss estimate may be considered as a lower estimate because our simulations were run for a moderate stellar activity level of UV radiation, and for the probabilities of the pre-dissociation, we took the minimal value of 0.1. Naturally, for higher XUV fluxes, the H_2 dissociation processes due to the forcing by stellar XUV radiation and the subsequent flux of photoelectrons will create a significantly higher contribution to the escaping flux of hydrogen atoms.

Conclusion

The photolysis of molecular and atomic hydrogen and helium by the XUV radiation of the host star leads to the production of the suprathermals – primary photoelectrons in the $H_2/H/He$ ionization and hydrogen atoms in the H_2 dissociation in the hydrogen-dominated atmospheres of exoplanets. These particles with an excess kinetic energies are an important source of thermal

(continued)

energy in the upper atmosphere of the exoplanet HD 209458b. However, in the contemporary aeronomical models the kinetics and transfer of hot hydrogen atoms and fresh photoelectrons were not calculated in detail, because they require the solving of the Boltzmann equation for a non-thermal population of these particles.

In this Chapter the important role of the suprathermal particles for aeronomy of the hydrogen-rich atmospheres of the exoplanets is pointed out and discussed. The numerical stochastic model which follows the kinetics and transport of suprathermals is described. This model was used to estimate the effect of the XUV radiation of the host star on the production of the suprathermals in the $H_2 \rightarrow H$ transition region in the upper atmosphere of the hydrogen-rich exoplanet. Partial deposition rates of the stellar XUV radiation due to the photolytic processes of Eqs. (6.7) and (6.8) in the $H_2 \rightarrow H$ transition region in the upper atmosphere of the hot Jupiter HD209458b were calculated using the Monte Carlo model developed by Shematovich (2008) and adapted for hydrogen atmospheres (Shematovich 2010). This allowed us to estimate the heating rate of the atmospheric gas by photoelectrons in the upper atmosphere of HD209458b. For the first time the heating efficiency η with and without taking into account the photoelectron impact processes in the $H_2 \rightarrow H$ transition region in the upper atmosphere of HD209458b was calculated.

Using the numerical stochastic model created by Shamatovich (2004) for a hot planetary corona the kinetics and transfer of suprathermal hydrogen atoms in the upper atmosphere and the emergent flux of atoms evaporating from the atmosphere were investigated. The latter is estimated as $5.8 \times 10^{12}\,\text{cm}^{-2}\,\text{s}^{-1}$ for a moderate stellar activity level of UV radiation, which leads to a planetary atmosphere evaporation rate of 5.8×10^{9} g/s due to the process of the dissociation of H_2. This additional non-thermal loss process should be taken into account in the current aeronomical models of the hydrogen-rich atmospheres of exoplanets.

Acknowledgements The authors acknowledge the support by the International Space Science Institute (ISSI) in Bern, Switzerland and the ISSI team *Characterizing stellar- and exoplanetary environments*. The authors also acknowledge the support by the RSCF project 14-12-01048. Finally, V.I. Shematovich and D. V. Bisikalo thank Lotfi Ben-Jaffel from the Institut Astrophysique de Paris (IAP) CNRS-UPMC, Paris and Tommi Koskinen from the Lunar and Planetary Laboratory University of Arizona, Tucson, USA for fruitful discussions.

References

Abgrall, H., Roueff, E., Liu, X., & Shemansky, D. E. (1997). *Astrophysics Journal, 481*, 557.

Ajello, J. M. Kanik, I., Ahmed, S. M., & Clarke, J. T. (1985). *Journal of Geophysical Research, 100*, 26411.

Ballester, G. E., Sing, D. K., & Herbert, F. (2007). *Nature, 445*, 511.
Ben-Jaffel, L. (2007). *Astrophysics Journal (Letters), 671*, L61.
Ben-Jaffel, L., & Sona Hosseini, S. (2010). *Astrophysics Journal, 709*, 1284.
Bird, G. A. (1994). *Molecular gas dynamics and direct simulation of gas flows*. New York: Oxford University Press.
Bisikalo, D. V., Shematovich, V. I., & Gerard, J. C., (1995). *Journal of Geophysical Research, 100*, 3715.
Bisikalo, D. V., Shematovich, V. I., Gérard, J. C., Gladstone, G. R., & Waite, J. H. (1996). *Journal of Geophysical Research, 101*, 21157.
Bisikalo, D. V., Kaygorodov, P. V., Ionov, D. E., Shematovich, V. I., Lammer, H., & Fossati, L. (2013a). *Astrophysics Journal, 764*, 19.
Bisikalo, D. V., Kaygorodov, P. V., Ionov, D. E., & Shematovich, V. I. (2013b). *Astronomy Reports, 57*, 715.
Bisikalo, D. V., Kaygorodov, P. V., Ionov, D. E., & Shematovich, V. I. (2014). In H. Lammer & M. L. Khodachenko (Eds.), *Characterizing stellar and exoplanetary environments* (pp. 81–104). Heidelberg/New York: Springer.
Bourrier, V., & Lecavelier des Etangs, A. (2013). *Astronomy and Astrophysics, 551*, A124.
Cecchi-Pestellini, C., Ciaravella, A., & Micela, G. (2006). *Astronomy and Astrophysics, 458*, L13.
Cecchi-Pestellini, C., Ciaravella, A., Micela, G., & Penz, T. (2009). *Astronomy and Astrophysics, 496*, 863.
Chamberlain, J. W., & Hunten, D. M. (1987). *Theory of planetary atmospheres. An introduction to their physics and chemistry*. Orlando: Academic Press.
Dalgarno, A., Yan, M., Liu, W. (1999). *Astrophysics Journal Supplement, 125*, 237.
Ehrenreich, D., Lecavelier des Etangs, A., Hébrard, G., Désert, J. M., Vidal-Madjar, A., McConnell, J. C., Parkinson, C. D., Ballester, G. E., & Ferlet, R. (2008). *Astronomy and Astrophysics, 483*, 933.
Ekenbäck, A., Holmström, M., Wurz, P., Grießmeier, J. M., Lammer, H., Selsis, F., & Penz, T. (2010). *Astrophysics Journal, 709*, 670.
Erkaev, N. V., Kulikov, Yu. N., Lammer, H., Selsis, F., Langmayr, D., Jaritz, G. F., & Biernat, H. K. (2007). *Astronomy and Astrophysics, 472*, 329.
Fossati, L., Haswell, C. A., Linsky, J. L., & Kislyakova, K. G. (2014). In H. Lammer & M. L. Khodachenko (Eds.), *Characterizing stellar and exoplanetary environments* (pp. 59–80). Heidelberg/New York: Springer.
Fox, J. L. & Hać, A. (1997). *Journal of Geophysical Research, 102*, 24005.
Fox, J. L., & Hać, A. B. (2009). *Icarus, 204*, 527.
Fox, J. L., Galand, M. I., & Johnson, R. E. (2008). *Space Science Reviews, 139*, 3.
García Muñoz, A. (2007). *Planetary and Space Science, 55*, 1426.
Garvey, R. H., & Green, A. E. S. (1976). *Physical Review A, 14*, 946.
Garvey, R. H., Porter, H. S., & Green, A. J. (1977). *Applied Physics, 48*, 4353.
Gerard, J. C., Richards, P. G., Shematovich, V. I., & Bisikalo, D. V. (1995). *Geophysical Research Letters, 22*, 279.
Gérard, J. C., Hubert, B., Bisikalo, D. V., & Shematovich, V. I. (2000). *Journal of Geophysical Research, 105*, 15795.
Glass-Maujean, M. (1986). *Physical Review A, 33*, 342.
Green, A. E. S., & Sawada, T. (1972). *Journal of Atmospheric and Terrestrial Physics, 34*, 1719.
Gröller, H., Shematovich, V. I., Lichtenegger, H. I. M., Lammer, H., Pfleger, M., Kulikov, Yu. N., Macher, W., Amerstorfer, U. V., & Biernat, H. K. (2010). *Journal of Geophysical Research, 115*, 12017.
Hodges, R. R. Jr., & Breig, E. L., (1991). *Journal of Geophysical Research, 96*, 7697.
Holmström, M., Ekenbäck, A., Selsis, F., Penz, T., Lammer, H., & Wurz, P. (2008). *Nature, 451*, 970.
Hubert, B., Gérard, J. C., Cotton, D. M., Bisikalo, D. V., & Shematovich, V. I. (1999). *Journal of Geophysical Research, 104*, 17139.

Hubert, B., Gérard, J. C., Killeen, T. L., Wu, Q., Bisikalo, D. V., & Shematovich, V. I. (2001). *Journal of Geophysical Research, 106*, 12753.
Huebner, W. F., Keady, J. J., & Lyon, S. P. (1992). *Astrophysics and Space Science, 195*, 1.
Ivanov, M. S., & Rogazinskij, S. V. (1988). Sov. J. Numer. Anal. Math. Modell., 453
Jackman, C. H., Garvey, R. H., & Green, A. E. S. (1977). *Journal of Geophysical Research, 82*, 5081.
Johnson, R. E., Combi, M. R., Fox, J. L., Ip, W. H., Leblanc, F., & McGrath, M. A., Shematovich, V. I., Strobel, D. F., & Waite, J. H. (2008). *Space Science Reviews, 139*, 355.
Kislyakova, K. G., Holmström, M., Lammer, H., Erkaev, N. V., (2014). In H. Lammer & M. L. Khodachenko (Eds.), *Characterizing stellar and exoplanetary environments* (pp. 137–150). Heidelberg/New York: Springer.
Koskinen, T. T., Yelle, R. V., Lavvas, P., & Lewis, N. K. (2010). *Astrophysics Journal, 723*, 116.
Koskinen, T. T., Harris, M. J., Yelle, R. V., & Lavvas, P. (2013). *Icarus, 226*, 1678.
Krestyanikova, M. A., & Shematovich, V. I. (2005). *Solar System Research, 39*, 22.
Krestyanikova, M. A., & Shematovich, V. I. (2006). *Solar System Research, 40*, 384.
Krstić, P. S., & Schultz, D. R. (1999a). *Physical Review A, 60*, 2118.
Krstić, P. S., & Schultz, D. R. (1999b). *Journal of Physics B: Atomic and Molecular Physics, 32*, 3485.
Lammer, H., Selsis, F., Ribas, I., Guinan, E. F., Bauer, S. J., & Weiss, W. W. W. (2003). *Astrophysics Journal (Letters), 598*, L121.
Lammer, H., Odert, P., Leitzinger, M., Khodachenko, M. L., Panchenko, M., Kulikov, Yu. N., Zhang, T. L., Lichtenegger, H. I. M., Erkaev, N. V., Wuchterl, G., Micela, G., Penz, T., Biernat, H. K., Weingrill, J., Steller, M., Ottacher, H., Hasiba, J., & Hanslmeier, A. (2009). *Astronomy and Astrophysics, 506*, 399.
Lammer, H., Kislyakova, K. G., Holmström, M., Khodachenko, M. L., & Grießmeier, J. M. (2011). *Astrophysics and Space Science, 335*, 9.
Lammer, H., Güdel, M., Kulikov, Yu. N., Ribas, I., Zaqarashvili, T. V., Khodachenko, M. L., Kislyakova, K. G., Gröller, H., Odert, P., Leitzinger, M., Fichtinger, B., Krauss, S., Hausleitner, W., Holmström, M., Sanz-Forcada, J., Lichtenegger, H. I. M., Hanslmeier, A., Shematovich, V. I., Bisikalo, D. V., Rauer, H., Fridlund, M. (2012). *Earth Planets Space, 64*, 179.
Lecavelier des Etangs, A., Vidal-Madjar, A., McConnell, J. C., & Hébrard, G. (2004). *Astronomy and Astrophysics, 418*, L1.
Lecavelier Des Etangs, A., Ehrenreich, D., Vidal-Madjar, A., Ballester, G. E., Désert, J. M., Ferlet, R., Hébrard, G., Sing, D. K., Tchakoumegni, K. O., & Udry, S. (2010). *Astronomy and Astrophysics, 514*, A72.
Lecavelier des Etangs, A., Bourrier, V., Wheatley, P. J., Dupuy, H., Ehrenreich, D., Vidal-Madjar, A., Hébrard, G., Ballester, G. E., Désert, J. M., Ferlet, R., Sing, & D. K. (2012). *Astronomy and Astrophysics, 543*, L4.
Linsky, J. L., Yang, H., France, K., Froning, C. S., Green, J. C., Stocke, J. T., & Osterman, S. N. (2010). *Astrophysics Journal, 717*, 1291.
Marov, M. Y., Shematovich, V. I., & Bisicalo, D. V. (1996). *Space Science Reviews , 76*, 1.
Nagy, A. F., Kim, J., & Cravens, T. E. (1990). *Annales Geophysicae, 8*, 251.
Opal, C. B., Peterson, W. K., & Beaty, E. C. (1971). *Journal of Chemical Physics, 55*, 4100.
Penz, T., Erkaev, N. V., Kulikov, Yu. N., Langmayr, D., Lammer, H., Micela, G., Cecchi-Pestellini, C., Biernat, H. K., Selsis, F., Barge, P., Deleuil, M., & Léger, A. (2008). *Planetary and Space Science, 56*, 1260.
Rjasanow, S., Schreiber, T., & Wagner, W. (1998). *Journal of Computational Physics, 145*, 382.
Shematovich, V. I. (2004). *Solar System Research, 38*, 28.
Shematovich, V. I. (2007). In Proceedings of 25th International Symposium on Rarefied Gas Dynamics, Siberian Branch of the Russian Academy of Sciences, 953.
Shematovich, V. I. (2008). In T. Abe (Ed.), American Institute of Physics Conference Series, (2008), 1084, 1047.
Shematovich, V. I. (2010). *Solar System Research, 44*, 96.

Shematovich, V.I., Bisikalo, D. V., & Gerard, J. C. (1994). *Journal of Geophysical Research, 99*, 23217.
Shematovich, V., Gérard, J. C., Bisikalo, D. V., & Hubert, B. (1999). *Journal of Geophysical Research, 104*, 4287.
Shematovich, V. I., Bisikalo, D. V., & Gérard, J. C. (2005). *Geophysical Research Letters, 32*, L02105.
Shizgal, B. D., & Arkos, G. G. (1996). *Reviews of Geophysics, 34*, 483.
Shyn, T. W., & Sharp, W. E. (1981). *Physical Review A, 24*, 1734.
van Kampen, A. G. (1984). *Stochastic processes in physics and chemistry*. Amsterdam: North-Holland.
van Zyl, B., & Stephen, T. M. (1994). *Physical Review A, 50*, 3164.
Vidal-Madjar, A., Lecavelier des Etangs, A., Désert, J. M., Ballester, G. E., Ferlet, R., Hébrard, G., & Mayor, M. (2003). *Nature, 422*, 143.
Vidal-Madjar, A., Désert, J. M., Lecavelier des Etangs, A., Hébrard, G., Ballester, G. E., Ehrenreich, D., Ferlet, R., McConnell, J. C., Mayor, M., & Parkinson, C. D. (2004). *Astrophysics Journal (Letters), 604*, L69.
Vidal-Madjar, A., Lecavelier des Etangs, A., Désert, J. M., Ballester, G. E., Ferlet, R., Hébrard, G., & Mayor, M. (2008). *Astrophysics Journal (Letters), 676*, L57.
Wayne, R. P. (1991). *Chemistry of atmospheres*. Oxford: Clarendon Press.
Yelle, R. V. (2004). *Icarus, 170*, 167.
Yelle, R. V. (2006). *Icarus, 183*, 508.
Yelle, R. V., Lammer, H., & Ip, W. H. (2008). *Space Science Reviews, 139*, 437.

Chapter 7
Stellar Driven Evolution of Hydrogen-Dominated Atmospheres from Earth-Like to Super-Earth-Type Exoplanets

Kristina G. Kislyakova, Mats Holmström, Helmut Lammer, and Nikolai V. Erkaev

Abstract In the present chapter we discuss the impact of a host stars radiation and plasma environment to the escape and evolution of hydrogen-dominated exoplanet atmospheres. We focus mainly on planets within the Earth- to super-Earth mass domain and consider both, thermal and nonthermal atmospheric escape processes. The type of thermal loss mechanism depends on the so-called escape parameter, which is the ratio of the gravitational energy of a particle to its thermal energy. For low values of this parameter a planetary atmosphere switches from classical Jeans to modified Jeans escape and finally to hydrodynamic blow off. During blow off the majority of the atmospheric particles dispose of enough energy to escape the planet's gravity field. This leads to extreme gas losses. It is shown that non-thermal losses for light species such as hydrogen never exceed blow off escape, but they are of significant importance for planets with relatively weak Jeans-type escape or heavier particles (e.g., O, C, N). From the diversity of non-thermal escape mechanisms, in the present chapter we focus on ion pick-up and discuss the importance of other loss mechanisms. The general conclusion of the chapter is, that escape processes strongly shape the evolution of exoplanet atmospheres and determine, if the planet loses its hydrogen and/or volatile-rich protoatmospheres or, on the contrary, remains

K.G. Kislyakova (✉) • H. Lammer
Austrian Academy of Sciences, Space Research Institute, Graz, A-8042, Austria
e-mail: kristina.kislyakova@oeaw.ac.at; helmut.lammer@oeaw.ac.at

M. Holmström
Swedish Institute of Space Physics, P.O. Box 812, SE-98128 Kiruna, Sweden
e-mail: matsh@irf.se

N.V. Erkaev
Institute for Computational Modelling, 660041 Krasnoyarsk 36, Russia

Russian Academy of Sciences, and Siberian Federal University, 660041 Krasnoyarsk, Russian Federation
e-mail: erkaev@icm.krasn.ru

H. Lammer, M. Khodachenko (eds.), *Characterizing Stellar and Exoplanetary Environments*, Astrophysics and Space Science Library 411,
DOI 10.1007/978-3-319-09749-7_7

as a mini-Neptune, which can probably not be considered as a potential habitat as we know it.

7.1 Introduction: Hydrogen-Rich Terrestrial Exoplanets

One of the most important questions still awaiting the solution in the present-day exoplanetary science is the determination of possible evolution scenarios of exoplanets, which could explain all currently observed types. All exoplanets as well as planets in the Solar System experience atmospheric mass losses. There exist a number of observations and measurements of losses from Solar System terrestrial planets (see, for example, Lundin 2011), and observations in the Lyman-α line for the hot Jupiters HD 209458b and HD 189733b (Vidal-Madjar et al. 2003; Lecavelier des Etangs et al. 2010) and a hot Neptune GJ 436b (Kulow et al. 2014), (see also Chap. 4, Fossati et al. 2014). Excess absorption in Lyman-α is usually explained as an observational evidence for mass loss (Bourrier and Lecavelier des Etangs 2013).

Here we focus on Earth-size to super-Earth type planets (we call them terrestrial in the present chapter), i.e., we exclude massive exoplanets from consideration. According to present-day modeling, terrestrial-sized planets should be common in the Universe (Broeg 2009). We make a short review about the atmospheric evolution of these moderate size planets and factors, which define it, as it is understood at present.

The atmosphere mass and composition of a terrestrial planet is defined, first, by the formation conditions and, second, by the following stellar-driven escape processes. Atmospheres may take origin in the primordial nebula, where all exoplanets are believed to be born (Ikoma and Genda 2006; Lammer et al. 2013a, 2014), or be later outgassed from the planetary interiors during solidification of magma oceans (Elkins-Tanton and Seager 2008). In the second case one speaks about a secondary atmosphere. Both types of atmosphere formation are believed to be important and work together on the atmosphere evolution.

Tables 7.1 and 7.2 show modeling results of modelled gravitational attraction and accumulation of nebula gas around an Earth-mass and a super-Earth core inside the habitable zone (HZ) of a solar-like host star as a function of protoplanetary nebula

Table 7.1 Captured hydrogen envelopes from a protoplanetary nebula around an Earth-size exoplanet within the HZ of a solar-like star with a protoplanet core mass of $1M_{\oplus}$ in units of EO_H ($1EO_H = 1.53 \times 10^{23}$ g), for nebula conditions with dust depletion factors f_d of 0.001, 0.01, and 0.1 and relative accretion rates $\frac{\dot{M}_{acc}}{M_{pl}}$ between 10^{-5} and 10^{-8} year^{-1} (Lammer et al. 2014)

$\frac{\dot{M}_{acc}}{M_{pl}}$ [year^{-1}]	$f_d = 0.001$	$f_d = 0.01$	$f_d = 0.1$
10^{-5}	9.608 EO_H	1.682 EO_H	0.316 EO_H
10^{-6}	44.6 EO_H	12.8 EO_H	2.313 EO_H
10^{-7}	210.0 EO_H	65.75 EO_H	16.8 EO_H
10^{-8}	1002 EO_H	332.6 EO_H	93.2 EO_H

Table 7.2 Captured hydrogen envelopes from a protoplanetary nebula around a super-Earth with a core size of $1.71R_{\oplus}$ and a core mass of $5M_{\oplus}$ in units of EO_H ($1EO_H = 1.53 \times 10^{23}$ g), for nebula conditions with dust depletion factors f_d of 0.001, 0.01, and 0.1 and relative accretion rates $\frac{\dot{M}_{acc}}{M_{pl}}$ between 10^{-5} and 10^{-8} year^{-1} (Lammer et al. 2014)

$\frac{\dot{M}_{acc}}{M_{pl}}$ [year^{-1}]	$f_d = 0.001$	$f_d = 0.01$	$f_d = 0.1$
10^{-5}	956.2 EO_H	299 EO_H	170 EO_H
10^{-6}	4653 EO_H	1600 EO_H	690 EO_H
10^{-7}	28620 EO_H	8117 EO_H	3250 EO_H
10^{-8}	105816 EO_H	103725 EO_H	17810 EO_H

parameters such as dust grain depletion factor f_d and protoplanetary luminosity during accretion. The results are based on a model that solves the hydrostatic structure equations for the protoplanetary nebula (Lammer et al. 2014). Depending on nebular properties, such as f_d, planetesimal accretion rates, and resulting luminosities one can see that an Earth-like rocky core can capture hydrogen from a few percent of the hydrogen amount in an Earth Ocean (EO_H) of up to several 10 s to even 1,000 EO_H (see also Lammer et al. 2014). Compared to a core with an Earth-mass, the super-Earth core with $5M_{\oplus}$ captures orders of magnitude more hydrogen. Even for a nebula condition with a high accretion rate and dust grain depletion factor the super-Earth would capture $\sim$170 EO_H. For low accretion rates an a $f_d = 0.001$ the planet would capture more than 10^5 EO_H. Thus, hydrogen envelopes that contain more than 1,000 EO_H may be common at super-Earths.

Recent findings of ESO's High Accuracy Radial velocity Planetary Searcher (HARPS) project and NASA's Kepler space observatory reveal from the radius-mass relation and the resulting density of discovered super-Earth, that these bodies probably have rocky cores, but are surrounded by significant H/He, H_2O envelopes, or both (e.g., http://www.exoplanet.eu). These findings demonstrate, that even if an exoplanet has a terrestrial size, its evolution may still differ much from the one of the Earth leaving the planet as a mini-Neptune type body. These discoveries raise the question about different evolution scenarios these exoworlds went and demonstrate, that not all of them could rid of their primordial dense envelopes. After the atmosphere formation is completed, its loss and evolution are defined by intensity of escape processes.

Several types of escape mechanisms from planetary atmospheres are known. The light gases like H and He are the most inclined to escape, however, heavier species can be lost too. It was shown by several authors such as Tian et al. (2008a,b), Lichtenegger et al. (2010), Lammer et al. (2013a), that nitrogen-dominated atmospheres are not easy to be kept during the first extreme phases of the star's activity after it's arrival at ZAMS due to extreme atmospheric expansion. Other species can be lost due to drag with the escaping hydrogen (this is considered as a possible explanation for Xe fractionation in the atmosphere of the Earth and was estimated, e.g., by Hunten et al. 1987). CO_2 atmosphere should not experience such strong expansion (Kulikov et al. 2006) and could be kept more easily, as one may see in the case of Venus.

Roche lobe effects can play a very important role for inflated close-in exoplanets, which fill their Roche lobe and consequently lose big amounts of mass (e.g. Erkaev et al. 2007; Lecavelier des Etangs et al. 2004). Close-in terrestrial planets lose their primordial hydrogen envelopes relatively easy (Leitzinger et al. 2011) and develop into hot bodies without atmosphere like CoRoT-7b. However, it is not so clear how mediate-sized planets in sub- to super-Earth mass and size domain can evolve in the HZs of their stars. Can they often evolve to Class I habitats with nitrogen atmosphere and life favorable conditions (Lammer et al. 2009a) which resemble the Earth, or do they usually evolve differently?

In this chapter we present the summary of our evolutionary studies for hydrogen-rich terrestrial-type exoplanets located in the HZ of its parent star. We consider thermal and nonthermal ion pick-up losses and discuss their possible influence on the planetary evolution. The main aim is to estimate the conditions under which these primordial hydrogen envelopes may be lost, leaving back an exoplanet with a possible Earth-type atmosphere.

7.2 Thermal Escape

On the basis of the sources of the host stars soft X-ray and EUV energy (XUV) input into the upper atmosphere, one can separate two main escape categories: thermal escape of neutral particles and non-thermal escape of neutrals and ions. Jeans escape is the classical thermal escape mechanism based on the fact that the atmospheric particles have velocities according to the Maxwell distribution. Figure 7.1 illustrates upper atmosphere processes and its response to solar/stellar XUV radiation (see also Chap. 6, Shematovich et al. 2014). Individual particles in the high tail of the distribution may reach escape velocity at the exobase altitude, where the mean free path is comparable to the scale height, so that they can escape from the planet's atmosphere. When the thermosphere temperature rises due to heating by the stellar XUV radiation, the bulk atmosphere starts hydrodynamically to expand with consequent adiabatic cooling. In such a case the velocity distribution at the exobase level is described by a shifted Maxwellian (e.g., Tian et al. 2008a, Erkaev et al. 2013). This regime can be called controlled hydrodynamic escape which resembles a strong Jeans-type escape but is still weaker compared to classical blow-off, where no control mechanism influences the escaping gas. If the XUV heating continues to increase so that the ratio between the gravitational and thermal particle energy becomes ≤ 1.5 than so-called blow off occurs leading to a stronger escape in comparison to the Jeans and even the controlled hydrodynamic escape. As it is shown by Erkaev et al. (2013), depending on the availability of possible IR-cooling molecules and the planets average density, hydrogen-rich super-Earths orbiting inside the HZ will reach hydrodynamic blow-off only for XUV fluxes several 10 times higher compared to today's Sun, which is the case for the early evolution of a Solar-type star (Güdel 2007). Most of their lifetime the upper atmospheres of these planets will experience non-hydrostatic conditions, but not blow off. In such

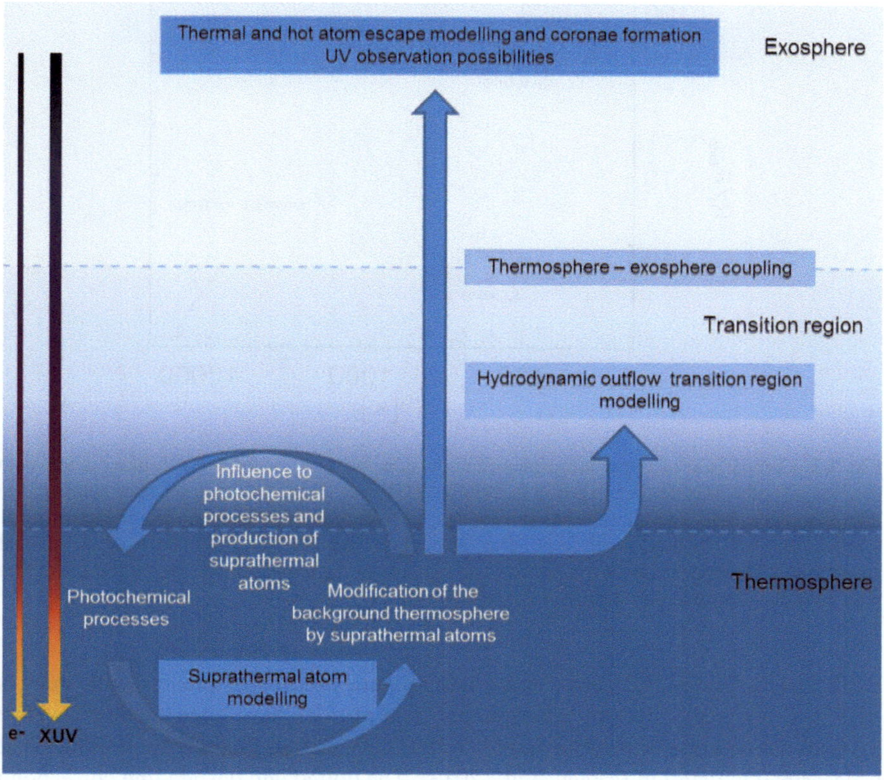

Fig. 7.1 Illustration of processes that occur in the upper atmosphere of planets. Heating by solar/stellar XUV radiation, photoelectrons, photochemical processes and the production of suprathermal atoms will modify the thermosphere structure and vice versa (see also Chap. 6, Shematovich et al. 2014). The temperature at which the transition from hydrostatic to hydrodynamic conditions occurs depends on the composition of the thermosphere and will also be affected by efficient non-thermal escape processes. The escape of suprathermal particles in turn depends on the physical structure of the thermosphere and its temperature

case the upper atmosphere expands hydrodynamically and the loss of the upward flowing gas results in controlled hydrodynamic escape. The blow off stage is more easily reached at less massive hydrogen-rich planets with mass equal to that of the Earth. These planets experience hydrodynamic blow off for much longer, and change from the blow off regime to the controlled hydrodynamic escape regime for XUV fluxes which are ≤ 10 times of today's Sun. Because of XUV heating and expansion of their upper atmospheres, both an exo-Earth and a super-Earth should produce extended exospheres or hydrogen coronae distributed above possible magnetic obstacles defined by intrinsic or induced magnetic fields. In such case the hydrogen-rich upper atmosphere will not be protected by possible magnetospheres like on present-day Earth, but could be eroded by the stellar wind plasma flow and lost from the planet in the form of ions (Khodachenko et al. 2007; Kislyakova

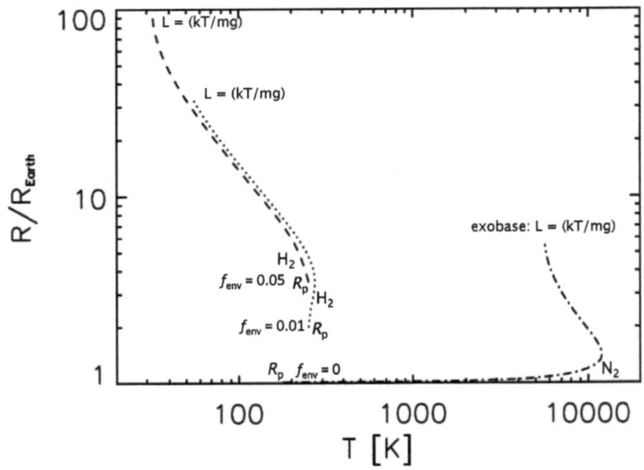

Fig. 7.2 Comparison of upper atmosphere temperature profiles for three Earth-like planets with core masses of 1 $M_\oplus$ that are exposed to a stellar XUV flux that is 10 times stronger compared to that of today's Sun. Two planets are surrounded by hydrogen envelopes f_{env} with $f_{env} = 0.01$ (*dotted line*) and 0.05 (*dashed line*). The third planet has an Earth-like nitrogen dominated atmosphere. The upper part of the profiles corresponds to the exobase level where the mean free path L of the atmosphere is similar as the scale height H

et al. 2013). Figure 7.2 shows three temperature profiles of planets with masses of $1M_\oplus$, but with different atmospheres. The profiles have been modeled with a hydrodynamic upper atmosphere and stellar XUV radiation absorption model that is described in detail in Erkaev et al. (2013). The dotted and dashed line correspond to Earth-mass planets with hydrogen envelopes f_{env}

$$f_{env} = \frac{M_{at}}{M_{at} + M_c}, \tag{7.1}$$

of 0.001 and 0.01, where M_{at} and M_c are the atmosphere and core masses and $R_p = R_c + z$ is the radius obtained from a transit. The third planet has an Earth-like nitrogen atmosphere with $f_{env} \sim 0$. In all three cases, the atmospheric masses are much lower than the core masses, so that $M_c \sim M_{pl} \sim M_\oplus$. However, according to Mordasini et al. (2012), Earth-mass planets with f_{env} of 0.001 and 0.01 have radii R_p that are much higher up in the atmosphere than their core radius R_c. The reason is that, in hydrogen atmospheres, visible light can not penetrate down to the surface (i.e., Jupiter, Saturn, Uranus, Neptune) because of Raleigh scattering (Lecavelier des Etangs et al. 2008). This is not the case for an Earth-like atmosphere. In a hydrogen dominated atmosphere, H_2 molecules are carriers of the Rayleigh scattering. The effective altitude at which the Rayleigh scattering of H2 molecules dominates depends on the mean density and as a consequence on the total pressure at the altitude. Because of this, the planetary radius R_p (see Fig. 7.2) corresponding to the atmospheric pressure P related to wavelength λ can then be written as

$$R_p = \frac{kT\mu g\tau^2}{P^2\sigma_{Rl}}, \tag{7.2}$$

with σ_{Rl} the λ dependent Rayleigh cross section, gravitational acceleration g, atmospheric temperature T, optical depth τ, the mean mass of atmospheric particles μ, and Boltzmann constant k. Therefore, for the two hydrogen-dominated protoplanets shown in Fig. 7.2, R_p will be larger in comparison to their core radii R_c and the nitrogen-rich planet case. This will result in the different average densities of the three planets shown in Fig. 7.2, although their masses are the same.

One can also see in Fig. 7.2, that if heated by the stellar XUV radiation that is 10 times higher compared to that of the present Sun, the temperature of the nitrogen atmosphere exceeds the one of the hydrogen atmospheres. The main reason is that the adiabatic cooling due to the hydrodynamic expansion is more effective in case of the hydrogen dominated upper atmosphere. One can also see from Fig. 7.2 that XUV exposed upper atmospheres can expand to several planetary radii. The profile curves end at the corresponding exobase levels where the mean free path L of the gas reaches the scale hight $H = kT/mg$, where m is the mean mass of the atmospheric species.

In this part of the chapter, calculations of the losses of captured hydrogen envelopes from protoplanets having masses in a range between Earth-like bodies and super-Earths with $5M_\oplus$ are presented by assuming that their rocky cores had formed before the nebula gas dissipated. In the thermal escape calculations we focused at bodies within the habitable zone (HZ) of a G star. These results are published in Lammer et al. (2014). This article is a continuation of the research performed by Erkaev et al. (2013), where the same methods were used to estimate the amount of hydrogen an Earth-type planet and a super-Earth ($M = 10M_\oplus$, $R = 2R_\oplus$) can lose in a HZ of a G star. The same code was also used by Lammer et al. (2013b) to estimate the thermal losses from 5 Kepler-11 super-Earths.

The model solves the system of the hydrodynamic equations for mass,

$$\frac{\partial \rho r^2}{\partial t} + \frac{\partial \rho v r^2}{\partial r} = 0, \tag{7.3}$$

momentum,

$$\frac{\partial \rho v r^2}{\partial t} + \frac{\partial \left[r^2(\rho v^2 + P)\right]}{\partial r} = -\rho g r^2 + 2Pr, \tag{7.4}$$

and energy conservation

$$\frac{\partial r^2\left[\frac{\rho v^2}{2} + \frac{P}{(\gamma-1)}\right]}{\partial t} + \frac{\partial v r^2\left[\frac{\rho v^2}{2} + \frac{\gamma P_p}{(\gamma-1)}\right]}{\partial r} = -\rho v r^2 g + q_{XUV} r^2. \tag{7.5}$$

Table 7.3 Thermal escape of hydrogen envelopes f_{env} in units of EO_H ($1EO_H = 1.53 \times 10^{23}$ g) from an Earth-size and mass planet inside the HZ according to a stellar XUV flux that is 100 times higher compared to that of the present Sun at 1 AU. Δt is the assumed exposure time, and z the envelope corresponding core-surface-radius distance (Mordasini et al. 2012; Lammer et al. 2014)

f_{env}	z [$R_\oplus$]	$\Delta t = 50$ Myr	$\Delta t = 100$ Myr	$\Delta t = 500$ Myr	$\Delta t = 1{,}000$ Myr
0.001	0.15	0.95 EO_H	1.9 EO_H	9.5 EO_H	19 EO_H
0.01	1	9.5 EO_H	19 EO_H	96.5 EO_H	193 EO_H

Table 7.4 Thermal escape of hydrogen envelopes f_{env} in units of EO_H ($1EO_H = 1.53 \times 10^{23}$ g) from a super-Earth with a core size of $1.71R_\oplus$ and a core mass of $5M_\oplus$ inside the HZ of its host star, according to a stellar XUV flux that is 100 times higher compared to that of the present Sun at 1 AU. Δt is the assumed exposure time, and z the envelope corresponding core-surface-radius distance (Mordasini et al. 2012; Lammer et al. 2014)

f_{env}	z [$R_\oplus$]	$\Delta t = 50$ Myr	$\Delta t = 100$ Myr	$\Delta t = 500$ Myr	$\Delta t = 1{,}000$ Myr
0.001	1.2	3.5 EO_H	6.9 EO_H	34.5 EO_H	69 EO_H
0.01	1.5	4.5 EO_H	9 EO_H	45 EO_H	90 EO_H
0.1	4	22.4 EO_H	44.7 EO_H	223.7 EO_H	447.5 EO_H

Here, r is the radial distance from the center of the protoplanetary core, ρ, P, T, v are the mass density, pressure, temperature and velocity of the nonhydrostatic outward flowing bulk atmosphere. γ is the polytropic index, g the gravitational acceleration and q_{XUV} is the XUV volume heating rate.

Tables 7.3 and 7.4 show the loss of protoplanetary nebula captured hydrogen envelopes due to hydrodynamic escape from an Earth-like core and a super-Earth with the size of a core size of $1.71R_\oplus$ and a core mass of $5M_\oplus$ inside the HZ of their host star in units of EO_H. The hydrogen envelopes are exposed to a XUV flux that is 100 times higher compared to that of today's Sun in 1 AU, a value expected during the early XUV saturation phase Δt of a young solar-like star (see also Chap. 1, Linsky and Güdel (2014)). Δt for G stars is expected to last about 100 Myr. It can be less for F stars but lasts longer for M dwarfs. The calculations have been performed in the domain from the lower thermosphere up to a critical point R_{crit}, where the Knudsen number equals to 0.1. The escape rates have been calculated with a heating efficiency of 15 % (see also Chap. 6, Shematovich et al. 2014). The escape of hydrogen from super-Earths can be large because of their large radii, bit it is too weak to evaporate the amount of captured hydrogen shown in Table 7.2, even if the exposure time equals 10^3 Myr. As it was shown by Erkaev et al. (2013) and Kislyakova et al. (2013), if an exoplanet located in the HZ does not lose the dense primordial hydrogen envelope during the first 90 Myr years of star's early XUV saturation period, it is unrealistic that the remaining gaseous envelope will be lost during the rest of the planet's lifetime. This can lead to the formation of a mini-Neptune instead of a super-Earth. Only close-in massive rocky planets such as CoRoT-7b (e.g., Leitzinger et al. 2011), that are exposed over a long time period to high XUV fluxes, or that grow to large massive bodies after the nebula evaporated, can lose their hydrogen envelopes completely (Lammer et al. 2014). From the results of these studies one finds that the nebula properties, protoplanetary growth time, planetary mass, size and the host stars radiation environment are the initial conditions, which define if the planets can evolve to the Earth-like class I habitats (Lammer et al. 2009a). Protoplanets with core masses that are $\leq 1 M_{\oplus}$ can lose their captured hydrogen envelopes during the active XUV saturation phase of their host stars inside the HZ of solar-like stars (Lammer et al. 2014), while rocky cores within the so-called super-Earth domain most likely can not get rid of their nebula captured hydrogen envelopes during their whole lifetime. Our results are in agreement with the suggestion that Solar System terrestrial planets, lost their nebula-based protoatmospheres during the XUV activity saturation phase of the young Sun. The results of these models are also in agreement with the discovery of low density hydrogen and/or volatile-rich super-Earths even at closer orbital distances <1 AU. These findings indicate that these planets could not lose their primordial atmospheres (see Fig. 13.1, Chap. 13, Fridlund et al. 2014). Furthermore, the results of these studies indicate that one should expect that many mini-Neptunes may populate the HZs of solar-like stars.

7.3 Ion Pick-Up

Besides thermal escape, various non-thermal escape mechanisms also contribute to the total mass loss and have to be accounted for. Non-thermal escape processes can be separated into ion escape and photochemical and kinetic processes that accelerate atoms beyond escape energy. Ions can escape from an upper atmosphere if the exosphere is not protected by a strong magnetic field and stretches above the magnetopause, or from the polar regions. In such a case, exospheric neutral atoms can interact with the stellar plasma environment. Also, planetary ions can be detached from an ionopause by plasma instabilities in the form of ionospheric clouds (Terada et al. 2002).

In this section we present our estimations of ion pick-up loss, which is one of the most effective among non-thermal escape processes (according to analysis of data for Venus and Mars obtained by the Analyzer of Space Plasma and Energetic Atoms (ASPERA) instruments on board Venus Express and Mars Express). We assume the exoplanets have no magnetic field to estimate the maximum possible losses. In our model, the ions are produces by charge-exchange with stellar protons, photoionization by stellar photons and electron impact ionization by electrons in the stellar wind.

Charge exchange reactions between stellar wind protons and neutral planetary particles consist of the transfer of an electron from a neutral atom to a proton leaving a cold atmospheric ion and an energetic neutral atom (ENA). This process is described by the following reaction: $H_{sw}^{+} + H_{pl} \rightarrow H_{pl}^{+} + H_{ENA}$.

We used the Direct Simulation Monte Carlo (DSMC) method to model the stellar wind – upper atmosphere interaction. The code starts at an inner boundary R_0 where the Knudsen number is 0.1. The details about the initial algorithm can be found in Holmström et al. (2008) and further developed versions in Kislyakova et al. (2013, 2014). Here we summarize only the main procedure.

The code includes two species: neutral hydrogen atoms and protons. The following processes/forces can act on neutral H atoms:

- Collision with a UV photon, which can occur if the particle is outside of the planet's shadow. Leads to an acceleration of the hydrogen atom away from the star. A UV photon is absorbed by a neutral hydrogen atom and then consequently reradiated in a random direction, leading to a radial velocity change. The UV collision rate is velocity dependent (depends also on the star Lyman-α flux and distance to the planet).
- Ionization by a stellar photon (photoionization) or by a stellar wind electron.
- Charge exchange between neutral hydrogen atoms and stellar wind protons. If a hydrogen atom is outside the planetary obstacle (magnetopause or ionopause) it can charge exchange with a stellar wind proton. The charge exchange cross-section is taken to equal $2 \times 10^{-19}\,\mathrm{m}^2$ (Lindsay and Stebbings 2005).
- Elastic collision with another hydrogen atom. Here the collision cross section was taken to be $10^{-21}\,\mathrm{m}^2$ (Izmodenov et al. 2000).

The coordinate system is centered at the center of the planet with the x-axis pointing towards the center of mass of the system, the y-axis pointing in the direction opposite to the planetary motion, and the z-axis pointing parallel to the vector Ω representing the orbital angular velocity of the planet. M_{st} is the mass of the planet's host star. The outer boundary of the simulation domain is the box $x_{min} \leq x \leq x_{max}$, $y_{min} \leq y \leq y_{max}$, and $z_{min} \leq z \leq z_{max}$. The inner boundary is a sphere of radius R_0.

Tidal potential, Coriolis and centrifugal forces, as well as the gravitation of the star and planet, acting on a hydrogen neutral atom are included in the following way (after Chandrasekhar 1963)

$$\frac{dv_i}{dt} = \frac{\partial}{\partial x_i}\left[\frac{1}{2}\Omega^2\left(x_1^2 + x_2^2\right) + \mu\left(x_1^2 - \frac{1}{2}x_2^2 - \frac{1}{2}x_3^2\right) + \right.$$
$$\left.\left(\frac{GM_{st}}{R^2} - \frac{M_{st}R}{M_p + M_{st}}\Omega^2\right)x_1\right] + +2\Omega\epsilon_{il3}v_l \tag{7.6}$$

Here $x_1 = x$, $x_2 = y$, $x_3 = z$, v_i are the components of the velocity vector of a particle, G is Newton's gravitational constant, R is the distance between the centers of mass, ϵ_{il3} is the Levi-Civita symbol, and $\mu = GM_{st}/R^3$. The first term in the right-hand side of Eq. 7.6 represents the centrifugal force, the second is the tidal-generating potential, the third corresponds to the gravitation of the planet's host star and the planet, while the last term stands for the Coriolis force. The self-gravitational potential of the particles is neglected.

Charge exchange takes place outside of an obstacle that corresponds to the magnetopause or ionopause boundary, which can be assumes as a surface described by

$$x = R_s \left(1 - \frac{y^2 + z^2}{R_t^2} \right). \tag{7.7}$$

Intensity of the stellar wind atmosphere erosion and, consequently, influence on the planetary evolution is different for different star ages and thus is strongly connected with the evolution of the star. It was discussed in Erkaev et al. (2013) and Kislyakova et al. (2013) that the stars of late spectral classes (red K and mostly M dwarfs) can keep high levels of the XUV radiation. The stellar wind evolution is more controversial (Wood et al. 2005), but higher XUV and X-ray heating intensify the non-thermal erosion of the atmosphere even if one considers the interaction with the constant stellar wind.

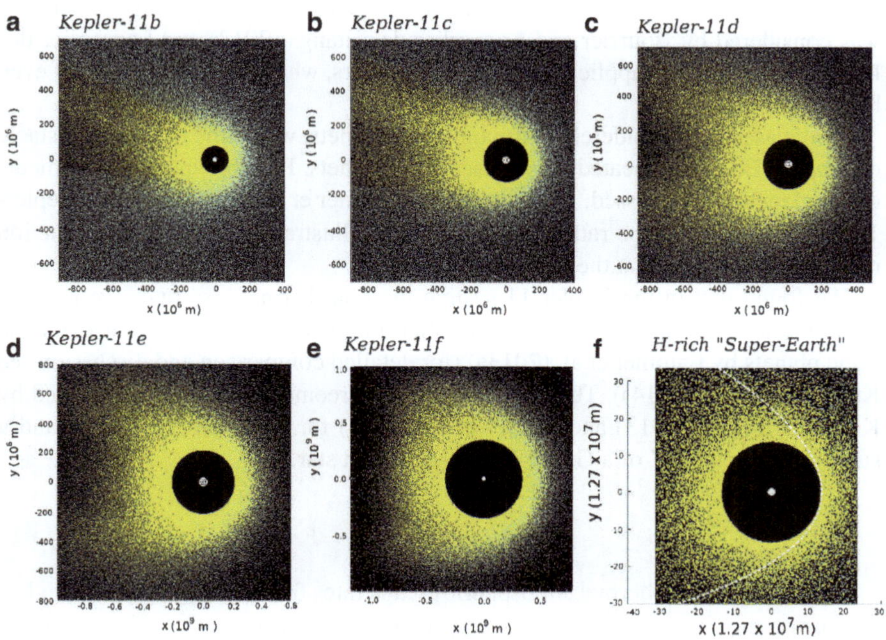

Fig. 7.3 **a–e**: Slices of modeled 3D atomic hydrogen coronae around the five Kepler-11 super-Earths for $-10^7 \leq z \leq 10^7$ m and heating efficiency $\eta = 40\,\%$. *Yellow* and *green dots* correspond to neutral hydrogen atoms and hydrogen ions, including stellar wind protons, respectively. The *white dot* in the center represents the planet. The *black empty area* around the planet corresponds to the XUV heated, hydrodynamically expanding thermosphere up to the height R_0 where Kn=0.1 (Kislyakova et al. 2013; 2014). **f**: Modeled atomic hydrogen coronae and stellar wind plasma interaction around a super-Earth hydrogen-rich planet inside an M star HZ at 0.24 AU. The XUV flux is 50 times higher than that of the present Sun, η=15 %; the *dashed line* denotes the planetary obstacle

Table 7.5 Ion pick-up loss rates and thermal escape rates for Kepler-11 planets for heating efficiency η (see Chap. 6, Shematovich et al. 2014) of 15 %. The values are given in [g·s^{-1}]. Thermal escape rates are taken from Table 3 in Lammer et al. (2013b)

Exoplanet	L_{ion}, η =15 %	L_{th}, η =15 %	L_{ion}, η =40 %	L_{th}, η =40 %
Kepler-11b	$\sim 1.17 \times 10^7$	$\sim 1.15 \times 10^8$	$\sim 1.3 \times 10^7$	$\sim 2.0 \times 10^8$
Kepler-11c	$\sim 1.07 \times 10^7$	$\sim 4.0 \times 10^7$	$\sim 1.37 \times 10^7$	$\sim 1.3 \times 10^8$
Kepler-11d	$\sim 1.47 \times 10^7$	$\sim 1.0 \times 10^8$	$\sim 2.33 \times 10^7$	$\sim 2.5 \times 10^8$
Kepler-11e	$\sim 1.84 \times 10^7$	$\sim 1.1 \times 10^8$	$\sim 3.34 \times 10^7$	$\sim 2.5 \times 10^8$
Kepler-11f	$\sim 6.0 \times 10^7$	$\sim 4.0 \times 10^8$	$\sim 6.8 \times 10^7$	$\sim 4.5 \times 10^8$

Figure 7.3 presents the results of the DSMC modeling of hydrogen coronae around five Kepler-11 super-Earths (Kislyakova et al. 2014) and around a test super-Earth with $M = 10M_{\oplus}$, $R = 2R_{\oplus}$ located in the HZ of an M dwarf (Kislyakova et al. 2013). As can be seen, in each case a huge hydrogen corona is formed composed of neutral hydrogen of planetary origin, ENAs and H atoms accelerated by the radiation pressure. Radiation pressure effects are of most importance in the vicinity of the host star (Kepler-11b and -c). This type of acceleration was considered by Bourrier and Lecavelier des Etangs (2013) and Lecavelier des Etangs et al. (2004) in application to the hot Jupiters, where these effects are of even higher importance.

In all cases, we considered non-magnetized planets, where the magnetic obstacle defined by Eq. 7.7 is located very close to the planets. Possible magnetic moments of exoplanets are discussed, for example, in Lammer et al. (2009b) and for Kepler-11b–f are believed to be rather weak. Table 7.5 illustrates average ion production rates, L_{ion}, depending on the heating efficiency η.

We found that in the Kepler-11 system the non-thermal loss rates are approximately one order of magnitude smaller than the thermal losses estimated for the same planets by Lammer et al. (2013a) (for detailed comparison and discussion see Kislyakova et al. (2014)). This ratio is in good agreement with results obtained by Kislyakova et al. (2013) and Erkaev et al. (2013) for an Earth-type planet and a super-Earth in the HZ of a GJ436-like M-type host star.

Conclusion

Thermal and non-thermal atmospheric escape rates from hydrogen-dominated upper atmospheres and their influence on atmospheric evolution of medium-size exoplanets are presented. The size, mass and distance to the planet's host star together with the stellar spectral type define in many aspects, if the planet may evolve into a terrestrial-type planet. The amount of initially accumulated gases, i.e. primordial nebula properties, plays also a big role (Lammer et al. 2014; Ikoma and Genda 2006). In general, close-in exoplanets lose their

(continued)

primordial envelopes and secondary atmospheres and reach the hydrodynamic blow off state much easier, and experience this escape condition longer (Lammer et al. 2013b), and they may lose their atmospheres also due to Roche-lobe overflow (Erkaev et al. 2007). At orbital distances of the HZ atmospheric evolution strongly depends also on the planet's mass and type of atmosphere. According to the results discussed in this chapter and in Lammer et al. (2014), Mars-sized bodies and planets with masses up to $<0.5 M_{\oplus}$ can lose a significant percentage or even the whole hydrogen-dominated protoatmosphere. On the other hand, so-called super-Earths will experience difficulties in losing their dense primordial envelopes composed mostly of light gases like hydrogen (Erkaev et al. 2013; Lammer et al. 2013b, 2014).

Non-thermal ion pick-up escape contributes to the total atmospheric losses as well, but although it is of most importance for small-sized planets like Mars (Lundin 2011), for bigger planets it makes up only several percent of total thermal and non-thermal losses and can not change the evolutionary scenarios significantly (Kislyakova et al. 2013, 2014).

Atmospheric evolution of an exoplanet is also deeply connected with the evolution of its host star. M dwarfs live longer and develop slower in comparison to Sun-type stars, which means that they stay longer in the very active XUV activity saturation phase (see also Chap. 1, Linsky and Güdel (2014). If an exoplanet orbits an M dwarf, it experiences severe stellar conditions (high levels of X-rays and EUV radiation, stronger stellar wind) much longer, leading to additional losses in comparison to an exoplanet of the same type orbiting a G star (Erkaev et al. 2013; Kislyakova et al. 2013).

Summarizing the above mentioned results, the evolution of a terrestrial-type nitrogen atmosphere requires several restrictions on an exoplanet and its host star. The star should be a long-lived object, which probably excludes the very early spectral classes; on the other hand, it should stay not too long in the highest XUV activity stage, when the planetary atmospheres undergo strong thermal and non-thermal atmospheric escape. The latter makes the fate of the exoplanets in M dwarf systems controversial, however, it is not excluded that the planets may evolve as a terrestrial-type body also orbiting these stars. As for the exoplanet itself, it has to be located in the right distance from its star, gain enough mass and volatiles during the formation. The volatile envelope should be thick enough to protect the atmospheres from the high stellar activity during the early age, but also thin enough to be lost. Otherwise the exoplanet would evolve into a mini-Neptune containing up to several percents of its weight as light gases.

Acknowledgements The authors acknowledge the support by the International Space Science Institute (ISSI) in Bern, Switzerland and the ISSI team *Characterizing stellar- and exoplanetary environments*. K. G. Kislyakova, and H. Lammer acknowledge support by the Austrian Research Foundation FWF NFN project S116 'Pathways to Habitability: From Disks to Active Stars, Planets and Life', and the related FWF NFN subproject, 'Particle/Radiative Interactions with Upper Atmospheres of Planetary Bodies Under Extreme Stellar Conditions.'

References

Bourrier, V., & Lecavelier des Etangs, A. (2013). *Astronomy and Astrophysics, 557*, A124.

Broeg, C. H. (2009). *Icarus, 204*, 15.

Chandrasekhar, S. (1963). *Astrophysical Journal, 138*, 1182.

Elkins-Tanton, L. T., & Seager, S. (2008). *Astrophysical Journal, 685*, 1237.

Erkaev, N. V., Kulikov, Y. N., Lammer, H., Selsis, F., Langmayr, D., Jaritz, G. F., & Biernat, H. K. (2007). *Astronomy and Astrophysics, 472*, 329.

Erkaev, N. V., Lammer, H., Odert, P., Kulikov, Y. N., Kislyakova, K. G., Khodachenko, M. L., Güdel, M., Hanslmeier, A., & Biernat, H. (2013). *Astrobiology, 11*, 1–19.

Fossati, L., Haswell, C. A., Linsky, J. L., & Kislyakova, K. G. (2014). In H. Lammer & M. L. Khodachenko (Eds.), *Characterizing stellar and exoplanetary environments* (pp. 59). Heidelberg/New York: Springer.

Fridlund, F., Rauer, H., & Erikson, A. (2014). In H. Lammer & M. L. Khodachenko (Eds.), *Characterizing stellar and exoplanetary environments* (pp. 253). Heidelberg/New York: Springer.

Güdel, M. (2007). *Living Reviews in Solar Physics, 4*, 3.

Holmström, M. Ekenbäck, A. Selsis, F. Penz, T. Lammer, H. & Wurz, P. (2008). *Nature, 451*, 970–972.

Hunten, D. M., Pepin, R. O., & Walker, J. C. G. (1987). *Icarus, 69*, 532–549.

Ikoma, M., & Genda, H. (2006). *Astrophysical Journal, 648*, 696.

Izmodenov, V. V., Malama, Y. G., Kalinin, A. P., Gruntman, M. Lallement, R., & Rodionova, I. P. (2000). *APSS, 274*, 71–76.

Khodachenko, M. L., Ribas, I., Lammer, H., Grießmeier, J.-M., Leitner, M., Selsis, F., Eiroa, C., Hanslmeier, A., Biernat, H. K., Farrugia, C. J., & Rucker, H. O. (2007). *Astrobiology, 7*, 167–184.

Kislyakova, K. G., Lammer, H., Holmström, M., Panchenko, M., Odert, P., Erkaev, N. V., Leitzinger, M., Khodachenko, M. L., Kulikov, Y. N., Güdel, M., Hanslmeier, A. (2013). *Astrobiology, 11*, 1030–1048.

Kislyakova, K. G., Johnstone, C. P., Odert, P., Erkaev, E. V., Lammer, H., Lüftinger, T., Holmström. M., Khodachenko, M. L., & Güdel, M. (2014). *Astronomy and Astrophysics, 562*, A116.

Kulikov, Yu. N., Lammer, H., Lichtenegger, H. I. M., Terada, N., Ribas, E., Kolb, C., Langmayr, D., Lundin, R., Guinan, E. F., Barabash, S., & Biernat, H. B. (2006). *PSS, 54*, 1425.

Kulow, J. R., France, K., Linsky, J., & Parke Loyd, R. O. (2014), *Astrophysical Journal, 786*, 132 (9pp)

Lammer, H., Bredehöft, J. H., Coustenis, A., Khodachenko, M. L., Kaltenegger, L., Grasset, O., Prieur, D., Raulin, F., Ehrenfreund, P., Yamauchi, M., Wahlund, J.-E., Grießmeier, J.-M., Stangl, G., Cockell, C. S., Kulikov, Y. N., Grenfell, J. L., & Rauer, H. (2009a). *Astronomy and Astrophysics Review, 17*, 181–249.

Lammer, H., Odert, P., Leitzinger, M., Khodachenko, M. L., Panchenko, M., Kulikov, Y. N., Zhang, T. L., Lichtenegger, H. I. M., Erkaev, N. V., Wuchterl, G., Micela, G., Penz, T., Biernat, H. K., Weingrill, J., Steller, M., Ottacher, H., Hasiba, J., & Hanslmeier, A. (2009b). *Astronomy and Astrophysics, 506*, 399.

Lammer, H., Kislyakova, K. G., Güdel, M., Holmström, M., Erkaev, N. V., Odert, P., & Khodachenko, M. L. (2013a). In J. M. Trigo-Rodriguez, F. Raulin, C. Muller & C. Nixon (Eds.), *The early evolution of the atmospheres of terrestrial planets*. Astrophysics and space science proceedings (p. 33). New York: Springer.

Lammer, H., Erkaev, N. V., Odert, P., Kislyakova, K. G., Leitzinger, M., & Khodachenko, M. L., (2013b). *Monthly Notices of the Royal Astronomical Society, 430*, 1247–1256.

Lammer, H., Erkaev, N. V., Odert, P., Kislyakova, K. G., Leitzinger, M., & Khodachenko, M. L. (2014). *Monthly Notices of the Royal Astronomical Society, 439*, 3225.

Lecavelier des Etangs, A., Vidal-Madjar, A., McConnell, J. C., & Hébrard, G. (2004). *Astronomy and Astrophysics, 418*, L1-L4.

Lecavelier des Etangs, A., Pont, F., Vidal-Madjar, A., & Sing, D. (2008). *Astronomy and Astrophysics, 481*, L83.

Lecavelier des Etangs, A., Ehrenreich, D., Vidal-Madjar, A., Ballester, G. E., Désert, J.-M., Ferlet, R., Hébrard, G., Sing, D. K., Tchakoumegni, K.-O., & Udry, S. (2010). *Astronomy and Astrophysics, 514*, A72.

Leitzinger, M., Odert, P., Kulikov, Y. N., Lammer, H., Wuchterl, G., Penz, T., Guarcello, M. G., Micela, G., Khodachenko, M. L. Weingrill, J. Hanslmeier, A. Biernat, H. K., & Schneider, J., (2011). *PSS, 59*, 1472.

Lichtenegger, H. I. M., Lammer, H., Grießmeier, J.-M., Kulikov, Y. N., von Paris, P., Hausleitner, W., Krauss, S., & Rauer, H. (2010). *Icarus, 210*, 1.

Lindsay, B. G. & Stebbings, R. F. (2005). *Journal of Geophysical Research, 110*, 12213.

Linsky, J. L., & Güdel, M., (2014). In H. Lammer & M. L. Khodachenko (Eds.), *Characterizing stellar and exoplanetary environments* (pp. 3). Heidelberg/New York: Springer.

Lundin, R. (2011). *SSR, 162*, 309.

Mordasini, C., Alibert, Y., Georgy, C., Dittkrist, K.-M., & Henning, T. (2012). *Astronomy and Astrophysics, 545*, A112.

Shematovich, V. I., Bisikalo, D. V., & Dmitry E. I. (2014). In H. Lammer & M. L. Khodachenko (Eds.), *Characterizing stellar and exoplanetary environments* (pp. 105). Heidelberg/New York: Springer.

Tian, F., Kasting, J. F., Liu, H.-L., & Roble, R. G. (2008a). *Journal of Geophysical Research, 113*, 5008.

Tian, F., Solomon, S. C., Quian, L., & Lei, J. (2008b). *Journal of Geophysical Research (Planets), 113*, 7005.

Terada, N., Machida, S., & Shinagawa, H. (2002). *Journal of Geophysical Research (Space Physics), 107*, 1471.

Vidal-Madjar, A., Lecavelier des Etangs, A., Désert, J.-M., Ballester, G. E., Ferlet, R., Hébrard, G., & Mayor, M. (2003). *Nature, 422*, 143.

Wood, B. E., Müller, H.-R., Zank, G. P., Linsky, J. L., & Redfield, S. (2005). *Astrophysical Journal Letters, 628*, L143.

Chapter 8
Interpretations of WASP-12b Near-UV Observations

Aline A. Vidotto, Dmitry V. Bisikalo, Luca Fossati, and Joe Llama

Abstract The near-UV observations of the hot-Jupiter WASP-12b obtained by Fossati et al. have revealed the presence of an asymmetric transit lightcurve that is both more pronounced in the near-UV and starts at an earlier time than the optical lightcurve. These features of the near-UV transit of WASP-12b have intrigued several modellers. In this Chapter, we review the different interpretations of the near-UV observations of the system.

8.1 Introduction: WASP-12b an Evaporating Hot Jupiter

In Chap. 4, Fossati et al. (2014) gives, an overview of the observations of exoplanet atmospheres and surrounding environments was presented. Here we focus on the particular case of WASP-12b, a hot-Jupiter transiting a late-F main-sequence star.

WASP-12b have been observed at near-UV wavelengths by the Hubble Space Telescope (HST) at a couple of occasions (Fossati et al. 2010b; Haswell et al. 2012). An intriguing feature of the near-UV transit of WASP-12b is that, when

A.A. Vidotto (✉)
SUPA, School of Physics and Astronomy, University of St Andrews, North Haugh, KY16 9SS, St Andrews, UK

Observatoire de Genève, Université de Genève, Ch. des Maillettes 51, Versoix, CH-1290, Switzerland
e-mail: Aline.Vidotto@st-andrews.ac.uk

D.V. Bisikalo
Russian Academy of Sciences, Institute of Astronomy, Moscow, Russian Federation
e-mail: bisikalo@inasan.ru

L. Fossati
Argelander-Institut für Astronomie der Universität Bonn, Auf dem Hügel 71, 53121, Bonn, Germany
e-mail: lfossati@astro.uni-bonn.de

J. Llama
Lowell Observatory, 1400 W Mars Hill Rd. Flagstaff. Arizona. 86001. USA
e-mail: joe.llama@lowell.edu

H. Lammer, M. Khodachenko (eds.), *Characterizing Stellar and Exoplanetary Environments*, Astrophysics and Space Science Library 411,
DOI 10.1007/978-3-319-09749-7_8

compared to the optical transit, it shows the presence of an asymmetric, deeper absorption that starts before the beginning of the transit in the optical but seems to end simultaneously with the optical transit. These asymmetric transit observations have been attributed to the presence of asymmetries in the exosphere of the planet (see also Chap. 5, Bisikalo et al. (2014)).

Fossati et al. (2010b) themselves speculated that the observed asymmetry could be due to the presence of a disc of previously stripped material, lost by the planet through Roche lobe overflow, following the predictions of Li et al. (2010).

Since then, the transit asymmetry observed by Fossati et al. (2010b) has raised the attention of several modelers. For example, Lai et al. (2010) studied in more details the mass transfer by Roche lobe overflow in the system, demonstrating that, as the overflowing gas passes through the inner Lagrangian point and flows along a narrow stream, it can form an accretion stream that can have a projected area comparable to that of the planet and a significant optical depth. They argue that this material could produce both the earlier ingress of WASP-12b as well as the deeper absorption associated to the near-UV transit.

Lai et al. (2010) also investigated the presence of a bow shock around the planet due to the interaction of the planet's magnetosphere with a stellar wind as the cause of the early ingress. Assuming a typical solar wind mass-loss rate and adopting solar wind properties, they derived the wind velocity at the distance of WASP-12b using a thermally-driven wind model (Parker 1958). They found that the wind is still subsonic at WASP-12b orbital distance, concluding that no bow shock is expected to form in the planet-wind interaction zone.

This last argument was later on disputed by Vidotto et al. (2010), who demonstrate that a bow shock can actually be formed around the planet if the relative azimuthal velocity between the planetary orbital motion and the ambient medium is taken into account. In their model, Vidotto et al. (2010) assumed that the bow shock is formed surrounding the planet's magnetosphere, whose pressure would be sufficient to hold the compressed material at the observed stand-off distance (derived from the time of the near-UV ingress).

More recently, a hydrodynamical simulation performed by Bisikalo et al. (2013b) further explored the idea of an outflow that emerged from the overfilling of the planet's Roche lobe. In their model, it is the ram pressure of this material that supports the bow shock, as opposed to the magnetic idea put forward by Vidotto et al. (2010).

At present, the data currently available of the system is insufficient to be able to distinguish between these different interpretations. In this Chapter, we focus on the latter two interpretations provided in the literature, namely that a bow shock is present surrounding the planet and that this bow shock could be either supported by the magnetic pressure of the planet (Vidotto et al. 2010) or by the ram pressure of the overflowing material (Bisikalo et al. 2013b).

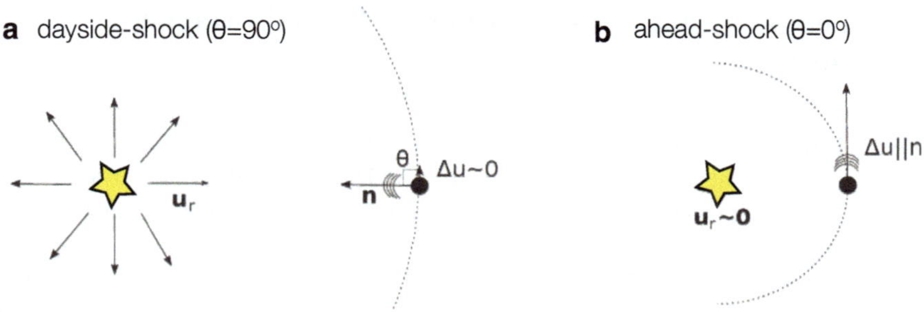

Fig. 8.1 Sketch of shock limits: (**a**) dayside-shock ($\theta = 90°$), (**b**) ahead-shock ($\theta = 0°$). *Arrows* radially leaving the star depict the stellar wind, *dashed semi-circles* represent the orbital path, θ is the deflection angle between the shock normal **n** and the relative azimuthal velocity of the planet $\Delta\mathbf{u}$ (After Vidotto et al. 2010)

8.1.1 The Bow Shock Model

A bow shock around a planet is formed when the relative motion between the planet and the stellar corona/wind is supersonic. The shock configuration depends on the direction of the flux of particles that arrives at the planet. We illustrate two different limits of the shock configuration in Fig. 8.1, where θ is the deflection angle between the azimuthal direction of the planetary motion and **n** is a vector that defines the outward direction of the shock. As seen from the planet, −**n** is the velocity of the impacting material. The first shock limit, a dayside-shock, occurs when the dominant flux of particles impacting on the planet arises from the (mainly radial) wind of its host star. For instance, the impact of the supersonic solar wind forms a bow shock at the dayside of Earth's magnetosphere (i.e., at the side that faces the Sun). This condition is illustrated in Fig. 8.1a and is met when $u_r > c_s$, where u_r and c_s are the local radial stellar wind velocity and sound speed, respectively.

A second shock limit, an ahead-shock (Fig. 8.1b), occurs when the dominant flux of particles impacting on the planet arises from the relative azimuthal velocity between the planetary orbital motion and the ambient plasma. This condition is especially important when the planet orbits at a close distance to the star, and therefore, possesses a high Keplerian velocity u_K. In this case, the velocity of the particles that the planet sees is supersonic if

$$\Delta u = |u_K - u_\varphi| > c_s, \tag{8.1}$$

where u_φ is the azimuthal velocity of the stellar corona. In general, we expect the bow shock to be formed at intermediate angles.

8.2 The Bow Shock Surrounding the Planet's Magnetic Obstacle

WASP-12b orbits a late-F main-sequence star with mass $M_* = 1.35\ M_\odot$ and radius $R_* = 1.57\ R_\odot$, at an orbital radius of $a = 3.15\ R_*$ (Hebb et al. 2009). Due to its close proximity to the star, the flux of coronal particles impacting on the planet should come mainly from the azimuthal direction, as the planet moves at a Keplerian orbital velocity of $u_K = (GM_*/a)^{1/2} \sim 230\,\mathrm{km\,s^{-1}}$ around the star. Therefore, a bow shock ahead of the of the planetary orbital motion is formed, where stellar coronal material is compressed. Vidotto et al. 2010's suggestion is that this material is able to absorb enough stellar radiation, causing the early-ingress observed in the near-UV lightcurve (see Fig. 8.2).

By measuring the phases at which the near-UV and the optical transits begin, one can derive the stand-off distance from the shock to the centre of the planet. In the geometrical consideration made next, it is assumed that the planet is fully superimposed on the disk of the central star, which is a good approximation for, e.g., small planets and transits with small impact parameters. Consider the sketches presented in Fig. 8.2, where d_{op} and d_{UV} are, respectively, the sky-projected

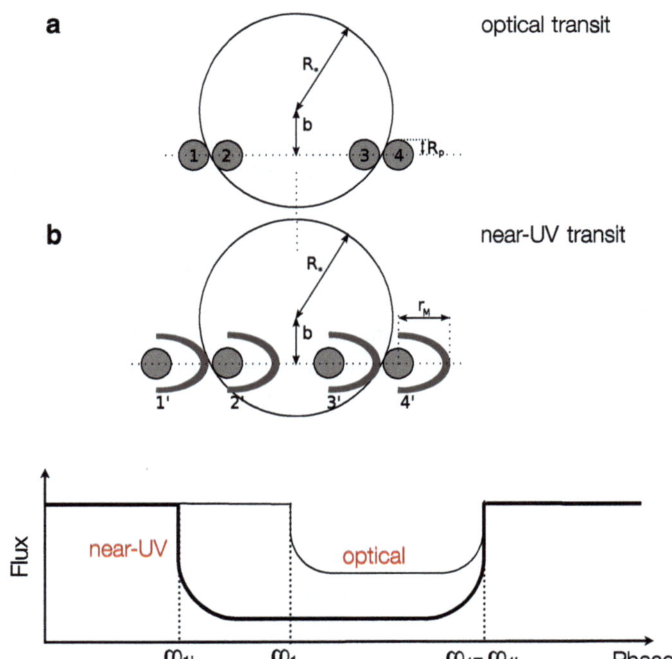

Fig. 8.2 Sketches of the lightcurves obtained through observations in the (**a**) optical and (**b**) near-UV, where the bow shock surrounding the planet's magnetosphere is also able to absorb stellar radiation (Figure adapted from Vidotto et al. 2011b)

distances that the planet (optical) and the system planet+magnetosphere (near-UV) travel from the beginning of the transit until the middle of the optical transit

$$d_{op} = (R_*^2 - b^2)^{1/2} + R_p \tag{8.2}$$

and

$$d_{UV} = (R_*^2 - b^2)^{1/2} + r_M, \tag{8.3}$$

where b is the impact parameter derived from transit observations, R_p is the planetary radius, and r_M is the distance from the shock nose (i.e., the head-on collision point) to the center of the planet. The start of the optical transit occurs at phase φ_1 (point 1 in Fig. 8.2), while the near-UV transit starts at phase $\varphi_{1'}$ (point $1'$ in Fig. 8.2). Taking the mid-transit phase at $\varphi = \varphi_m \equiv 1$, one notes that d_{op} is proportional to $(1 - \varphi_1)$, while d_{UV} is proportional to $(1 - \varphi_{1'})$. Using Eqs. (8.2) and (8.3), the stand-off distance r_M is derived from observed quantities

$$\frac{r_M}{R_p} = \frac{(1-\varphi_{1'})}{(1-\varphi_1)}\left[\sqrt{\left(\frac{R_*}{R_p}\right)^2 - \left(\frac{b}{R_p}\right)^2} + 1\right] - \sqrt{\left(\frac{R_*}{R_p}\right)^2 - \left(\frac{b}{R_p}\right)^2}. \tag{8.4}$$

Vidotto et al. (2010) assumed the stand-off distance to trace the extent of the planetary magnetosphere. At the magnetopause, pressure balance between the coronal total pressure and the planet total pressure requires that

$$\rho_c \Delta u^2 + \frac{[B_c(a)]^2}{8\pi} + p_c = \frac{[B_p(r_M)]^2}{8\pi} + p_p, \tag{8.5}$$

where ρ_c, p_c and $B_c(a)$ are the local coronal mass density, thermal pressure, and magnetic field intensity, and p_p and $B_p(r_M)$ are the planet thermal pressure and magnetic field intensity at r_M. In the case of a magnetized planet, the planet total pressure is usually dominated by the contribution caused by the planetary magnetic field (i.e., $p_p \sim 0$).

Vidotto et al. (2010) showed that, because WASP-12b is at a close distance to the star, the kinetic term of the coronal plasma may be neglected in Eq. (8.5). They also neglected the thermal pressure, so that Eq. (8.5) reduces to $B_c(a) \simeq B_p(r_M)$. Further assuming that stellar and planetary magnetic fields are dipolar, one has

$$B_p = B_* \left(\frac{R_*/a}{R_p/r_M}\right)^3, \tag{8.6}$$

where B_* and B_p are the magnetic field intensities at the stellar and planetary surfaces, respectively. Equation (8.6) shows that the planetary magnetic field can be obtained from *observationally-derived* quantities, namely the ratio a/R_* (from optical transit data), the normalised magnetospheric size r_M/R_p (from both near-UV and optical transits, Eq. 8.4) and the stellar magnetic field. For WASP-12, the stand-off distance obtained from the near-UV transit observation was found to be $r_M = 4.2\ R_p$ (Lai et al. 2010). Using the upper limit of $B_* < 10$ G for the stellar magnetic field derived by Fossati et al. (2010a), Vidotto et al. (2010) predicted, from Eq. (8.6), an upper limit for the planetary magnetic field of WASP-12b of

$$B_p < 24\ G.$$

8.2.1 Radiation Transfer Simulations of the Near-UV Transit

In order to test the hypothesis that the bow shock model was indeed able to cause the lightcurve asymmetry observed in WASP-12b, Llama et al. (2011) performed Monte Carlo radiation transfer simulations of the near-UV transit of WASP-12b. The characteristics of the stellar coronal plasma (density, velocity and temperature), modelled by Vidotto et al. (2010), were adopted in order to derive the density at the shock nose and the angle at which the shock is formed. As in Vidotto et al. (2010), Llama et al. (2011) assumed a shock in the adiabatic limit where the density behind such a shock increases by a factor of four with respect to the density ahead of the shock (stellar coronal material).

The characteristics of the local plasma surrounding the planet were derived based on simple models that are described next. Vidotto et al. (2010) adopted two scenarios. The first one considers the corona as a hydrostatic medium, so that it corotates with the star. The other one, considers that the corona is filled with an expanding, isothermal wind. In this case, the wind radial velocity u_r is derived from the integration of the differential equation along the radial coordinate r

$$u_r \frac{\partial u_r}{\partial r} = -\frac{1}{\rho_c}\frac{\partial p_c}{\partial r} - \frac{GM_*}{r^2}, \tag{8.7}$$

where G is the gravitational constant. The first prescription adopted has the advantage of having analytical solutions. The latter one, although lacking an analytical expression, can be easily integrated. In the first scenario, there is no radial velocity of the wind plasma, so that the shock forms ahead of the planet, while in the second scenario, the shock forms at an intermediate angle. These angles were used in the simulations of Llama et al. (2011).

To compute the near-UV light curve, the characterization of the three-dimensional geometry of the shock is required. Therefore, two unknowns of the shock geometry, its solid angle and its thickness, had to be specified. In order to tackle the influence of these two parameters, Llama et al. (2011) performed simulations for several shock geometries. As a result, they found that different sets of parameters could produce similar solutions.

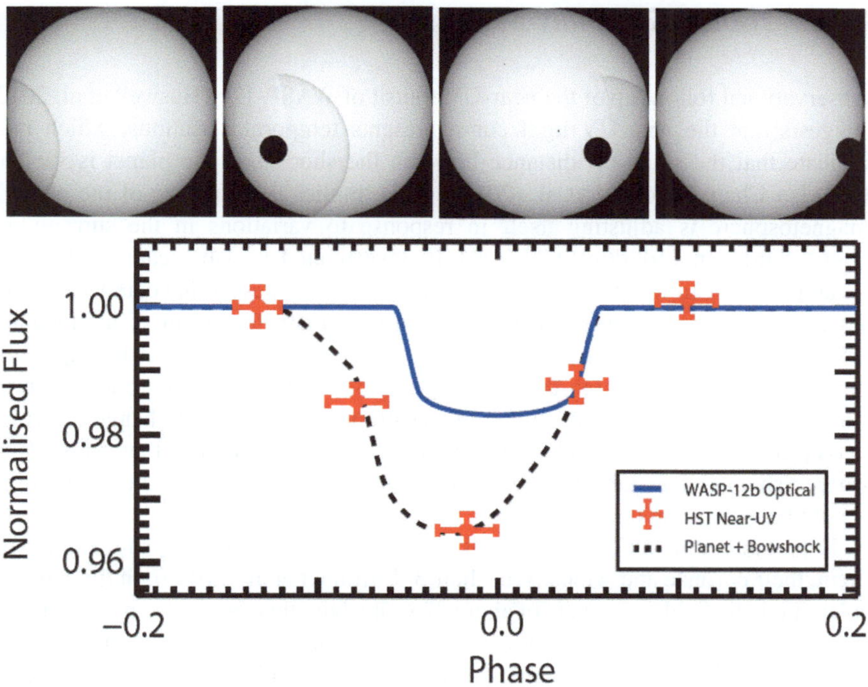

Fig. 8.3 Illustration of the near-UV transit of WASP-12b. The *solid line* shows the optical transit of WASP-12b, while the *dashed line* shows the simulated transit of a planet and its surrounding bow-shock in the near-UV. The near-UV transit observed with HST is shown in red (Figure based on the results of Llama et al. (2011))

To constrain model parameters, Llama et al. (2011) relied on the information present in the near-UV light curve by Fossati et al. (2010b). By analyzing different models that were equally able to provide a good fit to the HST data, they were able to place constraints on (1) the phase of the near-UV ingress ($\varphi_{1'}$) and (2) the optical depth of the shocked material, related to the extent of the shock and its thickness. From (1), they could constrain the projected stand-off distance as 5.5 R_p, slightly larger than the value derived by Lai et al. (2010) and used in Vidotto et al. (2010). From (2), they showed that the shocked material does not need to have a large optical depth to cause the amount of absorption observed in the near-UV HST light curve. Figure 8.3 illustrates one of their simulations that provided a good fit to the HST data. The simulations presented by Llama et al. (2011) support the hypothesis that a bow-shock could generate an early ingress of the transit, as the addition of a bow shock breaks the symmetry of the transit light curve. Nevertheless, the current data is not yet adequate to fully test this prediction. Near-UV observations of WASP-12b (Vidotto et al. 2011a) are desirable in order to test and constrain the models.

8.2.2 *Transit Variability*

Observational follow-up of the near-UV transit of WASP-12b (Haswell et al. 2012) suggests that the near-UV light curve presents temporal variations, which may indicate that the stand-off distance between the shock and the planet is varying (see also Chap. 4, Fossati et al. (2014)). This implies that the size of the planet's magnetosphere is adjusting itself in response to variations in the surrounding ambient medium. Vidotto et al. (2011b) investigated possible causes of transit variabilities due to variations of the coronal material. Processes that could lead to the variations in the surrounding medium of the planet are, for instance, an elongated orbit, a non-axisymmetric stellar corona, planetary obliquity (which may allow the planet to move through different regions of the host star's corona), intrinsic variations of the stellar magnetic field (resulting in stellar wind changes, coronal mass ejections, magnetic cycles). Vidotto et al. (2011b) concluded that, for systems where a shock is detectable through transit light curve observations, shock variations should be a common occurrence.

In particular, the magnetic fields of stars are observed to evolve and, along with them, their coronae and winds (see Chap. 3, Lüftinger et al. (2014)). In the Sun, the solar magnetic field has a cycle of about 22 years, but other systems can present even shorter cycles (Fares et al. 2009). Unfortunately, surface magnetic imaging is not available for WASP-12, implying that we do not have information of the topology of its magnetic field. Nevertheless, one can investigate how the evolution of stellar magnetic field can affect the variability of transits using targets whose magnetism is better constrained.

Recently, Llama et al. (2013) performed a detailed wind modelling of HD 189733, taking into consideration the large-scale surface maps derived for this star at different epochs by Fares et al. (2010). They used the results of the stellar wind model to derive the local stellar wind conditions throughout the orbital path of HD 189733b. The bow shock model, used to compute the geometry and thickness of the bow shock surrounding the planet, was improved as compared to the one adopted in Llama et al. (2011). Taking a magnetosphere whose characteristics are similar to that of Jupiter, Llama et al. (2013) performed simulated transit light curves in the optical and in the near-UV. They showed that depending on the nature of the stellar magnetic field, and hence its wind, both the transit duration and ingress time can vary when compared to optical lightcurves. As a consequence, any two consecutive near-UV transits might not be similar.

8.3 Gas Dynamic Simulation of the Interaction Between WASP-12b and Its Host Star

What happens if the exoplanet does not have its own proper magnetic field? Is it still possible to use a bow shock to explain the early ingress observed for WASP-12b? To answer this question Bisikalo et al. (2013b) performed pure 3D gas dynamic simulations of the interaction between the WASP-12b envelope and stellar wind to detail the flow pattern of the extended planet upper atmosphere. In particular, Bisikalo et al. (2013b) showed that the supersonic motion of the planet inside the stellar wind leads to the formation of a bow shock with a complex shape. Let us now consider the obtained solution in more details.

The first attempt to make gas dynamic simulations of the interaction between WASP-12b and its host star was made by Ionov et al. (2012). Assuming that the planet does not possess any magnetic field it was found that the bow shock forms immediately at the level of the upper atmosphere. Although the authors were able to qualitatively model the light curve, the shock results to be at a distance significantly different from the observed one. The distance between the contact discontinuity and the center of the planet can be defined analytically considering the dynamic pressure balance from both sides of the contact discontinuity (Baranov and Krasnobaev 1977). For the considered non magnetic case, this distance should not be greater than $\sim 1.88 R_{pl}$ and the wave front should be located at $\sim 2.23 R_{pl}$ from its center, half of that indicated by the observations. It is important to note that both early bow shock models (Vidotto et al. 2010; Ionov et al. 2012) do not account for the presence of the outflowing atmosphere. In presence of an overfilling Roche lobe the balance equations should take into account the dynamic and thermal pressure of the outflowing atmospheric gas (see Chap. 5, Bisikalo et al. (2014)). In this way, the positions of the bow shock and contact discontinuity will be defined by the gas properties. In case of an outflowing atmosphere, the gas from the vicinity of the L_1 point will form a stream with a slowly decreasing density, hence increasing the importance of the thermal pressure. More importantly, the stream from the L_1 point accelerates in the gravitational field of the star, increasing its radial velocity as $\sqrt{r}$. As a consequence, the dynamic pressure of the stream will increase ($\sim r$), while the importance of the magnetic field will decrease as $r^{-3}(B^2 \sim r^{-6})$. This means that for an outflowing atmosphere the role of the stream in the solution should be extremely relevant.

8.3.1 Model Description

The atmospheric parameters of WASP-12b are not well known, therefore Bisikalo et al. (2013b) assumed an isothermal upper atmosphere with a temperature, which is defined by the host star's deposition of XUV-energy, and is similar to that obtained for other hot-Jupiters: $T_{pl} = 10^4$ K (Yelle 2004; Koskinen et al. 2013). Note that the assumed temperature of the thermosphere does not correspond to the equilibrium temperature, which is about 2,500 K (Hebb et al. 2009), but to that of the lower thermosphere which is close to the visual radius and is heated by the stellar EUV radiation (Yelle 2004; Koskinen et al. 2013).

Bisikalo et al. (2013b) did not include stellar irradiation in the model, because this would affect just the boundary temperature value. In the simulation, Bisikalo et al. (2013b) attempted to recover the presence of an early ingress at the observed distance, by changing the boundary conditions to obtain a solution where the stream from the L_1 point is stopped by the stellar wind at the observed distance. On the other hand, the influence of the boundary temperature on the solution is rather important, because it defines the density of the atmospheric gas at the L_1 point. This is a key parameter because it is used to gather the Roche lobe overfilling and the power of the stream from the L_1 point; this consequently defines the solution. As a consequence, to develop a self-consistent gas dynamic model, it is necessary to include a more accurate description of the irradiation.

Bisikalo et al. (2013b) estimated the density of the lower thermosphere considering that the optical depth along the line of sight is $\tau = n_{pl} \times l_{pl} \times k_{pl} = 1$. Here, the distance l_{pl} corresponds to the path followed by the line of sight through the spherical layer $[R_{pl} - (R_{pl} + H_{pl})]$, where H_{pl} is the scale height. To avoid uncertainties due to the calculation of the opacities k_{pl}, Bisikalo et al. (2013b) assumed that the atmosphere of WASP-12b is hydrogen dominated, similarly to what Vidal-Madjar et al. (2003) considered for HD 209458b, for which the number density at the photometric radius is $\sim 2 \times 10^{10}\,\mathrm{cm}^{-3}$ (Murray-Clay et al. 2009). Taking into account that WASP-12b is ~ 2 times heavier and ~ 1.6 times larger (radius) than HD 209458b, they estimated the hydrogen number density at the photometric radius as $n_{pl} \approx 1.6 \times 10^{10}\,\mathrm{cm}^{-3}$, corresponding to a density of $\approx 2.7 \times 10^{-14}\,\mathrm{g/cm}^3$. They defined the obtained density at the photometric radius and the corresponding exobase at $1.55R_{pl}$. For comparison, Lai et al. (2010) defined the density at the 1 bar level, and their exobase at $1.59R_{pl}$, in agreement with the approach of Bisikalo et al. (2014).

Bisikalo et al. (2013b) considered a system configuration comparable to that of a binary system, consisting of the star with $M_* = 1.35\,M_\odot$ and WASP-12b with mass $M_{pl} = 1.3 \times 10^{-3}\,M_\odot$. Following the observations, Bisikalo et al. (2013b) assumed that the components of the binary system, having an orbital separation $A = 4.9\,R_\odot$, move in a circular orbit with a period $P_{orb} = 26\,\mathrm{h}$. The linear velocity of the planet in this system is $230\,\mathrm{km\,s}^{-1}$.

The flow is described by a 3D system of gravitational gas dynamic equations closed by the equation of state of a perfect neutral monatomic gas. In this model they neglected the non-adiabatic processes of radiative heating and cooling. An analysis of their solution shows that the density of the envelope is rather high, so everywhere in the envelope the Knudsen number[1] is <1. In the important regions of the solution (along the streams from the L_1 and L_2 points) the Knudsen number is always <0.1, allowing one to assume that a gas dynamic approach is valid for the problem considered here.

To solve the system of gas dynamic equations Bisikalo et al. (2013b) used a Roe-Osher TVD scheme of a high approximation order with the Einfeldt modification. This numerical method allows one to study flows with a significant

[1]The Knudsen number is defined as: $Kn = \lambda/H$, where λ is the mean free path, equal to $1/(n \times \sigma)$ where n is the number density and σ is the cross-section, $\sigma = 10^{-15}\,\mathrm{cm}^2$, and H is a typical scale of the solution $H = n/(dn/dr)$.

density contrast (Boyarchuk et al. 2002; Bisikalo et al. 2004, 2013a). Calculations have been performed in a rotating coordinate frame where the force field can be described by the Roche potential

$$\Phi = \frac{-GM_*}{\sqrt{x^2+y^2+z^2}} - \frac{GM_{pl}}{\sqrt{(x-A)^2+y^2+z^2}} - \frac{1}{2}\Omega^2\left[\left(x-\frac{AM_{pl}}{M_*+M_{pl}}\right)^2+y^2\right], \tag{8.8}$$

here Ω is the angular velocity of the system's rotation.

The origin of the coordinate system was placed at the center of the star; the X-axis is directed to the planet; the Z-axis coincides with the rotation axis of the system and is perpendicular to the orbital plane; the Y-axis finalizes the right-handed system. The calculations have been performed in a rectangular homogeneous grid. The size of the calculation domain is $(25 \times 20 \times 10)R_{pl}$ and the grid resolution is $464 \times 363 \times 182$ cells. The adopted size of the cell of 0.05 R_{pl} allows one to investigate all main flow pattern features in the planet's vicinity.

The boundary conditions have been set as follows. The upper atmosphere was assumed to be isothermal ($T_{pl} = 10^4$ K) and in hydrostatic equilibrium, i.e. the gas velocity in the upper atmosphere is zero. At the level $r = R_{pl}$, they set the value of the density at the visual radius equal to 2.7×10^{-14} g/cm^3. At the initial conditions, the density of the planet upper atmosphere was defined according to the barometric formula from the adopted boundary conditions up to the distance where the density becomes less than the wind density. Outside this region the computational domain was filled by gas of stellar wind. They treated the stellar wind as done by Vidotto et al. (2010). The particle number density in the stellar wind has been set equal to 5×10^6 cm^{-3} (Vidotto et al. 2010). The stellar wind parameters of WASP-12 are unknown, therefore Bisikalo et al. (2013b) used that of the solar wind. Namely, the temperature of the wind has been set equal to that of the Sun at the corresponding distance from the star $T = 10^6$ K (Withbroe 1988). The velocity of the wind was assumed to be 100 km s^{-1}, corresponding to the velocity of the solar wind at the distance of WASP-12b to the host star (Withbroe 1988). As the wind acceleration mechanism is still an open question, in the model (Bisikalo et al. 2013b) assumed the existence of a wind acceleration mechanism working similarly to light pressure. To take this into account Bisikalo et al. (2013b) modified the expression for the Roche potential, given in Eq. 8.8, as follows

$$\Phi = \frac{-\Gamma GM_*}{\sqrt{x^2+y^2+z^2}} - \frac{GM_{pl}}{\sqrt{(x-A)^2+y^2+z^2}} - \frac{1}{2}\Omega^2\left[\left(x-\frac{AM_{pl}}{M_*+M_{pl}}\right)^2+y^2\right], \tag{8.9}$$

where the Γ coefficient is used to account for the stellar wind acceleration and it is set to zero in the regions filled with wind material, and one in the rest of the calculation domain. This allows one to avoid the non-physical wind deceleration due to stellar gravity. The proper stellar wind velocity is subsonic with a Mach number $M = 0.85$. However, taking into account the supersonic orbital motion of the planet

($M = 1.97$) the total velocity of the planet with respect to the stellar wind is indeed supersonic with the rather large Mach number of $M = 2.14$.

8.3.2 The Flow Structure Around the Planet

The general morphology of the flow is shown in Fig. 8.4, indicating the density distribution and velocity vectors in the planet's envelope. The planet is depicted by the filled circle and it moves counterclockwise. The yellow solid lines denote the flow lines starting from the gas of the stellar wind. Besides, the Roche lobe, depicted by the white solid line, and Lagrangian points L_1 and L_2 are also shown.

Considering the density distribution shown in Fig. 8.4, Bisikalo et al. (2013b) noticed that the planet's envelope had a non-spherical complex shape. In addition to the upper atmosphere itself, one can see flows towards the L_1 and L_2 points. According to the conservation of angular momentum these streams are deflected in the direction of the planet's motion and against it, respectively. The supersonic motion of the planet and the presence of the stellar wind lead to the formation of a bow shock oriented according to the total vector of the wind material velocity with respect to the planet. The location of the shock and contact discontinuity can be easily determined from variations in the flow lines. It turns out that the head-on

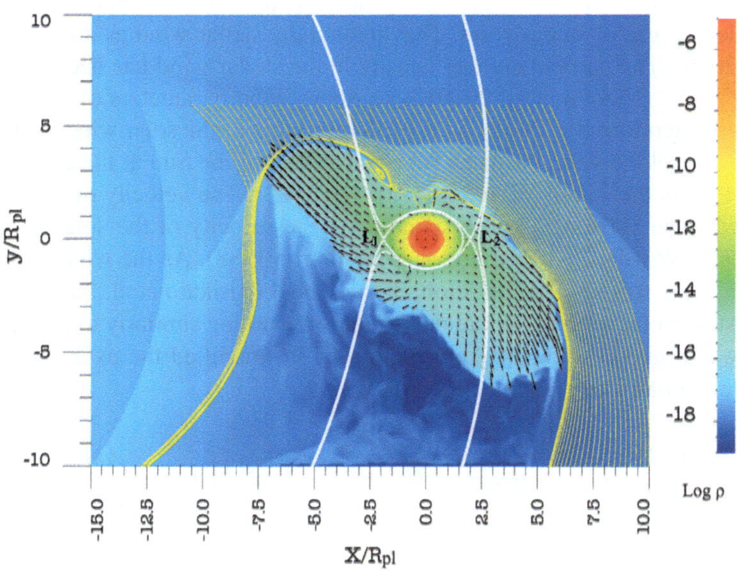

Fig. 8.4 Density distribution and velocity vectors in the envelope of WASP-12b. The planet is depicted by the filled circle and it moves counterclockwise. The *yellow solid lines* denote the flow lines starting in the gas of the stellar wind. The *white solid lines* denote the Roche equipotentials passing through the Lagrangian points L_1 and L_2 (Figure adapted from Bisikalo et al. 2013b)

collision point is located on the gas stream from the L_1 point. The distance between the planet center and the head-on collision point, as projected onto the stellar limb, is ~4.5 planet radii, in agreement with the estimation by Lai et al. (2010), who derived the position of the early-ingress at ~3.2 planet radii, from the planet surface.

Both shock and contact discontinuity have complex shapes. The asymmetric shape of the planet envelope, the upper atmosphere itself and two flows/prominences from L_1 and L_2, lead to the formation of a distinguishable double-peaked shock. The head-on collision point is at the peak of the prominence in the direction of L_1. However, when moving closer to the planet, the shock bends due to the presence of the planetary atmosphere, and shifts farther away from the planet, leading to the formation of a second hump in the shock. It is noticeable that the matter's flow, moving to the planet from the head-on collision point, undergoes strong disturbance when it gets into the cavity between the two humps in the shock waves. In particular, vortices form in this cavity. This effect smooths the contact discontinuity, mixing the matter of the stellar wind and upper atmosphere.

The flows towards the L_1 and L_2 points extend far away from the planet's Roche lobe. However, in their simulation the envelope is in a quasi-stationary regime, i.e. it is almost enclosed. The dynamic pressure of the stellar wind and flow, caused by the orbital motion of the planet, breaks the propagation of the streams from the L_1 and L_2 points, limiting the gaseous envelope of the planet by the bow shock and contact discontinuity. The numerical simulations briefly described here and presented in more detail in Bisikalo et al. (2013b) allow one to simulate the formation of an enclosed and steady-state gaseous envelope surrounding hot Jupiters which exceed the planet's Roche lobe, such as WASP-12b.

8.3.3 Early Ingress in Pure Gas Dynamic Model

Using 3D numerical simulations, Bisikalo et al. (2013b) investigated the flow pattern in the vicinity of the exoplanet WASP-12b, which is overfilling its Roche lobe. The results indicate that WASP-12b's envelope has a complex non-spherical shape. In addition to the central part of the spherical upper atmosphere two prominences, directed to the L_1 and L_2 Lagrangian points, develop in the system. These streams leave the planet and are deflected in the direction of the orbital motion and against it. Under the action of the dynamic pressure of the stellar wind these flows slow down and, then stop at distances of ~ 6 and ~ 4 planetary radii from its center, forming a stationary envelope.

The planet and its envelope move in the gas of the stellar wind with a supersonic velocity resulting in a Mach number $M = 2.14$. Thus, the dynamic pressure of the stellar wind not only works towards the formation of the stationary envelope, but leads also to the formation of a bow shock and contact discontinuity, which delimit the envelope. The wave has a complex double-peaked shape. The head-on collision point found in the calculations is at the peak of the stream from the L_1 point. The distance from the planet to the head-on collision point, as projected to the stellar limb, is of ~ 4.5 planetary radii, allowing us to explain the observed extent of the early ingress.

We should note one more important property of the described model of the planet's hydrogen envelope. Observations indicate (Fossati et al. 2010b) that the eclipse in the spectral bands where the early ingress has been observed is two times deeper than the eclipse of the planet itself (3.2 % versus 1.7 %). The possible explanation for this effect can also be the formation of a bow shock. Indeed, the motion of the planet inside the stellar wind is significantly supersonic and the formed bow shock is much hotter than the planet upper atmosphere. The heating of the gaseous envelope including the stream from the L_1 point leads to the excitation and broadening of additional lines in the UV bands. According to Vidal-Madjar et al. (2004) and Ben-Jaffel (2007) the depth of the transit in a given spectral band strongly depends on its width and presence of strong spectral lines. The larger the equivalent width of the lines in a given band, the deeper the transit in this band. Therefore, the additional heating of the planet envelope by the bow shock allows one to explain the fact that the eclipse, in the spectral bands where the early ingress has been observed, is two times deeper than the eclipse of the planet itself.

In summary, Bisikalo et al. (2013b) found that by using the pure gas dynamic simulations of the stellar wind interaction between WASP-12b and its host star, it is possible to obtain an alternative and self-consistent flow pattern which explains the existing observational data.

Conclusion

In this chapter, we presented a brief overview of possible interpretations of what is causing the intriguing early-ingress in the near-UV transit of the hot-Jupiter WASP-12b, as well as the deeper in-transit absorption when compared to the optical data, as observed by Fossati et al. (2010b) and Haswell et al. (2012). In particular, we focused on the interpretations that a bow shock, surrounding the planet at a certain stand-off distance, could be the cause of the near-UV transit features. In the first scenario we considered here, the bow shock forms surrounding the magnetosphere of the planet and may be used as a way to estimate planetary magnetic fields (Eq. 8.6). In the second scenario, the bow shock forms surrounding a dense exosphere, which is created by the overfilling of the Roche lobe. At present, the data currently available of the system is insufficient to be able to distinguish between these and other different interpretations. Repeated observations of the system is necessary in order to better constrain models, as well as for a deeper understanding of the structure and evolution of the stellar coronal material.

Acknowledgements The authors acknowledge the support by the International Space Science Institute (ISSI) in Bern, Switzerland and the ISSI team *Characterizing stellar- and exoplanetary environments*. A. A. Vidotto acknowledges also the support from a Royal Astronomical Society Fellowship and from an Ambizione fellowship of the Swiss National Science Foundation.

References

Baranov, V. B., & Krasnobaev, K. V. (1977). Hydrodynamic theory of a cosmic plasma, Moscow Izdatel Nauka.

Ben-Jaffel, L. (2007). *Astrophysical Journal, 671*, L61.

Bisikalo, D. V., Boyarchuk, A. A., Kaigorodov, P. V., Kuznetsov, O. A., & Matsuda, T. (2004). *Astronomy Reports, 48*, 449.

Bisikalo, D. V., Zhilkin, A. G., & Boyarchuk, A. A. (2013a). *Gas dynamic close binary stars (in Russian)*. Moscow: Physmatlit.

Bisikalo, D., Kaygorodov, P., Ionov D., Shematovich, V., Lammer, H., & Fossati, L. (2013b). *Astrophysical Journal, 764*, 19.

Bisikalo, D. V., Kaygorodov, P. V., Ionov, D. E., & Shematovich, V. I. (2014). In H. Lammer & M. L. Khodachenko (Eds.), *Characterizing stellar and exoplanetary environments* (pp. 81). Heidelberg/New York: Springer.

Boyarchuk, A. A., Bisikalo, D. V., Kuznetsov, O. A., & Chechetkin, V. M. (2002). Mass transfer in close binary stars, ESO.

Fares, R., Donati, J. F., Moutou, C., Bohlender, D., Catala, C., Deleuil, M., Shkolnik, E., Collier Cameron, A., Jardine, M. M., & Walker, G. A. H. (2009). *Monthly Notices of the Royal Astronomical Society, 398*, 1383.

Fares, R., Donati, J. F., Moutou, C., Jardine, M. M., Grießmeier, J. M., Zarka, P., Shkolnik, E. L., Bohlender, D., Catala, C., & Collier Cameron, A. (2010). *Monthly Notices of the Royal Astronomical Society, 406*, 409.

Fossati L., Bagnulo S., Elmasli. A, Haswell. C. A., Holmes S., Kochukhov O., Shkolnik E. L., Shulyak D. V., Bohlender D., Albayrak B., Froning C., & Hebb L. (2010a). *Astrophysical Journal, 720*, 872.

Fossati, L., Haswell, C. A., Froning, C. S., Hebb, L., Holmes, S., Kolb, U., Helling, C., Carter, A., Wheatley, P., Collier Cameron, A., Loeillet, B., Pollacco, D., Street, R., Stempels, H. C., Simpson, E., Udry, S., Joshi, Y. C., West, R. G., Skillen, I., & Wilson, D. (2010b). *Astrophysical Journal, 714*, L222.

Fossati, L., Haswell, C. A., Linsky, J. L., & Kislyakova, K. G. (2014). In H. Lammer, & M. L. Khodachenko (Eds.), *Characterizing stellar and exoplanetary environments* (pp. 59). Heidelberg/New York: Springer.

Haswell, C. A., Fossati, L., Ayres, T., France, K., Froning, C. S., Holmes, S., Kolb, U. C., Busuttil, R., Street, R. A., Hebb, L., Collier Cameron, A., Enoch, B., Burwitz, V., Rodriguez, J., West, R. G., Pollacco, D., Wheatley, P. J., & Carter, A. (2012). *Astrophysical Journal, 760*, 79.

Hebb, L., Collier-Cameron, A., Loeillet, B., Pollacco, D., Hébrard, G., Street, R. A., Bouchy, F., Stempels, H. C., Moutou, C., Simpson, E., Udry, S., Joshi, Y. C., West, R. G., Skillen, I., Wilson, D. M., McDonald, I., Gibson, N. P., Aigrain, S., Anderson, D. R., Benn, C. R., Christian, D. J., Enoch, B., Haswell, C. A., Hellier, C., Horne, K., Irwin, J., Lister, T. A., Maxted, P., Mayor, M., Norton, A. J., Parley, N., Pont, F., Queloz, D., Smalley, B., & Wheatley, P. J. (2009). *Astrophysical Journal, 693*, 1920.

Ionov, D. E., Bisikalo, D. V., Kaygorodov, P. V., & Shematovich, V. I. (2012). In M T Richards & I Hubeny (Eds.), *IAU Symposium, vol. 282*, Tatranska Lomnica, 545.

Koskinen, T. T., Harris, M. J., Yelle, R. V., & Lavvas, P. (2013). *Icarus, 226*, 1678.

Lai, D., Helling, C., & van den Heuvel, E. P. J. (2010). *Astrophysical Journal, 721*, 923.

Li, S. L., Miller, N., Lin, D. N. C., & Fortney, J. J. (2010). *Nature, 463*, 1054.

Llama, J., Wood, K., Jardine, M., Vidotto, A. A., Helling, C., Fossati, L., & Haswell, C. A. (2011). *Monthly Notices of the Royal Astronomical Society, 416*, L41.

Llama, J., Vidotto, A. A., Jardine, M., Wood, K., Fares, R., & Gombosi, T. I. (2013). *Monthly Notices of the Royal Astronomical Society, 436*, 2179.

Lüftinger, T., Vidotto, A. A., & Johnstone, C. P. (2014). In H. Lammer & M. L. Khodachenko (Eds.), *Characterizing stellar and exoplanetary environments* (pp. 37). Heidelberg/New York: Springer.

Murray-Clay, R. A., Chiang, E. I., & Murray, N. (2009). *Astrophysical Journal, 693*, 23.
Parker, E. N. (1958). *Astrophysical Journal, 128*, 664.
Vidal-Madjar, A., Lecavelier des Etangs, A., Désert, J. M., Ballester, G. E., Ferlet, R., Hébrard, G., & Mayor, M. (2003). *Nature, 422*, 143.
Vidal-Madjar, A., Désert, J. M., Lecavelier des Etangs, A., Hébrard, G., Ballester, G. E., Ehrenreich, D., Ferlet, R., McConnell, J. C., Mayor, M., & Parkinson, C. D. (2004). *Astrophysical Journal, 604*, L69.
Vidotto, A. A., Jardine, M., & Helling, C. (2010). *Astrophysical Journal, 722*, L168.
Vidotto, A. A., Jardine, M., & Helling, C. (2011a). *Monthly Notices of the Royal Astronomical Society, 411*, L46.
Vidotto, A. A., Jardine, M., & Helling, C. (2011b). *Monthly Notices of the Royal Astronomical Society, 414*, 1573.
Withbroe, G. L. (1988). *Astrophysical Journal, 325*, 442.
Yelle, R. V. (2004). *Icarus, 170*, 167.

Chapter 9
The Effects of Close-in Exoplanets on Their Host Stars

Eike W. Guenther and Stephan Geier

Abstract In analogy to the RS CVn stars, it is expected that close-in planets can induce stellar activity due to tidal and magnetic interaction between the star and the planet. The tidal and/or magnetic interaction between the star and the planet is expected to affect the magnetic field of the star which then cause structural changes of the stellar wind, the corona, the chromosphere, and also spots on the stellar surface. Observations of these effects are difficult and have led to mixed results that can at least partly be explained by the transient nature of the phenomena. The detection of induced activity has been claimed for a number of stars hosting close-in planets. However the most extreme object, WASP-18, does not show any such signs. The active regions, which have been claimed as being caused by induced activity are not located at the sub-planetary point but often $70° - 80°$ ahead of it. Some studies have claimed that stars with close-in, massive exoplanets are statistically more active, but others conclude that this correlation is just the product of a selection bias. Studies with the Kepler-satellite show that superflares exist, but they do not seem to be caused by the interaction between stars and planets. While it is still discussed whether induced activity has been detected for planet-hosting main-sequence stars, the engulfment of planets when the star evolves to a giant might have dramatic consequences.

9.1 Introduction: Stellar Activity Triggered by Hot Jupiters

Binary stars with orbital period of 15 days or less containing late-type stars usually show an enhanced level of stellar activity due to the interaction between the stars. Such binaries are called RS Canum Venaticorum stars, or RS CVn stars

E.W. Guenther (✉)
Thüringer Landessternwarte Tautenburg, Sternwarte 5, 07778 Tautenburg, Germany
e-mail: guenther@tls-tautenburg.de

S. Geier
European Southern Observatory, Karl-Schwarzschild-Str. 2, 85748 Garching bei München, Germany
e-mail: sgeier@eso.org

H. Lammer, M. Khodachenko (eds.), *Characterizing Stellar and Exoplanetary Environments*, Astrophysics and Space Science Library 411,
DOI 10.1007/978-3-319-09749-7_9

for short. After the discovery of close-in, massive planets orbiting late-type stars it was realized that the tidal and magnetic interaction between the star and the planet may also lead to an enhanced activity level of the star. As pointed out by Cuntz et al. (2000), the effects should be larger for systems with close-in planets than for systems with large separation, given that the tidal interaction scales with $M_p/M_* d^{-3}$, and the magnetic interaction with $B^{4/3}(B_p/B_*)^{1/3} d^{-2}$. Because hot Jupiters orbit within the Alfvén radius of the stars, a direct magnetic interaction with the stellar surface is possible (Preusse et al. 2005). The interaction may change the structure of the stellar wind, the corona, the chromosphere, the stellar spots and plage-regions of the star.

To some extent stars with close-in planets resemble the RS CVn stars. The signs of induced activity should thus be similar but of course of much smaller scale for planet-hosting stars. The properties of RS CVn stars thus give us a guideline what we should be looking for. For RS CVn stars the following properties are taken as evidence for induced activity (Hall 1992):

- Enhanced emission cores in Ca II and other chromospheric lines.
- An enhanced emission in the UV from the Chromosphere.
- An enhanced emission in the X-ray and radio regime from the Corona.
- An enhanced mass loss rate of the stellar wind.
- An enhanced coverage of star spots, and thus also an unusually large magnetic field strength.
- Enhanced Flare-like activity in the optical, X-ray and the radio regime.
- A reduced amount of differential rotation.

This article focuses on the observational aspects of the interaction between stars in planets. In Sect. 9.2 of this article, we will discuss the observational evidence in favor of and against induced activity. In Sect. 9.3 we will discuss the possible detection of bow shocks as indirect evidence for magnetic fields of planets (see also Chap. 8 (Vidotto et al. 2014), Chap. 10 (Alexeev et al. 2014), Chap. 11 (Grießmeier 2014) of this book). In Sects. 9.4 and 9.5 we will discuss the more dramatic effects that planets may have on stars, the change of the rotation rate of the stars, and the effects on the evolution by the engulfment of planets by the stars.

Prominent examples of stars with massive, close-in planets that will be mentioned several times in the article are:

- μ And, an F8V star with four know planets. The closest has $0.62 \pm 09\ M_{Jup}$, and a semi-major axis of 0.059 ± 001 AU (Butler et al. 1997).
- τ Boo, an F7 star with a planet that has a minimum mass of $5.95 \pm 0.28\ M_{Jup}$, and a semi-major axis of 0.046 AU (Butler et al. 1997).
- CoRoT-2b, a relatively young G7V star with a transiting planet of $3.31 \pm 0.16\ M_{Jup}$, and a semi-major axis of 0.0281 ± 0.0009 AU (Alonso et al. 2008).
- GJ 876, an M4V star with 4 know non-transiting planets. These have minimum masses of $2.2756 \pm 0.0045\ M_{Jup}$ (semi-major axis 0.208317 ± 0.00002 AU), $0.7142 \pm 0.0039\ M_{Jup}$ (semi-major axis 0.12959 ± 0.000024 AU), $0.021 \pm 0.001\ M_{Jup}$ (semi-major axis 0.0208066 ± 0.00000015 AU), $0.046 \pm 0.005\ M_{Jup}$ (semi-major axis 0.3343 ± 0.0013 AU) (Wright et al. 2011).

- HD179949, an F8V star with a planet which has a minimum mass of 0.95 $\pm$ 0.04 M_{Jup} and a semi-major axis of 0.045 $\pm$ 0.001 AU (Tinney et al. 2001).
- HD 189733, a K1-K2 star with a planet which has a minimum mass of 1.150 $\pm$ 0.028 M_{Jup} and a semi-major axis of 0.03142 $\pm$ 0.0038 AU (Southworth 2010).
- WASP-18, an F6 star with a transiting planet of 10.4 $\pm$ 4 M_{Jup}, and a semi-major axis of 0.02047 $\pm$ 0.0038 AU (Hellier et al. 2009).

9.2 Enhanced Chromospheric Activity and Spot Coverage Caused by Close-in Planets

9.2.1 The Ca II Lines

The easiest way to detect induced stellar activity is to search for periodic variations of the Ca II lines with the orbital period of the planet. Using the Ca II lines at 849.8, 854.2, and 866.2 nm Saar and Cuntz (2001) studied seven exoplanet host stars but did not find any periodic variations at the orbital period of the planets. In contrast to this, Shkolnik et al. (2003, 2005, 2008) and Walker et al. (2008) claimed to have detected a periodic enhancement of the variability of the emission (MAD, mean absolute deviation) of the Ca II H & K and other lines in a number of stars, most notably υ And, τ Boo and HD 189733. An interesting aspect is that the maximum emission did not occur at the sub-planetary point but at a longitude of 70°–80°(HD 179949; Fig. 9.1) and 160° ahead of it (υ And).

However, the signs of induced activity were only present 50–60 % of the time (effect seen in 2001, 2002, 2005, nothing seen in 2003, 2006). Walker et al. (2008) studied CaII-emission and the photometric variations of τ Boo. HD179949 was also the subject of a study by Scandariato et al. (2013) in which X-ray data and Ca II H&K spectra were obtained simultaneously. In contrast to the results obtained by Shkolnik et al. (2003), Shkolnik et al. (2005), and Shkolnik et al. (2008), no variability modulated at the orbital period of the planet was found. The υ And system also showed no sign of induced activity (Poppenhaeger et al. 2011b). No signs of planet induced activity was found in a study of seven stars with massive close-in planets using high-resolution spectra of the Ca II H & K and/or Hα by Lenz et al. (2011). No sign of induced stellar activity in the CaII H&K lines was found for WASP-18 in the 13 spectra obtained either (Miller et al. 2012). Because the phenomenon is transient, the negative results do not necessarily contradict the positive detections. However, the fact that nothing was found for the most extreme object known, WASP-18, casts doubts on previously claimed detections.

A statistical analyses of exoplanet host stars by Canto Martins et al. (2011) shows that exoplanet host stars in general do not have a higher activity level than stars without planets. However, they find some indications that stars with massive close-in planets seem to have an enhanced chromospheric activity. Hartman (2010) also claims to have found evidence that stars with close-in planets of high surface gravity orbit more active stars, but Poppenhaeger and Schmitt (2011a) argue that all these results are just due to selection biases.

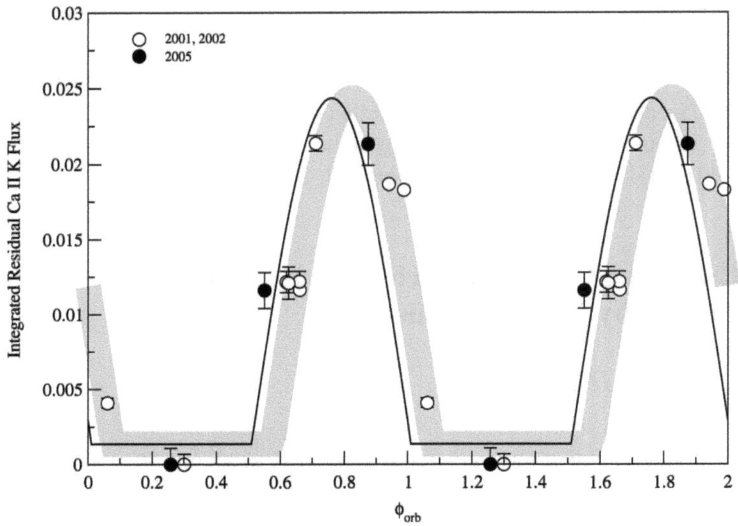

Fig. 9.1 Possible signs for induced activity: shown is the integrated flux of the Ca II K line residuals of HD 179949 as a function of orbital phase. The *grey line* is a best-fit spot model to the earlier data, whose thickness reflects the error in the phase shift. The *black line* is the same fit slightly shifted in phase by −0.07 to better fit the 2005 data. The effect was seen in 2001, 2002, and 2005 but not seen 2003, and 2006 (From Shkolnik et al. 2008)

9.2.2 The UV-Radiation from the Chromosphere

Enhanced chromospheric activity will lead to enhanced UV-radiation from the chromosphere. An increased depth of the transit has been observed in lines like Mg II in the near UV but this is a signature of an extended gas cloud around the planet rather than enhanced activity of the star (Haswell et al. 2012), and thus not subject of this article. France et al. (2012) observed the M4V star GJ 876 in the wavelength band from 115 to 314 nm with HST. This star has four planets. The most massive one has 2.2 M_{Jup} but it is already at 0.2 AU (period 61 days). The closest one has a semi-major axis of 0.02 AU (period 1.9 days) but has only 0.2 M_{Jup}. Induced activity thus is unlikely in this object, and the main purpose of this work was to find out how much UV-radiation the planet received in order to develop atmospheric models. It was found that the integrated Lyman-α flux is roughly equal to the rest of the integrated flux (115–121 + 122–314 nm), which means that this ratio is about 2,500 times larger than that for the Sun. A flare was also observed in the UV. In a second study, France et al. (2013) observed the UV radiation of five exoplanet host stars and one brown dwarf (GJ 581, M2.5V; GJ 876, M4V; GJ 436, M2.5V; GJ 832, M1.5V; GJ 667C, M1.5V; and GJ 1214, BD). In most of the stars the Lyman-α line fluxes comprise ∼37–75 % of the total 115–310 nm flux. The observed strong UV flux in these stars is possibly typical of these types of stars, rather than being a sign of enhanced activity. According to Knutson et al. (2010), planets orbiting stars that are less active often have strong high-altitude temperature inversion and water emission lines. The reason for this could be that the compounds responsible for the temperature inversions are destroyed by the UV-radiation in active stars.

Shkolnik (2013) made a statistical study of the UV-radiation of 272 planet host stars using data obtained with the NASA Galaxy Evolution Explorer (GALEX). She finds no clear correlation between the UV emission semi-major axis, or the mass of the planet, or the mass of the planet divided by the semi-major axis but points out that finding such a correlation is very difficult, because the FUV varies by a factor of 2–3 even for stars of the same T_{eff}. However, Shkolnik (2013) finds a tentative difference (at a level of 2.3σ) that stars with close-in planets are more FUV-active than stars with far-out planets.

9.2.3 *The Corona and the Stellar Wind*

Late-type stars (late F, G, K and M-star) with chromospheres also have coronae. If the chromospheric activity is enhanced by the interaction between the star and the planet, the coronae of these stars should also show an enhanced level of activity. Close-in planets orbit within the Alfvén radius of their host stars, which means that direct magnetic interaction between the star and the planet is possible.

Kashyap et al. (2008) claimed to have found that stars with close-in giant planets are on average a factor of about 4 brighter in the X-rays. Scharf (2010) examined the X-ray emission of 271 stars hosting planets and also finds a positive correlation between X-ray luminosity and the minimum mass of the most closely orbiting exoplanets. After analysing the X-ray emission of 72 stars with planets within 30 pc (Poppenhaeger et al. 2010; Poppenhaeger and Schmitt 2011a) conclude that there is no correlation between the X-ray flux and the mass or semi-major axis of the planet. They argue furthermore that the previous claims are due to selection biases. The reason is that low-mass planets with large orbital periods can only be detected with the radial velocity method, if the stars are very inactive. Since inactive stars are dim in X-rays, the sample is biased. Poppenhaeger et al. (2011b) also studied the υ And system in CaII emission and in X-rays. Contrary to the results published by Shkolnik et al. (2003), Shkolnik et al. (2005), and Shkolnik et al. (2008), they find that the CaII-emission is just modulated at the rotation period of the star and thus unrelated to the planet.

The HD 189733 system consists of a K-type star orbited by a transiting hot Jupiter and an M-type stellar companion. Interestingly, both Pillitteri et al. (2010) and Poppenhaeger et al. (2013) find a softening of the X-ray spectrum during the secondary transit of HD189733b. According to Poppenhaeger et al. (2013), the primary eclipse is also deeper in the X-ray than in the optical (6–8 % compared to 2.1 %) The authors argue that this means that the outer part of the planetary atmosphere is transparent at optical wavelengths, but opaque to X-rays. However, the X-ray emission of the corona is not coming from a homogeneous stellar disk but the emitting regions are very inhomogeneous and time-variable. For example, coronal helmets are much brighter than coronal wholes. If most of the emission would come from coronal helmets located at the equator and the planet orbits over the equator, we would observe a deep transit, which has nothing to do with the planet itself. The fact that the depth of the X-ray transit varies depending on which of the seven transit observations is averaged, already shows that the emission of the star is inhomogeneous. The radio emission from the coronae of planet hosting stars is discussed in Chap. 11 by Grießmeier (2014) in this book.

A change of the structure of the corona due to the presence of a planet will certainly affect the stellar wind. Although a number of theoretical studies show what can be expected, observational evidence is still missing (Cohen et al. 2010).

9.2.4 Magnetic Fields in the Photosphere and Stellar Spots

If the magnetic field of the star changes, this will affect the stellar wind, the corona and the chromosphere. The most direct way to study the magnetic interaction between the planet and the star would be to study the structure of the magnetic field of the star. This could be done be measuring the topology of the magnetic field at the level of the photosphere and then to calculate the field lines in the chromosphere and the corona.

In a pioneering study, Catala et al. (2007) mapped the magnetic field structure of τ Boo. The data is best reproduced with differential rotation leading to rotation periods between 3.0 and 3.7 days. In another study Donati et al. (2008) measured the differential rotation and found that the rotation ranges from 3.0 to 3.9 days over the stellar surface. Given that the orbital period of the planet is 3.3135 ± 0.0014 d, the orbital period of the planet is synchronized with the rotation of the star at a latitude of about 30°.

Photometric observation of τ Boo with the MOST satellite by Walker et al. (2008) in 2004 and 2005 showed a decrease in brightness by 1 mmag that repeats every 3.5 ± 0.7 days. This period is thus consistent with the orbital period of the planet. When the light-curve is phased to the orbital period of the planet, the dimming appears to be at a longitude of about 68°. This position of the spot is consistent with the CaII observations obtained by Shkolnik et al. (2008) in 2001, 2002, and 2005. The authors argue that the active region is likely to be related to the planet, because it has been observed over a time-span of 4 years, or 440 planetary orbits. They furthermore argue that the interaction must be magnetic, not tidal, because there is only one spot and this spot appears in advance of the sub-planet point. In the case of tidal interaction, two tidal bulges would be expected, like in the Earth Moon-system.

Additional studies of the magnetic field topology of τ Boo by Fares et al. (2009), Fares et al. (2010), Fares et al. (2012), and Fares et al. (2013) show that the rotation rate of the star is equal to the orbital period of the planet at a latitude of 40°. The magnetic field detected is quite weak, only 5–10 G. An interesting result of the studies of the magnetic field of τ Boo is that the length of the magnetic cycle of

2 years is unusually short. The authors argue that the short cycle could be caused by the interaction with the planet. However, no evidence for enhancement in the activity in CaII H&K or Hα caused by the planet was found. The same is true for HD 189733 and HD 179949 (Moutou et al. 2007; Fares et al. 2010, 2012, 2013).

Fossati et al. (2013) measured also the magnetic field strength and topology of the old and inactive solar-like planet-hosting stars HD 70642, HD 117207, and HD 154088. Evidence for magnetic planet-star interaction was found in none of them. Thus, the only sign of induced activity is the short length of the magnetic cycle of τ Boo. Neither the field strength, nor the topology of the field of planet-host stars are any different from those of stars without planets (see also Chap. 3, Fig. 3.2 (Lüftinger et al. 2014)).

A problem of most previous studies of the interaction between stars and planets is that they focused on inactive stars. This is because most radial-velocity surveys are biased towards inactive stars, because active stars rotate more rapidly making radial-velocity surveys more difficult. Transit surveys are in principle not biased towards slowly rotating stars. It is thus no surprise that the first planet orbiting a relatively young ($\sim$100 Myr), active star was found in a transit survey: CoRoT-2b (Alonso et al. 2008). The orbital period of the planet is 1.7429964 days. An indirect way to study the magnetic fields at photospheric levels is to study the spots. Using the superb light-curves obtained by CoRoT, a number of studies of stellar spots and activity have been obtained. Pagano et al. (2009) found two active longitudes separated by about 180° and rotating with periods of 4.5221 and 4.5543 days, respectively. They also find that the total spotted area oscillates with a period of about 28.9 days. The authors argue that the variance of the stellar flux is modulated with the planet orbital period, which may indicate that stellar activity is induced by magnetic star-planet interaction. Lanza (2011) also finds some evidence in the light-curve of CoRoT-6b that the passage of the planet over the stellar active regions seem to be associated with triggering of new magnetic flux emergence. This finding is remarkable, because CoRoT-6b has a relatively long orbital period of 8.9 days (Fridlund et al. 2010). In contrast to this, no evidence for induced activity was found in a photometric study of the young active K dwarf HD 63454 which is hosting a Jovian planet in a 2.82 day period orbit (Kane et al. 2011).

9.2.5 Flares

Because hot Jupiters orbit within the Alfvén radius of the stars, field-lines can connect the planets with the star, which means that giant flares could occur (Preusse et al. 2005).

The idea that close-in giant planets could cause giant flares was first discussed by Cuntz et al. (2000), Rubenstein and Schaefer (2000), Schaefer et al. (2000). Using a more detailed model Lamza (2009) also predicts that large flares extending over most of the stellar Corona may sometimes occur in stars with close-in planets.

Pillitteri et al. (2010) in fact observed a large flare in the HD 189733 system that seems to come from active regions displaced by 75°–78° in respect to the sub-planetary point. The size of the flaring region was estimated to be of the order of the radius of the star. The HD 189733 system consists of a K-type star orbited by a transiting Hot Jupiter and an M-type stellar companion. Also Bentley et al. (2009) reported on the observation of a flare during the transit of OGLE-TR-10b.

Using all kind of astronomical data Rubenstein and Schaefer (2000) and Schaefer et al. (2000) came up with a list of nine cases in which normal F8 and G8 main-sequence stars seem to have had a flare that was 10^2–10^7 times more energetic than the largest solar flare, corresponding to energies of 10^{33}–10^{38} ergs. The nine stars were all single (or in very wide binaries), slow rotators, and not of young age. From this data, they estimate that normal, solar-like stars should have one super-flare every 600 years. Since it seems unlikely that the Sun had such a flare in the past two millennia, they came up with the idea that superflares might be caused by magnetic reconnection between fields of the primary star and a close-in Jovian planet. However, up to now no massive, close-in, planets were found around any of these stars.

In this discussion Rubenstein and Schaefer (2000) and Schaefer et al. (2000) used the argument that the Sun did not have giant flares in the last 2,000 years. But how certain is that? Using radiocarbon data Solanki et al. (2004) reconstructed the sunspot numbers over the last 11,000 years. They found that about 90 % of this time the Sun was even less active than today. However, Schrijver et al. (2012) recently showed that estimating the solar activity using indirect methods is very difficult, and argued that the most reliable source of information are the observations of spots obtained in the last 400 years, from this data. They concluded that solar flares in the past four centuries did not substantially exceeded the level of the largest flares observed during the space era. A review on the activity of the Sun in the past and its effects on the climate of the Earth has recently been published by Solanki et al. (2013).

However, Usoskin et al. (2013) pointed out that an event observed in the year 775 could have been a giant flare on the Sun. Another hint that the Sun may have had giant flares in the past came from the discovery of the Apollo 11 mission. In this mission it was found that small lunar craters frequently contain lumps of soil whose upper surfaces are coated with glassy glaze, and a detailed analysis showed that this "glass" could not be older than about 20,000 years. Gold (1969) and Zook et al. (1977) then came up with the idea that the "glass" could have been created by an event in which the lunar surface was heated up to the melting point by a giant solar flare. Using dynamo models Shibata et al. (2013) showed that the Sun could have one superflare with energy of 10^{34} erg to occur once every 800 years, and a superflare with 10^{35} erg once every 5,000 years. However, Kovaltsov and Usoskin (2013) reconstructed the occurrence rate of extreme events related to solar energetic particles from an analysis of cosmogenic radionuclides measured in lunar rocks over the last 1 Myr. From this data, they concluded that events only four times larger than the largest events observed during the space age appear only once every 10,000 years. This implies that events of about $2 \cdot 10^{33}$ erg should appear only once every 10,000 years. It thus seems that giant flares are rare on the Sun.

If solar-like stars would have one super-flare every few hundred years, then there should be many super-flares in the Kepler-data. Compared to earlier studies the Kepler data has the great advantage that it is very homogeneous and has a superb photometric quality. In a first study Maehara et al. (2012) analyzed the light-curves of 83,000 stars using 120 days of Kepler-data. In total they found 365 superflares with an energy release between 10^{29} and 10^{32} ergs on a timescale of hours. Using 500 days of Kepler-data Shibayama et al. (2013) improved the statistics. They found 1,547 superflares originating from 279 G-type dwarfs. The analysis of the data showed that the occurrence rate (dN/dE) of superflares versus flare energy (E) shows a power-law distribution with $dN/dE \sim E^{-\alpha}$, where $\alpha \sim 2$. The power-law index thus is quite close to that of solar flares. The occurrence rate of superflares with energies in the range between 10^{34} and 10^{35} ergs for solar-like stars (T_{eff} ~5,600–6,000 K, rotation period >10 days) is once every 800–5,000 years. They also found that stars with a high rate of superflares usually have also large spots. In another study Notsu et al. (2013) showed that the energy of superflares is related to the coverage of the star by spots. Already in the first study Maehara et al. (2012) pointed out that no hot transiting Jupiter was found in any of the stars that had superflares. However, in order to show that superflares are not related to close-in planets it remains to be shown that these do not have hot Jupiters at all, not only transiting ones.

9.2.6 The Solar System

Given that the Sun does not have a planet with the mass of Jupiter at a distance of 0.1 AU or closer, we may think that we have not to discuss the solar system. Surprisingly, even for the solar system it has been claimed that the planets have an effect on the activity cycle of the Sun. For example, Abreu et al. (2012) claimed that they had found a relation between the variations of cosmogenic isotopes and the torques caused by the planets. However, in a subsequent article Cameron and Schüssler (2013) showed that the claimed correlations are spurious.

9.3 Bow Shocks

Observations of the transit of WASP-12 in the UV by Vidotto et al. (2010) showed an ingress that was much earlier than in the optical regime. They interpret this as evidence for a shock-front ahead of the planetary orbital motion. They furthermore conclude that the presence of such a shock front can be taken as indirect evidence for a magnetic field of the planet. In the case of WASP-12b they estimate the magnetic field strength to be about $B_p = 24\,G$ (see Chap. 4 (Fossati et al. 2014) and Chap. 8 (Vidotto et al. 2014) for details).

9.4 Can Planets Affect Stellar Rotation?

If the orbital period of the planet is shorter than the rotation rate of the star, the planet moves inwards and the rotation of the star is accelerated. In extreme cases, the planet might even be engulfed by the star, which we will discuss in the next section. Likewise, if the orbital period of the planet is longer than the rotation rate of the star, the rotation of the star is reduced and the planet moves outwards. The rotation rate of the star could also be changed if planets either enhances, or reduces the stellar wind. A large number of studies of the tidal interaction between stars and close-in planets have been carried out in which the fate of these planets is discussed (e.g. Pätzold and Rauer 2002; Jiang et al. 2003; Mardling and Lin 2004; Pätzold et al. 2004; Adams and Laughlin 2006; Carone and Pätzold 2007; Jiang et al. 2007; Levrard et al. 2009; Matsumura et al. 2010; Deleuil et al. 2012). The rotation of stars should thus be affected by planets, but is there also evidence for this?

Lamza (2009) finds that G- and K-type stars with $P_{rot} > 10$ days that have close-in planets rotate faster than planets of the same age without planets. However, Gonzalez (2011) disputes the differences. Cohen et al. (2010) find that angular momentum loss of the star due to the stellar wind is reduced by a factor of 4 due to the interaction of the stellar wind with the magnetic field of a close-in planet. This means that the reduction in wind-driven angular momentum loss can compete with, and perhaps dominate, spin-up of the star due to tidal interaction.

The projected angle between the spin-axis of the star and the orbital spin-axis of the planet has been measured for many systems using the Rossiter-McLaughlin effect. The measurements show that many systems are misaligned. As pointed out by Winn et al. (2010), cool stars usually have aligned spin-axis, whereas planets orbiting hot stars are often misaligned. This result was confirmed independently by Schlaufman (2010) who finds that the transition is at $T_{eff} = 6{,}250$ K. This is also the effective temperature where the outer stellar convective layer disappears. The basic idea is that convective envelopes of cool stars are massive enough to generate tidal forces that align the spin axis of the orbit of the planet with the stellar spin. Hotter stars have either a very thin convective envelope or none at all. The initial spin-orbit angle thus is preserved. If this idea is correct, it would mean that the spin axis of stars and planets were originally misaligned. In cool stars they become aligned after a few years.

An alternative scenario for the misaligned systems is that the orbits of the planets are tilted by the interaction with a third, relatively massive object, for example another star, a brown dwarf or another massive planet. In our solar system we know that the orbits of some minor planets are strongly tilted in respect to the ecliptic because of the interaction with massive planets like Jupiter. This mechanism is known as Kozai mechanism (Kozai 1962). Fabrycky and Tremaine (2007) as well as Nagasawa et al. (2008) showed that the Kozai mechanism could also work for extrasolar planets and it could explain why there are misaligned systems.

The Kepler satellite has now found a large number of transiting multiple-planet systems. This means that the planets of these systems all have to orbit roughly in the same plane. A particularly interesting case is the Kepler-30-system that contains three planets, that all orbit in the same plane and where it could be shown that this plane is aligned with the equator of the star (Sanchis-Ojeda et al. 2012). The first system containing several planets in which the plane in which the planets orbit is tilted with respect to the stellar equator is Kepler-56. As shown by Huber et al. (2013), the two planets of this system are coplanar. But using asteroseismology, they could demonstrate that the spin-axis of the star is misaligned with respect to the spin-axis of the orbit of the planets. If confirmed, this discovery rules out planet-planet interaction as the cause for misaligned systems. However, this discovery does not necessarily imply that the spin axis of the convection zone of the star has been tilted by the tidal interaction with a planet because the spin axis could have been tilted by chaotic accretion.

9.5 The Engulfment of Planets

Pätzold and Rauer (2002) argued that very massive, close-in planets are rare, because such planets would have spiraled into the host stars due to tidal interaction. If this hypothesis were correct, it would mean that there could be many stars, which have incorporated a planet. Jiang et al. (2007) interpret the weak positive-mass period relation for exoplanets found as an indication that massive, close-in planets might be missing because the host star engulfs them. A planet that is engulfed by the star would change the abundances of the star, because a planet adds heavy elements to the outer layers of the star. According to Fossati et al. (2010) WASP-12 is an ideal object to look for pollution signatures in the stellar atmosphere. After analysing the abundances as a function of the condensation temperature and comparing them with those published for planet-hosting and non-planet-hosting stars, they find hints of atmospheric pollution in WASP-12's photosphere. However, the discovery of WASP-18b, a planet with an orbital period of 0.94 days and 10 M_{Jup} shows that there are at least a few very massive close-in planets (Hellier et al. 2009). The idea that massive planets do not exist because they are engulfed thus seems now less likely than when it was originally proposed.

Once a star becomes a giant, close-in planets are certainly engulfed. However, which planets are engulfed, whether planets can survive the engulfment to reappear at later stages and how this affects the stars is still a very active field of research. As pointed out by Kunitomo et al. (2011), in contrast to main sequence stars, no planet of a giant star in the mass-range between $\sim 1.5 - 3\,M_{sun}$ has been found within 0.6 AU yet. The only exception is HD 102956 but this star is a subgiant, not a giant (Johnson et al. 2010). In their study Kunitomo et al. (2011) determine the critical semi-major axis (or the survival limit) inside which planets are eventually engulfed by their host stars and find that the critical semi-major axis is quite sensitive to stellar mass in the range between 1.7 and 2.1 M_{sun}. They conclude that all planets orbiting giants with spectral types G and K that have been detected are beyond the survival limit. All known planets orbiting stars with a mass higher than 2.1 M_{sun} are orbiting far beyond the survival limit. This implies that engulfment by host stars may not be the main reason for the observed lack of short-period giant planets.

One possible explanation why massive stars seem to have only planets at large orbital distances is that the lifetime of the disks of more massive stars is simply too short for the planets to form and to migrate inward (Currie 2009). Villaver and Livio (2007) calculated the survival of gas planets around stars in a mass-range between one and five M_{sun} as these stars evolve off the main sequence. They show that planets with masses smaller than Jupiter do not survive the planetary nebula phase if located initially at orbital distances smaller than 3–5 AU. Planets more massive than two Jupiter masses around low-mass stars ($M_* \leq 1 M_{sun}$ on the main sequence) survive the planetary nebula stage down to orbital distances of $\sim$3 AU. They finally conclude that planets around white dwarfs with masses $M_{WD} > 0.7\,M_{sun}$ should typically have orbital distances larger than 15 AU.

An interesting question is what happens to a planet after it is engulfed and how this affects the subsequent evolution of the host star. It is widely accepted that the closest binary stars with separations of the order of one solar radii can only be formed after a common-envelope (CE) phase, where a companion star is engulfed by a red-giant (Ivanova et al. 2013). In those cases the companion transfers enough of its orbital energy and angular momentum to eject the envelope and expose the close binary remnant. At the outset it was unclear, whether substellar companions like planets or brown dwarfs can do that as well.

Soker (1998) studied the engulfment of substellar objects by red giants and found that objects with masses higher than $10\,M_{Jup}$ might be able to eject a common envelope and survive as close companions to the hot, helium-burning cores of the red giants. The discovery of close ($< 1\,R_{sun}$) brown dwarf companions with masses of about $60\,M_{Jup}$ around such so-called hot subdwarf (sdB) stars (SDSSJ082053.53+00 0843.4, Geier et al. 2011b; SDSSJ162256.66+473051.1, Schaffenroth et al. 2014) provided observational evidence for this scenario. Substellar objects with masses lower than 10 M_{Jup} like hot Jupiter planets were predicted to either be evaporated in the common envelope or merge with the red-giant core during or shortly after the CE-phase (Politano et al. 2008). Two single, fast rotating hot subdwarf stars have been proposed as the results of such a CE-merger (EC 22081-1916, Geier et al. 2011a; SB290,Geier et al. 2011a, 2013).

The fact that only about half of the observed sdB stars are in close binaries and that planets in wider orbits around sdB stars detected by the timing method seem to be quite common (Silvotti et al. 2007; Beuermann et al. 2012) might be an indication for past interactions with even smaller, now evaporated objects that helped to form single sdBs. Charpinet et al. (2011) detected reflections from two nearly Earth-sized bodies orbiting the pulsating sdB star KIC05807616 at distances of 0.0060 and 0.0076 AU. Those objects are interpreted as being the dense cores of either one (Bear and Soker 2012) or two evaporated giant planets (Charpinet et al. 2011) that were transported closer to the star during the engulfment and have triggered the mass loss necessary to form the sdB.

Nelemans and Tauris (1998) invoked engulfed substellar companions to explain the existence of apparently single low-mass ($< 0.4\,M_{Sun}$) helium white dwarfs (He-WDs). In the classic picture such He-WDs cannot be formed by single-star evolution, but are the cores of red giants that lost their envelopes after a common envelope phase even before helium-burning was ignited. Although most of these objects are indeed found in close binaries (Brown et al. 2013), some seem to be single stars (Maxted and Marsh 1998). Subsequently, a close ($< 0.4\,R_{Sun}$) brown dwarf companion (53 M_{Jup}) has been discovered orbiting such a low-mass white dwarf (WD0137-349; Maxted et al. 2006) providing evidence that the scenario proposed by Nelemans and Tauris (1998) might be feasible.

The calculations by Livio and Soker (2002) show that the engulfment of planets may lead to a significant enhancement of the stellar mass-loss rate. They also point out that about 3.5 % of the giant stars are expected to be affected. In another theoretical study, Massarotti (2008) investigates the increase in the rotation rate of post-main-sequence stars as they expand and ingest orbiting planets. He finds that about 1 % of the horizontal branch stars with solar metallicity and mass should show abnormal rotation, which could have been caused by the engulfment of planets.

According to Siess and Livio (1999a), the effects of the engulfment of brown dwarfs and massive planets can be quite dramatic. The calculations show that for high accretion rates ($\dot{M}_{acc} = 10^{-4}\,M_{Sun}\mathrm{year}^{-1}$) , the star expands substantially which gives rise to hot bottom burning at the base of the convective envelope. The accretion of brown dwarfs and planets can induce the ejection of shells around giant stars, increase their surface lithium abundance and lead to significant spin-up. In a second study Siess and Livio (1999b) focus on asymptotic giant branch (AGB) stars. They find again that the accretion of brown dwarfs and massive planets leads to a substantial expansion of the star, and hot bottom burning can be activated. Possible observational signatures of the engulfment of planets include the ejection of a shell, an increase of the ^7Li surface abundance, an enrichment of heavy elements, an increase of the rotation speed of the star, and the generation of magnetic fields.

The abundance patterns of giant stars were studied by Pasquini et al. (2007). They find that in contrast to stars on the main sequence, giant stars hosting planets do not show enhanced metallicity. They argue that if the enhancement observed in main-sequence stars is due to the engulfment of planets, giant stars would not show enhanced metallicity, because the convective layers of giant stars are so large.[1]

A study of the abundance pattern of giant stars by Maldonado et al. (2013) shows that the situation is probably more complicated than previously thought. These authors find a very strong relation between the metallicity distribution and the stellar mass. For giant stars with $M_* \leq 1.5\,M_{sun}$ there is no difference between stars with and without planets but for giant stars with $M_* > 1.5\,M_{sun}$ they find a significant difference. The metallicity is also enhanced for planet-hosting subgiant stars. As

[1] There is a long list of articles about the abundance patterns of planet host stars in which it is also discussed whether these are caused by the engulfment of planets (Perryman 2011). However, very few of them discuss the giant stars.

Pasquini et al. (2007) they also argue that the pollution scenario explains why main sequence and sub-giant stars with planets are metal rich and also why giant stars with $M_* \leq 1.5\, M_{sun}$ hosting planets are not enriched because the pollution is diluted when the star becomes a giant. But the pollution scenario does not explain why planet-hosting giant stars with $M_* > 1.5\, M_{sun}$ are metal rich.

An interesting question is what happens to the planet after it is engulfed. As pointed out by Charpinet et al. (2011), the discovery of post-red-giant host stars with giant planets orbiting as close as 0.116 AU shows that these objects must have survived engulfment. Charpinet et al. (2011) also found two nearly Earth-sized bodies orbiting the post-red-giant, hot B Subdwarf star at distances of 0.0060 and 0.0076 AU. They interpret these planets as being the dense cores of evaporated giant planets that were transported closer to the star during the engulfment and which triggered the mass loss necessary for the formation of the hot subdwarf B star.

Another interesting aspect is that the shapes of planetary nebulae (PN) are believed to be affected if the primary is either a close binary, or if it has a substellar companion (Soker 1997; de Marco and Soker 2011). Soker (1997) originally invoked a planetary role in shaping PNs because there are not enough close binaries to shape the large fraction of nonspherical PNs. The shaping of PNs by planets is only possible if these planets have survived interactions with red giant branch stars. However, in order to explain why 20 % of the PNs are highly asymmetric although only about 4 % of the main sequence stars have close-in, massive planets, de Marco and Soker (2011) had to assume that only 20 % of the stars develop a PN, and these stars have to be the ones with close-in planets. They argue that this makes sense because close-in planets enhance the mass-loss rate of stars, and only stars with enhanced mass-loss rates will develop PNs.

Conclusion
Thus, while it is not yet clear how much close-in planets affect main sequence stars, the engulfment of planets at the later stages of the evolution might have very dramatic effects. Significant fractions of the populations of evolved stars with unclear origins like sdBs, He-WDs and maybe also some more exotic post-AGB objects may have been formed by such star-planet interactions.

Acknowledgements The authors acknowledge the support by the International Space Science Institute (ISSI) in Bern, Switzerland and the ISSI team *Characterizing stellar- and exoplanetary environments*.

References

Abreu, J.A., Beer, J., Ferriz-Mas, A., McCracken, K.G., & Steinhilber, F. (2012). *Astronomy and Astrophysics, 548*, A88.

Adams, F.C., & Laughlin, G. (2006). *Astrophysical Journal, 649*, 1004.

Alexeev, I. I., Grygoryan, M. S., Belenkaya, E. S., Kalegaev, V. V., & Khodachenko, M. L. (2014). In H. Lammer & M. L. Khodachenko (Eds.), *Characterizing stellar and exoplanetary environments* (pp. 189). Heidelberg/New York: Springer.

Alonso, R., Auvergne, M., Baglin, A., Ollivier, M., Moutou, C., Rouan, D., Deeg, H. J., & The CoRoT Team (2008). *Astronomy and Astrophysics, 482*, L21.

Bear, E., & Soker, N. (2012). *Astrophysical Journal, 749*, L14.

Bentley, S. J., Hellier, C., Maxted, P. F. L., Dhillon, V. S., Marsh, T. R., Copperwheat, C. M., & Littlefair, S. P (2009). *Astronomy and Astrophysics, 505*, 901.

Beuermann, K., Dreizler, S., Hessman, F. V., & Deller, J. (2012). *Astronomy and Astrophysics, 543*, 138.

Brown, W. R., Kilic, M., Allende Prieto, C., Gianninas, A., & Kenyon, S. J. (2013). *Astrophysical Journal, 769*, 66.

Butler, R.P., Marcy, G.W., Williams, E., Hauser, H., & Shirts, P. (1997). *Astrophysical Journal Letters, 474*, L115.

Cameron, R.H., & Schüssler M. (2013). *Astronomy and Astrophysics, 557*, A83.

Canto Martins, B.L., Das Chagas, M.L., Alves, S., Leão,, I. C., de Souza Neto, L. P., & de Medeiros, J. R. (2011). *Astronomy and Astrophysics, 530*, A73.

Carone, L., & Pätzold, M. (2007). *Planetary and Space Science, 55*, 643.

Catala, C., Donati, J.-F., Shkolnik, E., Bohlender, D., & Alecian, E. (2007). *Monthly Notices of the Royal Astronomical Society, 374*, L42.

Charpinet, S., Fontaine, G., Brassard, P., Green, E. M., Van Grootel, V., Randall, S. K., Silvotti, R., Baran, A. S., Østensen, R. H., Kawaler, S. D., & Telting, J. H. (2011). *Nature, 480*, 496.

Cohen, O., Drake, J.J., Kashyap, V. L., Sokolov, I.V., & Gombosi, T.I. (2010). *Astrophysical Journal Letters, 723*, L64.

Cuntz, M., Saar, S. H., & Musielak, Z. E. (2000). *Astrophysical Journal Letters, 533*, L151.

Currie, T. (2009). *Astrophysical Journal Letters, 694*, L171.

Deleuil, M., & The CoRoT Team (2012). *Astronomy and Astrophysics, 538*, A145.

de Marco, O., & Soker, N. (2011). *Publications of the Astronomical Society of the Pacific, 123*, 402.

Donati, J.-F., Moutou, C.,Farés, R., Bohlender, D., Catala, C., Deleuil, M., Shkolnik, E., Collier Cameron, A., Jardine, M. M., & Walker, G. A. H. (2008). *Monthly Notices of the Royal Astronomical Society, 385*, 1179.

Fabrycky, D., & Tremaine, S. (2007). *Astrophysical Journal, 669*, 1298.

Fares, R., Donati, J.-F., Moutou, C., Bohlender, D., Catala, C., Deleuil, M., Shkolnik, E., Collier Cameron, A., Jardine, M. M., & Walker, G. A. H. (2009). *Monthly Notices of the Royal Astronomical Society, 398*, 1383.

Fares, R., Donati, J.-F., Moutou, C., Jardine, M. M., Grießmeier, J.-M., Zarka, P., Shkolnik, E. L., Bohlender, D., Catala, C., & Collier Cameron, A. (2010). *Monthly Notices of the Royal Astronomical Society, 406*, 409.

Fares, R., Donati, J.-F., Moutou, C., Jardine, M., Cameron, A. C., Lanza, A. F., Bohlender, D., Dieters, S., Martínez Fiorenzano, A. F., Maggio, A., Pagano, I., & Shkolnik, E. L. (2012). *Monthly Notices of the Royal Astronomical Society, 423*, 1006.

Fares, R., Moutou, C., Donati, J.-F., Catala, C., Shkolnik, E. L., Jardine, M. M., Cameron, A. C., & Deleuil, M. (2013). *Monthly Notices of the Royal Astronomical Society, 435*, 1451.

Fossati, L., Bagnulo, S., Elmasli, A., Haswell, C. A., Holmes, S., Kochukhov, O., Shkolnik, E. L., Shulyak, D. V., Bohlender, D., Albayrak, B., Froning, C., & Hebb, L. (2010). *Astrophysical Journal, 720*, 872.

Fossati, L., Kochukhov, O., Jenkins, J. S., Stancliffe, R. J., Haswell, C. A., Elmasli, A., & Nickson, E. (2013). *Astronomy and Astrophysics, 551*, A85.

Fossati, L., Haswell, C. A., Linsky, J. L., & Kislyakova, K. G. (2014). In H. Lammer, M. L. Khodachenko (Eds.), *Characterizing stellar and exoplanetary environments* (pp. 59). Heidelberg/New York: Springer.

France, K., Linsky, J.L., Tian, F., Froning, C.S., & Roberge, A. (2012). *Astrophysical Journal Letters, 750*, L32.

France, K., Froning, C. S., Linsky, J. L., Roberge, A., Stocke, J. T., Tian, F., Bushinsky, R., Désert, J.-M., Mauas, P. Vieytes, M., & Walkowicz, L. M. (2013). *Astrophysical Journal, 763*, 149.

Fridlund, M., Hébrard , G., Alonso, R., Deleuil, M., Gandolfi, D., Gillon, M., Bruntt, H., & The CoRoT Team (2010). *Astronomy and Astrophysics, 512*, A14.

Geier, S., Classen, L., & Heber, U. (2011a). *Astrophysical Journal, 733*, L13.

Geier, S., Schaffenroth, V., Drechsel, H., & The MUCHFUSS Team (2011b). *Astrophysical Journal, 731*, L22.

Geier, S., Heber, U., Heuser, C., Classen, L., O'Toole, S. J., & Edelmann, H. (2013). *Astronomy and Astrophysics, 551*, L4.

Gold, Th. (1969). *Science 165*, 1345.

Gonzalez, G. (2011). *Monthly Notices of the Royal Astronomical Society, 416*, L80.

Grießmeier, J.-M. (2014). In H. Lammer & M. L. Khodachenko (Eds.), *Characterizing stellar and exoplanetary environments* (pp. 213). Heidelberg/New York: Springer.

Hall, D.S. (1992). Binary stars, RS Canum venaticorum type. In Maran (Ed.), *The astronomy and astrophysics encyclopaedia* (pp. 74). Cambridge: Cambridge University Press.

Hartman, J.D. (2010). *Astrophysical Journal Letters, 717*, L138.

Haswell, C.A., Fossati, L., Ayres, T., France, K., Froning, C. S., Holmes, S., Kolb, U. C., Busuttil, R., Street, R. A., Hebb, L., Collier Cameron, A., Enoch, B., Burwitz, V., Rodriguez, J., West, R. G., Pollacco, D., Wheatley, P. J., & Carter, A. (2012). *Astrophysical Journal, 760*, 79.

Hellier, C., Anderson, D.R., Collier Cameron, A., , Gillon, M., Hebb, L., Maxted, P. F. L., Queloz, D., Smalley, B., Triaud, A. H. M. J., West, R. G., Wilson, D. M., Bentley, S. J., Enoch, B., Horne, K., Irwin, J., Lister, T. A., Mayor, M., Parley, N., Pepe, F., Pollacco, D. L., Segransan, D., Udry, S., & Wheatley, P. J. (2009). *Nature, 460*, 1098.

Huber, D., Carter, J.A., Barbieri, M., Miglio, A., Deck, K.M., Fabrycky, D. C., Montet, B. T., Buchhave, L. A., Chaplin, W. J., Hekker, S., Montalbán, J., Sanchis-Ojeda, R., Basu, S., Bedding, T. R., Campante, T. L., Christensen-Dalsgaard, Jørgen, Elsworth, Y. P., Stello, D., Arentoft, T., Ford, E. B., Gilliland, R. L., Handberg, R., Howard, A.W., Isaacson, H., Johnson, J. A., Karoff, Ch., Kawaler, St. D., Kjeldsen, H., Latham, D. W., Lund, M. N., Lundkvist, M., Marcy, G. W., Metcalfe, T. S., Silva Aguirre, V., & Winn, J. N. (2013). *Science, 342*, 331.

Ivanova, N., Justham, C., Chen, X., De Marco, O., Fryer, C. L., Gaburov, E., Ge, H., Glebbeek, E., Han, Z., Li, X.-D., Lu, G., Marsh, T., Podsiadlowski, P., Potter, A., Soker, N., Taam, R., Tauris, T. M., van den Heuvel, E. P. J., & Webbink, R. F. (2013). *Annual Review of Astronomy and Astrophysics, 21*, 59.

Jiang, I.-G., Ip, W.-H., & Yeh, L.-C. (2003). *Astrophysical Journal, 582*, 449.

Jiang, I.-G., Yeh, L.-C., Chang, Y.-C., & Hung, W.-L. (2007). *Astronomical Journal, 134*, 2061.

Johnson, J.A., Bowler, B.P., Howard, A.W., Henry, G. W., Marcy, G. W., Isaacson, H., Brewer, J. M., Fischer, D. A., Morton, T. D., & Crepp, J. R. (2010). *Astrophysical Journal Letters, 721*, L153.

Kane, S.R., Dragomir, D., Ciardi, D.R., Lee, J.-W., Lo Curto, G., Lovis, Ch., Naef, D., Mahadevan, S., Pilyavsky, G., Udry, St., Wang, X., & Wright, J. (2011). *Astrophysical Journal, 737*, 58.

Kashyap, V. L., Drake, J. J., & Saar, S. H. (2008). *Astrophysical Journal, 687*, 1339.

Knutson, H.A., Howard, A.W., & Isaacson, H. (2010). *Astrophysical Journal, 720*, 1569.

Kovaltsov, G.A., & Usoskin, I.G. (2013). *Solar Physics, 182*.

Kozai, Y. (1962). *Astronomical Journal, 67*, 591.

Kunitomo, M., Ikoma, M., Sato, B., Katsuta, Y., & Ida, S. (2011).*Astrophysical Journal, 737*, 66.

Lanza, A. F. (2009). *Astronomy and Astrophysics, 505*, 339.

Lanza, A. F. (2011). *Astrophysics and Space Science, 336*, 303.

Lenz, L.F., Reiners, A., & Kürster, M. (2011). In *16th cambridge workshop on cool stars, stellar systems, and the Sun*, Seattle (Vol. 448, pp. 1173).

Levrard, B., Winisdoerffer, C., & Chabrier, G. (2009). *Astrophysical Journal Letters, 692*, L9.

Livio, M., & Soker, N. (2002). *Astrophysical Journal Letters, 571*, L161.

Lüftinger, T., Vidotto, A. A., & Johnstone, C. P. (2014). In H. Lammer & M. L. Khodachenko (Eds.), *Characterizing stellar and exoplanetary environments* (pp. 37). Heidelberg/New York: Springer.

Maehara, H., Shibayama, T., Notsu, S., Notsu, Y., Nagao, T., Kusaba, S., Honda, S., Nogami, D., & Shibata, K. (2012). *Nature, 485*, 478.

Maldonado, J., Villaver, E., & Eiroa, C. (2013). *Astronomy and Astrophysics, 554*, A84.

Mardling, R. A., & Lin, D. N. C. (2004). *Astrophysical Journal, 614*, 955.

Massarotti, A. (2008). *Astronomical Journal, 135*, 2287.

Matsumura, S., Peale, S. J., & Rasio, F. A. (2010). *Astrophysical Journal, 725*, 1995.

Maxted, P. F. L., & Marsh, T. R. (1998). *Monthly Notices of the Royal Astronomical Society, 296*, 34.

Maxted, P. F. L., Napiwotzki, R., Dobbie, P., & Burleigh, M. R. (2006). *Nature, 442*, 543.

Miller, B. P., Gallo, E., Wright, J. T., & Dupree, A. K. (2012). *Astrophysical Journal, 754*, 137.

Nagasawa, M., Ida, S., & Bessho, T. (2008). *Astrophysical Journal, 678*, 498.

Nelemans, G., & Tauris, T. (1998). *Astronomy and Astrophysics, 335*, L85.

Notsu, Y., Shibayama, T., Maehara, H., Notsu, S., Nagao, T., Honda, S., Ishii, T. T., Nogami, D., & Shibata, K. (2013). *Astrophysical Journal, 771*, 12.

Moutou, C., Donati, J.-F., Savalle, R., Hussain, G., Alecian, E., Bouchy, F., Catala, C., Collier Cameron, A., Udry, S., & Vidal-Madjar, A. (2007). *Astronomy and Astrophysics, 473*, 651.

Pagano, I., Lanza, A.F., Leto, G., Messina, S., Barge, P., & Baglin, A. (2009). *Earth Moon and Planets, 105*, 373.

Pasquini, L., Döllinger M.P., Weiss, A., Girardi, L., Chavero, C., Hatzes, A. P., da Silva, L., & Setiawan, J. (2007). *Astronomy and Astrophysics, 473*, 979.

Pätzold, M., & Rauer, H. (2002). *Astrophysical Journal Letters, 568*, L117.

Pätzold, M., Carone, L., & Rauer, H. (2004). *Astronomy and Astrophysics, 427*, 1075.

Perryman, M. (2011). *The exoplanet handbook*. Cambridge: Cambridge university press

Pillitteri, I., Wolk, S.J., Cohen, O., Kashyap, V., Knutson, H., Lisse, C. M., & Henry, G. W. (2010). *Astrophysical Journal, 722*, 1216.

Politano, M., Taam, R.E., van der Sluys, M., & Willems, B. (2008). *Astrophysical Journal Letters, 687*, L99.

Poppenhaeger, K., Robrade, J., & Schmitt, J. H. M. M. (2010). *Astronomy and Astrophysics, 515*, A98.

Poppenhaeger, K., & Schmitt, J. H. M. M. (2011a). *Astrophysical Journal, 735*, 59.

Poppenhaeger, K., Lenz, L.F., Reiners, A., Schmitt, J. H. M. M., & Shkolnik, E. (2011b). *Astronomy and Astrophysics, 528*, A58.

Poppenhaeger, K., Schmitt, J. H. M. M., & Wolk, S. J. (2013). *Astrophysical Journal, 773*, 62.

Preusse, S., Kopp, A., Büchner, J., & Motschmann, U. (2005). *Astronomy and Astrophysics, 434*, 1191.

Rubenstein, E. P., & Schaefer, B. E. (2000). *Astrophysical Journal, 529*, 1031.

Saar, S. H., & Cuntz, M. (2001). *Monthly Notices of the Royal Astronomical Society, 325*, 55.

Sanchis-Ojeda, R., Fabrycky, D. C., Winn, J. N., Barclay, Th., Clarke, B. D., Ford, E. B., Fortney, J. J., Geary, J. C., Holman, M. J., Howard, A. W., Jenkins, J. M., Koch, D., Lissauer, J. J., Marcy, G. W., Mullally, F., Ragozzine, D., Seader, Sh. E., Still, M., & Thompson, S. E. (2012). *Nature, 487*, 449.

Scandariato, G., Maggio, A., Lanza, A. F., Pagano, I., Fares, R., Shkolnik, E. L., Bohlender, D., Cameron, A. C., Dieters, S., Donati, J.-F., Martínez Fiorenzano, A. F., Jardine, M., & Moutou, C. (2013). *Astronomy and Astrophysics, 552*, A7.

Schaefer, B. E., King, J. R., Deliyannis, C. P. (2000). *Astrophysical Journal, 529*, 1026.

Schaffenroth, V., Geier, S., Heber, U., Kupfer, T., Ziegerer, E., Heuser, C., Classen, L., Cordes, O. (2014), *Astronomy and Astrophysics, 564*, A98

Scharf, C. A. (2010). *Astrophysical Journal, 722*, 1547.

Schlaufman, K. C. (2010). *Astrophysical Journal, 719*, 602.

Schrijver, C.J., Beer, J., Baltensperger, U., Cliver, E. W., Güdel, M., Hudson, H. S., McCracken, K. G., Osten, R. A., Peter, T., Soderblom, D. R., Usoskin, I. G., & Wolff, E. W. (2012). *Journal of Geophysical Research (Space Physics), 117*, 8103.

Shibata, K., Isobe, H., Hillier, A., Choudhuri, A. R., Maehara, H., Ishii, T. T., Shibayama, T., Notsu, S., Notsu, Y., Nagao, T., Honda, S. & Nogami, D. (2013). *Publications of the Astronomical Society of Japan, 65*, 49.
Shibayama, T., Maehara, H., Notsu, S., Nagao, T., Honda, S., Ishii, T.T., Nogami, D., & Shibata, K. (2013). *Astrophysical Journal Supplement, 209*, 5.
Shkolnik, E., Walker, G. A. H., & Bohlender, D. A. (2003). *Astrophysical Journal, 597*, 1092.
Shkolnik, E., Walker, G. A. H., Bohlender, D. A., Gu, P.-G., & Kürster, M. (2005). *Astrophysical Journal, 622*, 1075.
Shkolnik, E., Bohlender, D. A., Walker, G. A. H., & Collier Cameron, A. (2008). *Astrophysical Journal, 676*, 628.
Shkolnik, E. L. (2013). *Astrophysical Journal, 766*, 9.
Silvotti, R., Schuh, S., Janulis, R., Solheim, J.-E., Bernabei, S., Østensen, R., Oswalt, T. D., Bruni, I., Gualandi, R., Bonanno, A., Vauclair, G., Reed, M., Chen, C.-W., Leibowitz, E., Paparo, M., Baran, A., Charpinet, S., Dolez, N., Kawaler, S., Kurtz, D., Moskalik, P., Riddle, R., & Zola, S. (2007). *Nature, 449*, 189.
Soker, N. (1997). *Astrophysical Journal Supplement, 112*, 487.
Soker, N. (1998). *Astronomical Journal, 116*, 1308.
Solanki, S.K., Usoskin, I.G., Kromer, B., Schüssler, M., & Beer, J. (2004). *Nature, 431*, 1084.
Solanki, S. K., Krivova, N. A., & Haigh, J. D. (2013). *Annual Review of Astronomy and Astrophysics, 51*, 311.
Southworth, J. (2010). *Monthly Notices of the Royal Astronomical Society, 408*, 1689.
Siess, L., & Livio, M. (1999a). *Monthly Notices of the Royal Astronomical Society, 308*, 1133.
Siess, L., & Livio, M. (1999b). *Monthly Notices of the Royal Astronomical Society, 304*, 925.
Tinney, C.G., Butler, R.P., Marcy, G.W., Jones, H. R. A., Penny, A. J., Vogt, St. S., Apps, K., & Henry, G. W. (2001). *Astrophysical Journal, 551*, 507.
Usoskin, I.G., Kromer, B., Ludlow, F., Beer, J., Friedrich, M., Kovaltsov, G. A., Solanki, S. K., & Wacker, L. (2013). *Astronomy and Astrophysics, 552*, L3.
Vidotto, A. A., Jardine, M., & Helling, C. (2010). *Astrophysical Journal Letters, 722*, L168.
Vidotto, A. A., Bisikalo, D. V., Fossati, L., & Llama, J., (2014). In H. Lammer & M. L. Khodachenko (Eds.), *Characterizing stellar and exoplanetary environments* (pp. 153). Heidelberg/New York: Springer.
Villaver, E., & Livio, M. (2007). *Astrophysical Journal, 661*, 1192.
Walker, G. A. H., Croll, B., Matthews, J. M., Kuschnig, R., Huber, D., Weiss, W. W., Shkolnik, E., Rucinski, S. M., Guenther, D. B., Moffat, A. F. J., & Sasselov, D. (2008). *Astronomy and Astrophysics, 482*, 691.
Winn, J.N., Fabrycky, D., Albrecht, S., & Johnson, J. A. (2010). *Astrophysical Journal Letters, 718*, L145.
Wright, J.T., Fakhouri, O., Marcy, G.W., Han, E., Feng, Y., Johnson, John Asher, Howard, A. W., Fischer, D. A., Valenti, J. A., Anderson, J., & Piskunov, N. (2011). *Publications of the Astronomical Society of the Pacific, 123*, 412.
Zook, H. A., Hartung, J. B., & Storzer, D. (1977). *Icarus, 32*, 106.

Part III
Exoplanet and Astrophysical Magnetic Fields

Not much is known about magnetic fields of exoplanets. So far magnetosphere observations and models known from Solar System planets can be used and applied in a comparative way to exoplanets. In Solar System planets direct numerical simulations of planetary dynamos have worked out well in reproducing the observed magnetic fields. In the following articles the authors give an overview on the fundamental properties of planetary magnetism. The hypotheses and main results of planetary dynamo modeling are reviewed in a comparative way to exoplanets.

Chapter 10
Magnetosphere Environment from Solar System Planets/Moons to Exoplanets

Igor I. Alexeev, Maria S. Grygoryan, Elena S. Belenkaya, Vladimir V. Kalegaev, and Maxim Khodachenko

Abstract First we discuss the solar wind plasma interaction with the Solar System planets that have intrinsic magnetic fields: Mercury, Earth, Jupiter, and Saturn are discussed. As a result of such an interaction cavities, which are free from the solar wind plasma and occupied by the planetary magnetic field are created. These cavities are usually called magnetospheres are surrounded and bound by the magnetopause. The magnetopause preserved the planetary magnetic field penetration into the magnetosheath so that its impossible for the magnetosheath plasma flow to penetrate into the magnetosphere. The magnetosheath are placed between the bow shock and the magnetopause. The bow shock forms a boundary against the unshocked super Alvénic plasma flow. As demonstrated by the analysis of the Mercury, Earth, Jupiter, and Saturn magnetopauses, these surfaces can be well described by a paraboloid of revolution with different subsolar distances and flaring angles. Based on this fact an universal model of the planetary magnetosphere can be constructed. We chose the planets in the inner magnetospheres of which the magnetic field vectors have been measured by orbiting spacecraft magnetometers. The proposed models describe the basic physical processes that are responsible for the structure and dynamics of the planetary magnetospheres. Additionally to the inner planetary field the different magnetospheric sources of magnetic field are included in the model. Finally, we discuss how these magnetosphere models can be applied to exoplanets in a comparative way.

I.I. Alexeev (✉) • M.S. Grygoryan • E.S. Belenkaya • V.V. Kalegaev
Scobeltsyn Institute of Nuclear Physics, Lomonosov Moscow State University, Leninskie Gory, 1(2), Moscow, 119992, Russian Federation
e-mail: alexeev@dec1.sinp.msu.ru; grg@dec1.sinp.msu.ru; elena@dec1.sinp.msu.ru; klg@dec1.sinp.msu.ru

M. Khodachenko
Scobeltsyn Institute of Nuclear Physics, Lomonosov Moscow State University, Leninskie Gory, 1(2), Moscow, 119992, Russian Federation

Space Research Institute, Austrian Academy of Sciences, Graz, A-8042, Austria
e-mail: maxim.khodachenko@oeaw.ac.at

H. Lammer, M. Khodachenko (eds.), *Characterizing Stellar and Exoplanetary Environments*, Astrophysics and Space Science Library 411,
DOI 10.1007/978-3-319-09749-7_10

10.1 Introduction: Magnetospheres

The magnetopause of a planet is a boundary formed by the shocked solar wind plasma between a planetary magnetic field and a solar wind plasma flowing past. The shape and location of a planetary magnetopause can be determined by balancing the solar wind dynamic ram pressure with the magnetic and thermal pressures originating inside the boundary. In this review we discuss several studies (Slavin et al. 1985; Alexeev et al. 2008; Shue et al. 1998; Joy et al. 2002; Kivelson and Southwood 2003; Kanani et al. 2010; Arridge et al. 2006) related to planetary magnetopause shapes and the applications of an universal model for planetary magnetopauses.

Using a Newtonian form of the pressure balance equation, the stand-off distance, R_{ss}, is estimated as a size-pressure dependence described by a power law $R_{ss} \sim p_{sw}^{\alpha_m}$, here p_{sw} is the solar (stellar) wind dynamic pressure. This exponent is consistent with that one derived from the numerical magnetohydrodynamic simulations. The power low index α is estimated as $\alpha_E = -1/6.6$ for Earth (Shue et al. 1998), $\alpha_J = -1/4$ for Jupiter (Alexeev and Belenkaya 2005; Huddleston et al. 1998), and $\alpha_S = -1/5$ for Saturn (Kanani et al. 2010; Belenkaya et al. 2006a,b). A similar function for Mercury is not determined till now. Several spacecrafts launched over the past 55 years measured directly the magnetic field in the magnetospheres of the Solar System planets. It was found that Venus and Mars do not have magnetic fields, while Mercury, Jupiter, Saturn, Uranus, and Neptune- have their intrinsic magnetic fields like the Earth. The magnetic field of a planet deflects the incident supersonic plasma flow of the solar wind and forms the magnetosphere, a region or cavity that almost free from the solar wind plasma. Voyager 2 was the only spacecraft which flyby at Uranus and Neptune, so that one has now enough data so far to check models on the observed structures of their magnetospheres (Herbert 2009). However, here we confine ourselves to the magnetized planets Mercury, Earth, Jupiter, and Saturn. All of these planets have been visited by orbiting spacecraft with magnetometers onboard. On the other hand it is natural that Earth's magnetosphere has been studied by using numerous spacecrafts in much more detail in the past years than the magnetospheres of distant planets.

10.2 Magnetospheres of the Earth, Jupiter, and Saturn

A magnetospheric magnetic field paraboloid model originally developed by Alexeev (1986) and references therein for Earth's magnetosphere, was later further developed for Mercury (Alexeev et al. 2008), Jupiter (Alexeev and Belenkaya 2005; Belenkaya 2004), and Saturn (Alexeev et al. 2006). The existing paraboloid magnetosphere model of the Earth's magnetosphere has been incorporated by different magnetopause flaring (Belenkaya et al. 2005) for Mercury, Jupiter, and Saturn (Arridge et al. 2006). The planetary magnetosphere, as well as the terrestrial one, undergoes expansion and compression, particularly, due to a change in the solar wind dynamic pressure. The concept of the dependence of the Earth's magnetopause location on the dynamic pressure of the solar wind was first introduced by Chapman and Ferraro (1931).

10.2.1 Paraboloid Magnetosphere Model: General Issues

The name of the model follows from its key simplifying assumption that the planetary magnetopause may be represented as a paraboloid of revolution, elongated in the direction of the stellar wind flow. The paraboloid magnetosphere model calculates the magnetic field generated by a variety of current systems located on the boundaries and within the boundaries of a planetary magnetosphere. The main contributors to the magnetic field in paraboloid magnetosphere model, in the most general case, are:

- The planetary intrinsic magnetic dipole field;
- The magnetic field of a current disk (magnetodisk) around the planet;
- The magnetopause current which provides confinement of the dipole and magnetodisk fields inside the magnetopause;
- The cross-tail currents and their closure currents on the magnetopause;
- The interplanetary magnetic field (IMF), which partially penetrates into the magnetosphere as a result of reconnection with its magnetic field.

The overall magnetic field produced by the magnetopause currents, magnetotail and magnetodisk current systems is calculated using a method developed by Alexeev (1978). It ensures that the magnetic fields of various magnetospheric sources are confined within the area delimited by the paraboloidal shape magnetopause. This is achieved by the implementation of the appropriate shielding potential at the magnetopause border.

The paraboloid magnetosphere model is formulated in the planetary-dipole-centered stellar-magnetospheric coordinate system (PSM), with the planetary magnetic dipole moment, **M** located in the XZ plane and the X-axis pointed towards the Sun (star). In the most general case, the following parameters characterize the structure of the planetary magnetosphere in the paraboloid magnetosphere model (Alexeev and Bobrovnikov 1997; Alexeev et al. 2003; Belenkaya et al. 2005; Alexeev and Belenkaya 2005):

- The distances R_{ss}, R_{bs} from the center of planet to the subsolar points on the magnetopause and on the bow shock, respectively;
- The distance R_2 from the center of planet to the inner edge of the magnetospheric tail current sheet;
- The outer and inner edges of the magnetodisk relative the center of planet, e.g., R_{D1} and R_{D2}, respectively;
- The value of the planetary dipole magnetic field $B_{d0} = B_d\ (r = r_p, z = 0)$ in the equatorial plane at the surface of planet, r_p is the planet radius;
- The value of magnetic field $B_{DC} = B_{MD}\ (r = R_{D1},\ z = 0)$ produced by the magnetodisk at its outer edge;

- The value of magnetic field B_{t0} produced by the current sheet alone at the inner edge of the magnetotail current sheet (e.g., for $r = R_2$), where parabolic coordinates are equal to $\{\alpha = \alpha_0 = \sqrt{s^2 + (1+s^2)\frac{R_2}{R_{ss}}},\ \beta = 0\,, \phi = \pm\frac{\pi}{2}\}$;
- The magnetic dipole tilt angle ψ - relative Z-axis;
- The vector of the dipole offset relative to the planet center $\mathbf{d}_{dip} = \{dx,\ dy,\ dz\}$;
- The displacement z_0 of the magnetotail current sheet relative to the magnetic equatorial plane $z = 0$; and
- The portion, $\mathbf{b}$, of the IMF, $\mathbf{B}$, penetrating into magnetosphere, given as $\mathbf{b} = k_r\mathbf{B}$, where k_r is the reconnection efficiency coefficient (Slavin and Holtzer 1979).

In the present paper we consider a simplified version of the paraboloid magnetosphere model, assuming an orthogonal orientation of the planetary magnetic dipole relative the stellar wind flow (e.g., $\psi = 0$), no displacement of magnetotail current sheet (e.g., $z_0 = 0$) and without taking into account the IMF.

Here we investigate various flaring conditions of the planetary magnetopause and generalize the geodipole screening current field for the case of Mercury, Jupiter, and Saturn. From the known boundary conditions, a solution of the Laplace equation for the scalar potential of the magnetopause current magnetic field will be obtained.

Based on the knowledge of the Earth's magnetosphere where flaring changes significantly with variations in the solar wind parameters and interplanetary magnetic field, we can suggest that other planetary magnetospheres will show the same behavior (till now there are not enough measurements to make a definite conclusion about changes in the magnetopause flaring angle). In the Earth's magnetosphere such behavior is caused by the dependence of the current sources inside the magnetosphere on the solar wind density, velocity, and magnetic field. Moreover, own magnetospheric dynamics may also play an essential role. In the Jupiter's case the inner magnetospheric sources are more intense in comparison with that of Earth's ones. Thus, the changes in the currents inside the magnetosphere should result in significant changes in the magnetopause flaring.

The result of calculations for different flaring angles for the planetary magnetospheres which include the dipole field and the field of the magnetopause currents screening it, is shown in Fig. 10.1. The magnetic field in the model is calculated using the separation of variables in paraboloid coordinates while solving the Laplace

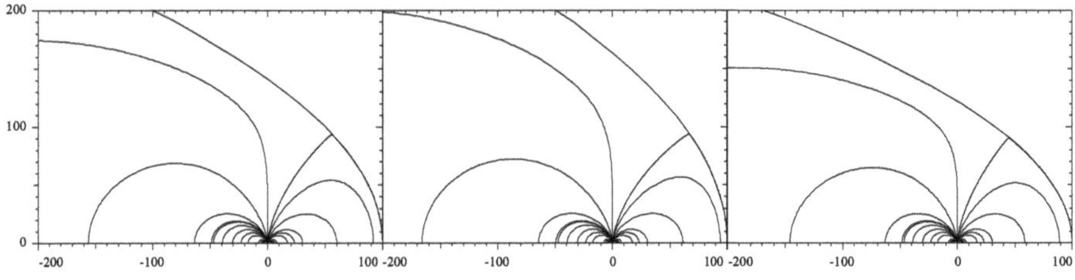

Fig. 10.1 The planetary dipole magnetic field screened by the magnetopause currents for different values of the magnetopause flaring. From *left* to *right*: Earth case $s = 1$, the axis tips scale is 0.1 R_E; Saturn case $s = 0.67$, the axis tips scale is 0.22 R_S; and Jupiter case $s = 1.25$, the axis tips scale is 1 R_J

equation (see Alexeev and Shabansky 1972, Greene and Miller 1994). The most simple way to take into account possible changes in the magnetopause flaring in the paraboloid model is to make changes in the coordinate system (Greene and Miller 1994). The coordinate surfaces are confocal paraboloids of revolution around the x axis. The relation between dimensionless parabolic (α, β, φ) and solar-magnetospheric coordinates (x, y, z) and the inverse transformation are defined by

$$x = \frac{R_1}{2}\left(\beta^2 - \alpha^2 + s^2\right), \quad y = R_1\alpha\beta\sin\varphi, \quad z = R_1\alpha\beta\cos\varphi,$$
$$\alpha = \sqrt{\tfrac{R_f - x}{R_1} + \tfrac{s^2}{2}}, \qquad \beta = \sqrt{\tfrac{R_f + x}{R_1} - \tfrac{s^2}{2}}, \quad \tan\phi = \tfrac{y}{z}. \tag{10.1}$$

Here the distance between the point (x, y, z) and the focus of the coordinate surfaces placed at $\left(x = \frac{s^2 R_1}{2}, 0, 0\right)$ is $R_f = \sqrt{\left(x - \frac{s^2 R_1}{2}\right) + y^2 + z^2} = \frac{R_1}{2}\left(\alpha^2 + \beta^2\right)$. The X axis is directed toward the Sun, the XZ surface contains the planetary dipole, φ is the azimuth angle around the X axis, s is a dimensionless constant that is a measure of the magnetopause expansion, and R_1 is the radius of the magnetopause curvature in the subsolar point. This radius is determined by the subsolar magnetopause distance R_{ss} and a flaring parameter s as $R_1 = 2\,R_{ss}\left(1 + s^2\right)^{-1}$:

$$s = \sqrt{\frac{4R_{ss}^2}{{R_T}^2} - 1}\,, \quad R_{ss} = \frac{R_1}{2}\left(1 + s^2\right), \text{ and } \quad \frac{R_T}{R_{ss}} = \frac{2}{\sqrt{1 + s^2}}, \tag{10.2}$$

which represents the dimensionless flaring of the paraboloid of revolution considered as the magnetopause when $\beta = 1$. The coordinates α, β, φ are the parabolic coordinates connected with the solar-magnetospheric coordinates x, y, z by expressions given in Eq. 10.1. The flaring parameter s is determined by $R_T = y|_{x=0}$ which is the distance from the planet's center to the magnetopause at $x = 0$ in the dawn–dusk direction. For $s = 1$ we have an ordinary (with average flaring) terrestrial magnetosphere in the paraboloid model. For this case the dimension scale R_1 is equal to R_{ss}, and $R_T = \sqrt{2}R_{ss}$. For $s > 1$ $(s < 1)$ the flaring angle is smaller (higher). The equation of the magnetopause (the surface $\beta = 1$) is given by the paraboloid of revolution in the Cartesian coordinate system

$$\frac{x_{mp}}{R_{ss}} + \frac{y_{mp}^2 + z_{mp}^2}{R_T^2} = 1\,. \tag{10.3}$$

Here x_{mp}, y_{mp}, z_{mp} are the Decart coordinates of the magnetopause. The transformations of the Decart coordinates to parabolic ones as well as definition of the dimensional scales R_{ss} (R_{ss} have been given above by Eq. 10.2). The appropriateness of such a shape for the forward magnetopause is supported by the boundary fit of Russell (1977) and Slavin et al. 2009. In particular, based on Mariner 10 observations, Russell, (1977) found that a near-paraboloidal shape of the real planetary magnetopause is characterized by an eccentricity of 0.8 whereas a true parabola has an eccentricity of 1.0. The paraboloid model of the magnetosphere is based on the assumption that a magnetospheric field can be described by a planetary dipole supplemented by two current systems. One of them corresponds to the magnetopause currents that shield the dipole field and the magnetic field component normal to the magnetopause becomes equal to zero. Another current system corresponds to the tail currents. These create a magnetic field tangential to the magnetopause and form two bundles of field lines coming from opposite directions. The current layer in the magnetospheric tail separates the northern and southern tail lobes, in which opposite bundle of magnetic field lines are almost parallel to each other. The tail currents form a theta shaped current placed in the planes perpendicular to Earth–Sun line in the nightside sector of the magnetosphere. The described model of the magnetosphere received the name "paraboloid" due to the shape of the surface, a paraboloid of revolution, approximating the magnetopause (Alexeev 1986).

The dipole field is compressed by the solar wind flow past the magnetosphere. As a consequence one can expected that the dayside field lines shifted closer to the Earth compared to an undisturbed dipole field. The magnetic field of the shielded dipole is bigger than the dipole field. As demonstrated by our calculations it is true till some distance in the night side (L_n) (see Fig. 10.2). Tailward from the point

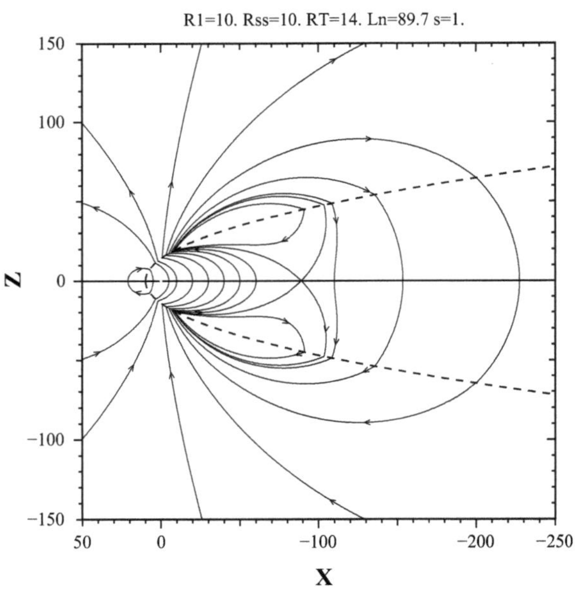

Fig. 10.2 In the noon-midnight plane of the Earth's magnetosphere the field lines of the magnetopause (Chapman – Ferraro) currents are shown. The *dashed curved* marked the magnetopause section (a parabola $\beta = 1.0$). One can see the subsolar distance $R_{ss} = 10.$, The terminator magnetopause radius $R_T = \frac{2R_{ss}}{\sqrt{1+s^2}} = 14$. The flaring parameters s are equal to 1.0. For the Chapman–Ferraro current field the position of the X-type neutral line at the noon–midnight plane marked by the cross at the $x = -89.7\, z = 0$. Distance from this point to the Earth is L_n. All distance are scaled to Earth's radius $R_E = 6400$ km

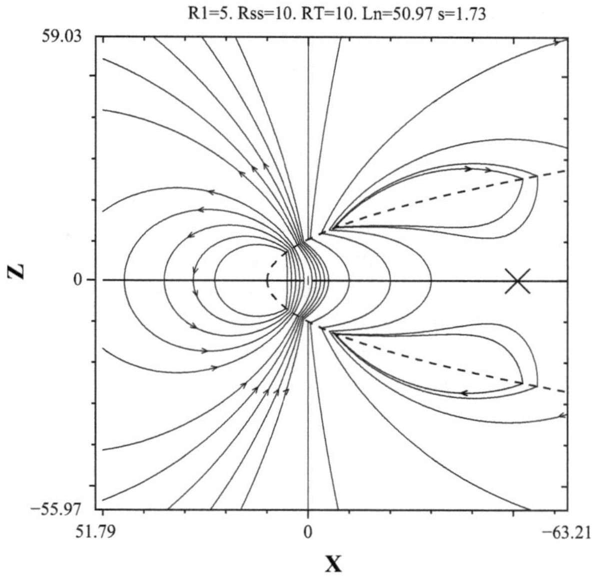

Fig. 10.3 Similar as in Fig. 10.2 but for different magnetopause flaring parameter value, for Earth's magnetosphere in the noon-midnight plane of the Earth's magnetosphere the field lines of the magnetopause (Chapman – Ferraro) currents are shown. The flaring parameters s is equal to $\sqrt{3} = 1.73$.The *dashed curved* marked the magnetopause section (a parabola $\beta = 1.0$). One can see that the subsolar distance $R_{ss} = 10.$ (as in Fig. 10.2 case), but the terminator magnetopause radius, R_T, is about 1.5 tomes smaller, $R_T = \frac{2R_{ss}}{\sqrt{1+s^2}} = R_{ss} = 10$. The position of the X-type neutral line at Sun-planet lines marked by cross and it placed at $x = -L_n = -51.0$ (Compare with $L_n = 89.7$ for $s = 1$)

$x = -L_n$ the magnetopause current field change its signum and the total (dipole + shield currents) field value is smaller than the dipole field. The $X = -89.7$ is the inflection point for tailward dependence of the total field on the distance to the Earth. At smaller distance the magnetic field B_z component decreased more slowly compare to the dipole $\left(\frac{R_E}{r}\right)^3$ dependence, but at greater distance the B_z component decreased exponentially $\left(\text{proportional to } e^{-\frac{|x|}{L_n}}\right)$ (Alexeev et al. 1998a). In some sense this distance can be used for the estimation of the downtail distance of the magnetosphere where the distant neutral point for the total magnetospheric field including the tail current system and the penetrated IMF can exist. Comparison of the two Figs. 10.2 and 10.3 are shown with a decreasing magnetopause flaring angle and the downtail magnetospheric size decreased also. For rough estimations the L_n/R_{ss} can be taken proportional to R_T/R_{ss} and equal to $\frac{9.0}{s}$.

10.2.2 Paraboloid Model of Mercury's Magnetosphere

Before we will go toward the Mercury's paraboloid magnetospheric model, we will introduce in the model IMF distortion in the course of the solar wind flow past the magnetosphere. We define three regions of space: (1) the undisturbed solar wind, (2) the magnetosheath, and (3) the magnetosphere. The interplanetary magnetic field represented by sum of two uniform vectors in the solar wind (region (1)) ($\mathbf{B}_1 = \mathbf{B}_0 + \mathbf{b} = const.$). Here $\mathbf{B}_0$ is the shield part of the IMF, which do not penetrate the magnetosphere. This field compressed and disturbed in the magnetosheath formed a magnetic barrier upstream the magnetopause. Rest part of the IMF $\mathbf{b}$ without distortion penetrated into the magnetosphere. The total magnetic field $\mathbf{B}_1$ and $\mathbf{B}_2$ satisfied the continuity conditions for the normal to magnetopause (bow shock) magnetic field components. The specified boundary conditions at the bow shock and the magnetopause are:

$$B_{1\beta} = B_{2\beta}|_{\beta=\beta_{bs}}, \qquad B_{2\beta}|_{\beta=1} = b_\beta, \tag{10.4}$$

Here $\mathbf{B}_2$ is the disturbed IMF inside the magnetosheath. The coordinate surface $\beta = 1$ is the surface defining the magnetopause, and $\beta = \beta_{bs}$ is the surface defining the bow shock. We use $\beta_{bs} = \sqrt{1.8} = 1.34$, corresponding to a distance between bow shock and magnetopause along x axis (Δ) of $0.4R_{ss}$ (compare $0.3R_{ss}$ obtained by Fairfield (1971).

We allow for a rotational diskontinuity magnetopause by allowing a fraction of the magnetosheath magnetic field at the magnetopause, $\mathbf{b}$, to penetrate that boundary. For the conductive plasma flow the reconnection efficiency coefficient, $k_{r\perp}$ is proportional to the magnetic Reynolds number (Alexeev 1986) $k_{r\perp} = 0.9R_m^{-\frac{1}{4}}$. Here the magnetic Reynolds number is given by $R_m = \mu_0\sigma V R_{ss}$, where R_{ss} is a characteristically dimension scale, σ is the anomalous plasma conductivity at the magnetopause, μ_0 is the vacuum permeability, and V is the solar wind velocity.

Alexeev and Kalegaev (1995) considered the flow of a conductive fluid past the magnetosphere (e.g., Alexeev 1986) and derived an expression for the portion of the magnetosheath magnetic field penetrating the magnetopause $\mathbf{b}$, as

$$b_x = \frac{2.5}{\sqrt{\pi}} k_{r\perp}^2 B_{0x}, \; b_y = k_{r\perp} B_{0y}, \; b_z = k_{r\perp} B_{0z}. \tag{10.5}$$

For typical solar wind plasma condition $R_m = 10^4$, in which case $|\mathbf{b}|$ is about 10 times smaller than $|B_0|$. R_m was estimated by Alexeev (1986) on the basis of the observed characteristic thickness of MHD diskontinuities in the solar wind and threshold values of the current density for the ionacoustic instability. The Mercury total magnetic field strength calculated by the paraboloid model are shown in the Fig. 10.4 (noon-midnight plane) and in the Fig. 10.5 (equatorial plane).

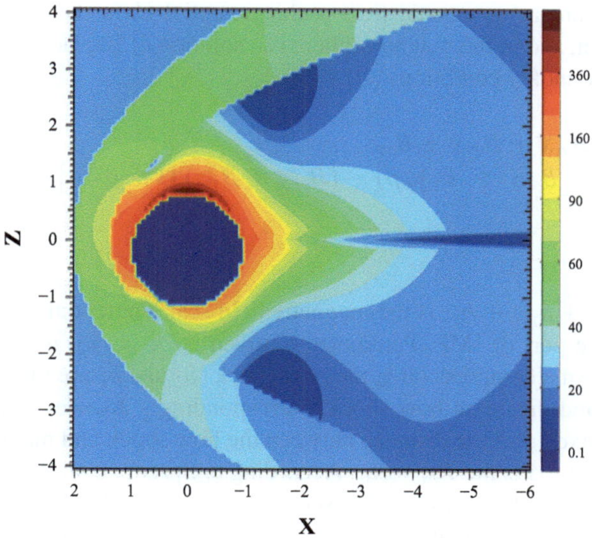

Fig. 10.4 For Mercury magnetospheric field the countours of the magnetic field strenght equal to constant in the noon – midnight plane are shown. All planetary magnetopauses are fixed by the subsolar magnetopause distance $R_{ss} = 1.4\ R_M$, the terminator magnetopause radius $R_T = 2.0\ R_M$, and the tailward distance to the inflection point is $L_n = 12.6\ R_M$. The Mercury radius is $R_M = 2439$ km. The flaring parameters s are equal to 1.0

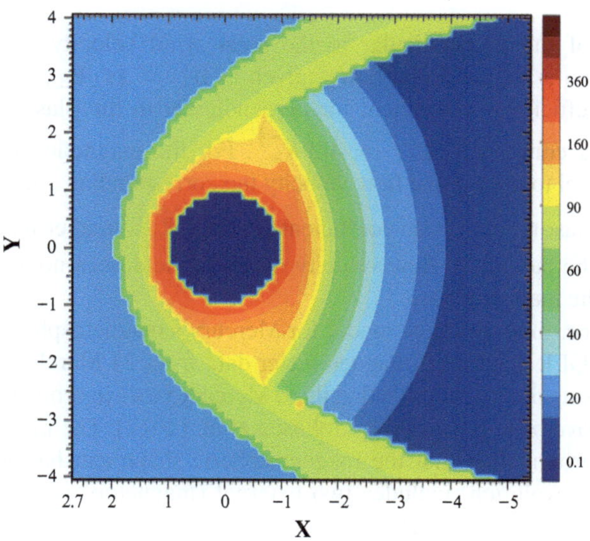

Fig. 10.5 For Mercury magnetospheric field the contours of the magnetic field strength equal to constant in the equatorial plane are shown. All planetary magnetopauses are fixed by the subsolar magnetopause distance $R_{ss} = 1.4\ R_M$, the terminator magnetopause radius $R_T = 2.0\ R_M$, and the night tail length $L_n = 12.6\ R_M$ the Mercury radius is $R_M = 2439$ km. The flaring parameters s are equal to 1.0

The interplanetary magnetic field scalar potential U in the solar wind, in the magnetosheath, and in the magnetosphere is a solution of Laplace's equation which satisfies the boundary conditions (10.4):

$$\begin{aligned} U_1 &= u + B_{ox}x + B_{oy}y + B_{oz}z && \beta > \beta_{bs}, \\ U_2 &= u + k\left[B_{ox}\left(x - R_1 \ln\beta\right) + \left(B_{oy}y + B_{oz}z\right)\left(1+\beta^{-2}\right)\right] && \beta_{bs} > \beta > 1, \\ U_3 &= u && \beta < 1, \end{aligned} \tag{10.6}$$

here $u = b_x x + b_y y + b_z z$ is the magnetic field scalar potential of the penetrating magnetosphere part of IMF. Parameter k, given by $k = \rho_2/\rho_1$, deternimes the solar wind plasma compression at the bow shock (ρ_2 and ρ_1 are the plasma dencity downstream and upstream bow shock, correspondingly Alexeev et al. 2003). The coefficient k fixed also the distance between the bow shock and magnetopause $k = 1 + R_1/2\Delta$, here Δ is the $R_{bs} - R_{ss}$. According to Spreiter et al. (1966), $k = 2.125$ for $\Delta = 0.4R_1$ and $R_1/2\Delta = 1.125$ (see also Landau and Lifshitz 1966).

In the parabolic coordinates the magnetosheath magnetic field components at the magnetopause (given by $\mathbf{B} = -\nabla\mathbf{U}$) are (see Alexeev et al. 1998b)

$$\begin{aligned} B_{2\alpha} &= k\tfrac{R_1}{R_f}\left[-\alpha B_{0x} + 2\left(B_{0y}\sin\varphi + B_{0z}\cos\varphi\right)\right] + b_\alpha. \\ B_{2\beta} &= b_\beta, \\ B_{2\varphi} &= 2k\left(B_{0y}\cos\varphi - B_{0z}\sin\varphi\right) + b_\varphi, . \end{aligned} \tag{10.7}$$

The portion of the perpendicular to the solar wind velocity vector IMF, $\mathbf{B}_\perp$, penetrating into a magnetosphere, $\mathbf{b}_\perp$, given as $\mathbf{b}_\perp = k_{r\perp}\mathbf{B}_\perp$, where $k_{r\perp}$ is the reconnection efficiency coefficient for perpendicular to the plasma flow velocity magnetic field component, $k_{r\perp} = 0.9R_m^{-\frac{1}{4}}$. For the magnetic field component, which is parallel to the plasma flow velocity the corresponding coefficient is $k_{r\parallel} = \frac{2}{\sqrt{\pi}}R_m^{-\frac{1}{2}}$. The last reconnection coefficient is about $R_m^{-\frac{1}{4}}$ smaller than $k_{r\perp}$. Reason for it is that the parallel to flow velocity magnetic field does not compressed and drapping by the magnetopause.

The baseline, time-averaged model for Mercury's magnetosphere, derived from the MESSENGER spacecraft magnetometer data from 24 March to 12 December 2011, comprising the spacecraft's first three Mercury years in orbit around the innermost planet have been presented by Johnson et al. (2012). The model, constructed under the approximation that the magnetospheric shape can be represented as a paraboloid of revolution, includes two external (magnetopause and magnetotail) current systems and an internal (dipole) field and allows for reconnection. We take advantage of the geometry of the orbital magnetometer data to estimate all but one of the model parameters, and their ranges, directly from the observations. These parameters are then used as a priori constraints in the paraboloid magnetospheric model of Alexeev et al. (2003), and the sole remaining parameter, the dipole moment, is estimated as 190 nT R_M^3 from a grid search. We verify that the best-fit dipole moment is insensitive to changes in the other parameters within their determined ranges. The model provides an excellent first-order fit to the MESSENGER observations, with a root-mean-square misfit of less than 20 nT globally. The results show that the magnetopause field strength ranges from 10 to 50 % of the dipole field strength at observation locations on the dayside and at nightside latitudes north of 60° N. Globally, the residual signatures observed to date are dominated by the results of magnetospheric processes, confirming the dynamic nature of Mercury's magnetosphere.

The orbit tracks for the second year of MESSENGER observations with the corresponding magnetic field observations demonstrates the overall success of the model. Residuals (data minus model) are typically less than 50 nT in magnitude compared with a maximum signal in the data of ~500 nT. Model predictions for the B_X, B_Y, and B_Z components for MESSENGER dawn-dusk and midnight-noon orbits provide an excellent fit to the observations.

The magnitudes of the magnetopause, tail, and dipole fields predicted at the planetary surface have been calculated by Johnson et al. (2012). The strong north-south asymmetry in the internal field imposed by the large northward offset of the dipole is evident, with the field strength at high northern latitudes over 700 nT, approximately three times that at the corresponding southern latitudes. Magnetopause fields are strongest on the dayside as expected. In addition, the offset dipole results in a magnetopause surface that is closer to the planetary surface at southern latitudes than at corresponding northern latitudes. This north-south difference is greatest at low to middle latitudes at local noon, and thus the strongest magnetopause fields are found at low southern dayside latitudes, where the magnetopause field strength reaches almost 80 nT, about 40 % of the dipole field strength at those locations.

10.2.3 Jupiter's Magnetosphere

The dependence of the Jupiter's magnetospheric boundary location (in particular, the distance of the Jovian subsolar magnetopause, R_{ss}) on the solar wind dynamic ram pressure, p_{sw}, was investigated in Huddleston et al. (1998). In good agreement with results of Slavin et al. (1985), it occurred that R_{ss} depends on p_{sw} in the power -0.22 ± 0.04. Results presented in Huddleston et al. (1998) were summarized by the following empirical formula (Cowley and Bunce 2003)

$$R_{ss} = 35.5 R_J / {p_{sw}}^{0.22} (\text{nPa}) . \tag{10.8}$$

According to Alexeev and Belenkaya (2005) the Jupiter's magnetospheric size is proportional to $p_{sw}^{-0.23}$. All these expressions show more strong dependence than for the dipole magnetosphere, when R_{ss} is proportional to $p_{sw}^{-1/6}$ (as in the case of the Earth). Huddleston et al. 1998 noted that in the Jovian magnetosphere, the hot internal plasma gives significant contribution to the pressure balance which determines the location of the magnetopause. In addition, the activity of the volcanoes of Io is episodic, so this process can also influence on the Jovian magnetospheric size. According to observations by spacecraft, in the Jovian environment, R_{ss} changed from 45 R_J to 110 R_J. However, for each value of R_{ss}, different distances from the planetary center to the magnetopause in the direction perpendicular to Jupiter–Sun line (X axis) could exist.

The paraboloid model of the Earth's magnetosphere has a modular structure. The scaling relations allow to adapt the magnetopause and the tail current systems developed for the Earth's magnetosphere to the case of Mercury, Jupiter, and Saturn. However, for the currents caused by the rapid planetary rotation (magnetodisk for Jupiter and Saturn) there is no analogy in the terrestrial magnetosphere. The magnetodisk is the main source of Jupiter's magnetospheric magnetic field. Its effective magnetic moment prevails Jupiter's dipole magnetic moment of ~2.6 times (Belenkaya 2004; Alexeev and Belenkaya 2005).

The dimensions of planetary magnetospheres may vary by a factor of several thousand (see Table 10.1), but different planets have similarly shaped magnetopauses. The front part of the magnetosphere coincides with the paraboloid of revolution, which has a symmetry axis that is a line joining the planet and the Sun. The magnetic field at the subsolar point can be determined from the balance of the solar wind plasma dynamic pressure and the pressure of the magnetic field inside the magnetosphere. This field does not depend on the value of the planetary dipole and is determined unambiguously by the solar wind dynamic presssure. The planetary dipole determines the size of the magnetosphere, R_{ss}. For Jupiter's and Saturn's magnetospheres the equatorial magnetodisks are formed by the rotational uploading of the satellite's plasma. As a result, R_{ss} is determined not by the planetary dipole only, but by some "effective" dipole which is bigger than the planetary one.

The magnetic field lines for the dipole and screening magnetopause currents without the tail current field of Mercury and Earth are shown in Fig. 10.6 in the noon – midnight plane for the general magnetospheric model. All planetary magnetopauses are scaled to R_{ss}. Circles in Fig. 10.6 show the Mercury (light black circle), the Earth (blue circle), the Saturn (green circle), and Jupiter (red circle). Because the paraboloids of the revolution in the chosen coordinate system are the same for both Mercury and Earth (blue dashed curve), the magnetic field lines (black curves) are coincided. The yellow curves are marked by the magnetic field lines which go to the cusp (the neutral points).

The magnetopause of Jupiter (dashed red curve) is more compressed to the $X-$axis comparable to the Earth's magnetopause. The magnetopause of Saturn (dot-and-dashed green curve) is more expanded comparable to the Earth's case. The best fit magnetopause obtained by Shue et al. (1998) is shown in Fig. 10.6 by the dashed purple curve.

The strength of the currents at the magnetopause and those in the tail depends on the magnetic moment of the "effective" planetary dipole and on the character size of the magnetosphere. The comparison of the results obtained in the model with the

Table 10.1 Solar System Planetary Magnetosphere Parameters

Parameters (units)	Equation	Mercury	Earth	Jupiter	Saturn
Heliocentric distance (AU)	r_0	0.38	1	5.2	9.5
Equatorial radius ($R_E = 6,371$ km)	r_P	0.38	1	11.2	9.45
Magnetic moment ($T \cdot km^3$)	$B_0 \cdot r_P^3$	2.8 10^{-6}	0.008	150.	4.6
Dipole tilt angle (degrees)	ψ	0°	10.5°	10°	0°
Equatorial magnetic field (μT)	B_0	0.196	30.	420.	20.
Dipole hemisphere magnetic flux (GWb)	$2\pi B_0 r_P^2$	0.0072	7.7	13, 450	456.
Open field line magnetic flux (GWb)	$\pi B_0 r_P^2 \frac{r_P}{R_{ss}}$	0.0024	0.42	450	11.4
Polar oval radius (degrees)	$\sin\theta = \sqrt{\frac{r_P}{R_{ss}}}$	55°	20°	15°	13°
Average IMF magnitude (nT)	$\frac{5\sqrt{1+r_0^2}}{\sqrt{2}\cdot r_0^2}$	10.2	5	1	0.5
Nominal Parker spiral angle ϕ	$\tan\phi = \frac{1}{r_0}$	20.8°	45°	80°	85°
Solar wind ram pressure (nPa)	$\frac{1.7}{r_0^2}$	11.8	1.7	0.07	0.015
Subsolar magnetic field magnitude (nT)	$B_{ss} = \frac{74.5}{r_0}$	196	74.5	14.3	7.8
Subsolar magnetopause distance (R_P)	R_{ss}	1.4 R_M	10 R_E	70 R_J	22 R_S
Magnetopause flaring parameter s	$\frac{\sqrt{4R_{ss}^2-R_T^2}}{R_T}$	1.	1.	1.25	0.66
Nose magnetopause curvature radius (R_P)	$\frac{2R_{ss}}{1+s^2}$	1.4 R_M	10 R_E	55 R_J	31 R_S
Terminator magnetopause radius (R_p)	$\frac{2\,R_{ss}}{\sqrt{1+s^2}}$	2. R_M	14 R_E	112 R_J	37 R_S
Distance to the inflection point (R_P)	$\frac{8.97}{s}\,R_{ss}$	12.6 R_M	89.7 R_E	502 R_J	220 R_S
Night tail length (R_P)	16.2 R_{ss}	22.7 R_M	162. R_E	1300 R_J	400 R_S

observational data for Mercury shows that the two above mentioned current systems suffice to describe their magnetospheric field. For Earth, Jupiter, and Saturn, the ring current in the inner magnetosphere is significant. Different planets have ring currents of different natures. Near Earth, the ring current is formed by trapped particles; ring current enhancement mostly determines the depression of the equatorial field during geomagnetic storms. In the case of Jupiter, this current is included into a

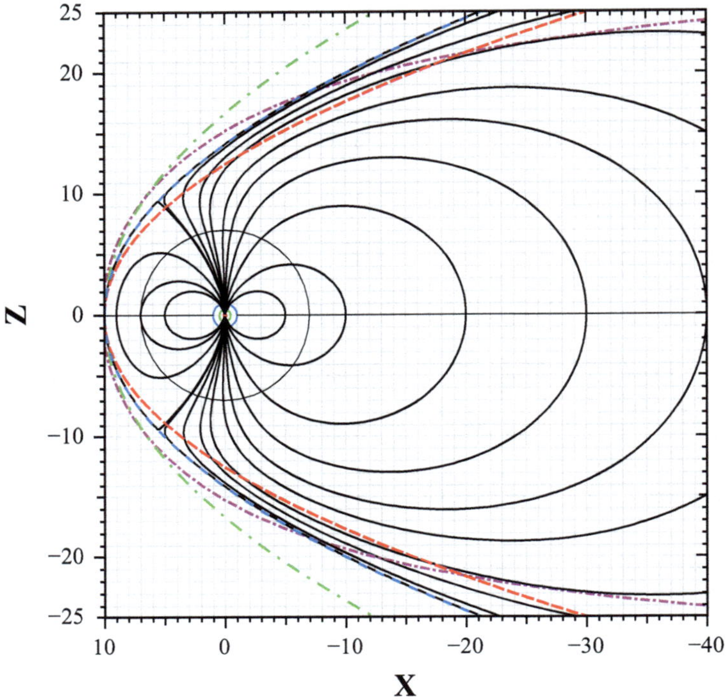

Fig. 10.6 Mercury's and Earth's magnetospheric field lines are shown for the dipole and screening magnetopause currents in the noon-midnight plane for a general magnetospheric model. All planetary magnetopauses are scaled to R_{ss} and one tip on *horizontal axis* is 0.1 R_{ss} ($R_{ss} = 10.$). *Biggest circle* shows Mercury with a radius, $R_M = R_{ss}/1.4 = 7$. Earth's surface with a radius $R_E = R_{ss}/10 = 1$, the Saturn's surface with radius $R_S = R_{ss}/22 = 0.45$, and Jupiter with radius $10./70. = 0.14$ are shown. The flaring parameters s are equal to 1.0 for Mercury and Earth, and to 1.25 and 0.66 for Jupiter and Saturn, correspondingly

plasma disk that is formed by the ionized eruptions from Io's volcanoes. The fast rotation of Jupiter drives and accelerates cool plasma, which is thrown into the outer magnetosphere. As the magnetic moment of the plasma disk is more than twice exceeds the magnetic moment of Jupiter, the size of the Jovian magnetosphere is almost two times bigger than the size of the magnetosphere formed around the planetary dipole only. The pressure of the magnetospheric Jovian plasma is equal to the magnetic field pressure near the noon magnetopause. Saturn and Jupiter rotate at similar angular rates. The sizes of both planets are also close to each other. The planetary magnetic field of Saturn, however, is almost 20 times weaker than that of Jupiter. Therefore, the "dipole enhancement effect" is considerably weaker for Saturn than for Jupiter, and the plasma disk of Saturn enhances its effective magnetic moment by a factor of 1.2–1.5.

10.3 Paraboloid Model Application to Hot Jupiter Magnetospheres

The paraboloid model approach to the study of planetary magnetospheres has been elaborated in details during the last decades in variety of papers devoted to the Solar System planets (Alexeev et al. 2003; Alexeev and Belenkaya 2005; Alexeev et al. 2010). These generalized models have been successfully tested against the magnetic field measurements performed by various spacecraft (Mercury: MESSENGER; Jupiter: Ulysses and Saturn: Pioneer 11, Cassini), showing good agreement between the paraboloid magnetosphere model predictions and the in-situ measured values.

Therefore, the goal is the application of the model to the case of a close-orbit giant exoplanet, in order to reveal a more realistic configuration of the magnetic field around such a planet and to estimate the size of the planetary magnetic obstacle (see also Chap. 11 (Grießmeier 2014)). The typical characteristics of hot Jupiter and their surrounding stellar wind plasma environment addressed in the previous sections will be used as input parameters for the quantitative characterization of the magnetosphere of an exoplanet on the basis of the paraboloid magnetosphere model (Khodachenko et al. 2012). We will repeat be low only some basic points of the paraboloid magnetosphere model concept and demonstrate the importance of magnetodisks in the scaling of magnetosphere of hot Jupiter. This study sheds more light on the problem of hot Jupiter – stellar wind interaction and magnetospheric protection of planetary atmospheres against stellar wind erosion.

10.3.1 Magnetodisks Are Key Elements of Hot Jupiter Magnetospheres

According to the model, a magnetodisk is placed in the equatorial plane at the interval of radial distances $[R_{D2}, R_{D1}]$ from the center. The formation of magnetodisk may be justified in the following way. It is well known that the field of a rotating planetary magnetic dipole can drive the inner magnetospheric plasma to rigid corotation with a planet only inside of the so-called Alfvénic surface, where the strength of magnetic field is high enough (Mestel 1968; Vasyliunas 1983). The equatorial boundary of the Alfvénic surface, R_A, is determined from the equality of energy densities of the plasma rotational motion $\varepsilon_p = \rho_A \omega_p^2 R_A^2/2$ and of the dipole magnetic field $\varepsilon_B = M_d^2/2\mu_0 R_A^6$, where $M_d = \mathscr{M}\frac{\mu_0}{4\pi}$ (Mestel 1968; Coroniti and Kennel 1977). Beyond the Alfvénic surface ($r > R_A$), i.e. in the area where the rotating planetary dipole magnetic field becomes too weak to drive the plasma in rigid co-rotation, a centrifugal outflow of the sub-corotating material begins. Plasma goes away with a radial speed V_{esc}. Therefore, the inner edge of the disk may be taken as approximately coinciding with the Alfvénic surface radius, i.e. R_{D2}.

One of the first cases when a magnetodisk model similar to that has been discussed, is the study by Mestel (1968). The author considers there the escape of plasma from a rotating magnetized star and proposes to distinguish two zones in the stellar wind flow. First, a dead zone where the roughly dipolar field of a star is strong enough to force the plasma flow to follow the field and to keep the gas co-rotating with a star, and second, a wind zone where the gas flow drags the field to follow the flow, resulting in the formation of an open-field lines region. By this, the Alfvénic surface, which crosses the equatorial plane at $r = R_A$, separates these two zones. Already in this work, two possibilities for the creation of an outflowing material flux were addressed. One is related to the pressure gradient driven thermal wind, caused by the high temperature of the expanding stellar corona, and another – due to the magnetically-controlled centrifugal forces which drive a centrifugal wind. The last becomes important if the coronal temperature is too low for driving a thermal wind. In that sense the situation is completely similar to the hot Jupiter case considered here, when the thermal expansion of the heated and ionized planetary upper atmosphere takes place under the conditions of a rotating planetary intrinsic magnetic dipole field.

The outflowing plasma, moving along the field lines inside the Alfvénic surface ($r < R_A$) is concentrated near the equatorial plane and provides the material source for creation of magnetodisk. The plasma, escaping along the filed lines, penetrating beyond the Alfvénic surface, deforms the original planetary magnetic dipole field, resulting in the radial stretching of the field lines (Mestel 1968) and the creation of a thin disk-type current sheet in the equatorial region. The situation with the magnetodisk formation and confinement is essentially non-stationary, characterized by continuous load of plasma to the disk, as well as to the entire magnetosphere, and simultaneous loss of the expanding material from the system by the non-thermal mechanisms at the boundary of magnetosphere. This comprises the major specifics of the magnetodisk beyond R_A. In such a dynamical situation the usual force balance approach is inapplicable. Analytic solutions for the similar case kinematic model of a stationary electromagnetic field and electric current environment around a rotating magnetized sphere (with a dipole-type magnetic field) in the presence of an axisymmetric radial outflow of plasma have been considered in Alexeev et al. (1982). Magnetic field configuration obtained there clearly indicates the formation of a thin equatorial current disk.

The ring electric current of the disk is determined by the magnetic flux above (and below) the disk (Alexeev and Belenkaya 2005). This flux is a part of the total magnetic flux of the planetary dipole which corresponds to the dipolar field lines extended outside the Alfvénic surface, i.e., the lines which in the case of an undisturbed magnetic dipole would cross the equatorial plane beyond the Alfvénic surface radius R_A. Contrary to the dipole field case, these lines are elongated almost parallel to the equatorial plane and have, in the case of a highly conducting plasma, a component $B_\theta \sim 0$. In the considered disk model the magnetic flux above (and below) the disk is assumed to be conserved, i.e. independent of the distance. This is true for a very high ($\to \infty$) conductivity of the disk plasma. Under this conditions the pressure of the magnetic field outside the disk may be considered to be the same as the pressure of plasma at the disk center.

The density of plasma at the inner edge of the disk ρ_A may be estimated from the planetary thermal mass loss dM_p^{th}/dt. Let us assume that the portion $\gamma dM_p^{th}/dt$ of the total thermally escaping material takes part in the formation of magnetodisk. In this case $\gamma dM_p^{th}/dt = 2\pi R_A^2 \delta\theta\ \rho_A V_{\rm esc}$ where $\delta\theta$ is the angular thickness of the disk. Using this expression one can write:

$$\rho_A = \frac{\gamma (dM_p^{th}/dt)}{2\pi R_A^2 \delta\theta\ V_{\rm esc}} \tag{10.9}$$

From the equality of the above defined energy densities of the plasma rotational motion and the dipole magnetic field one obtains the expression for the equatorial radius of the Alfvénic surface, r_A or the inner radius of the disk R_{D2}, measured in the planetary radii r_p

$$r_A = \frac{R_{D2}}{r_p} = \left(\frac{2\pi R_A^3 \delta\theta\ B_{d0}^2}{\mu_0\, \omega_p\, \gamma\, (dM_p^{th}/dt)} \right)^{1/6} \tag{10.10}$$

where $r_A = R_A/r_p$ is the dimensionless Alfvénic radius and $B_{d0} = B_d\ (r = r_p,\ z = 0)$ is the value of the planetary dipole magnetic field at the surface of the planet in the equatorial plane. When obtaining Eq. (10.10) we took into account that $M_d = B_{d0} r_p^3$.

Here, we are ignored a partional ionization of the escaping upper atmospheric material. It folows to the recent aeronomical calculations of the upper atmospheric structure of giant exoplanets at close orbits (e.g., Yelle 2004, García Muñoz 2007, Koskinen et al. 2010) with inclusion of the major photo-chemistry processes of atmospheric species, the expanding material at the heights above $3r_p$ is mostly ionized with the ratio of ion to neutral number densities about 10 (Yelle 2004). The coefficient γ has to reflect a fact that not all the escaping and ionized material contributes the creation of magnetodisk. There is a part of plasma flow which is lost along the open field lines. Since motion of the escaping plasma happens mostly along the field lines, we may assume that the escaping material flux is proportional to the magnetic field flux. The total flux of the non-disturbed magnetic dipole field, which would cross the equatorial plane beyond the planetary radius r_p is $F_0 = \int_0^{2\pi} \{\int_{r_p}^{\infty} B_z(r, z = 0) r\, dr\} d\phi = 2\pi \int_{r_p}^{\infty} \left(B_{d0} r_p^3 / r^3 \right) r\, dr = 2\pi B_{d0}\, r_p^2$. Then, the portion of the dipole flux beyond the Alfvénic radius defined in the similar way, is $(r_p/R_A) F_0$.

According to the paraboloid magnetosphere model calculations, this flux is divided in equal parts between the magnetic flux through the magnetodisk and the flux of open field lines going to the external magnetosphere (Alexeev and Belenkaya 2005). Therefore, the part of plasma flow lost along the open field lines, which does not take part in the formation of magnetodisk, is the fraction $(2r_A)^{-1}$ of the total escaping ionized material flow $\left(\frac{dM_p^{th}}{dt} \right)$. As it can be seen in Fig. 10.7 $2R_A \gg 1$, it is

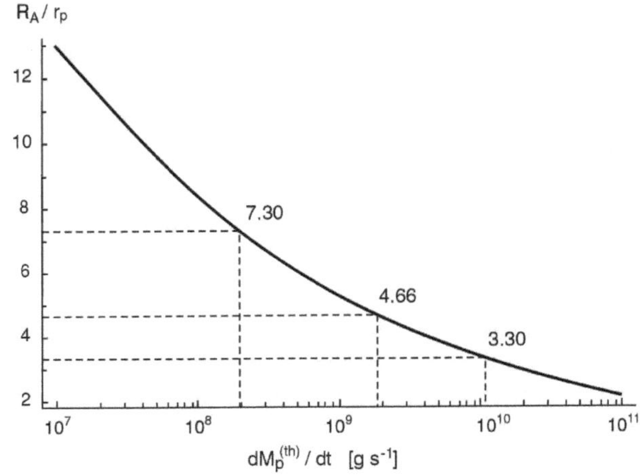

Fig. 10.7 Equatorial dimensionless Alfénic surface radius r_A of a hot Jupiter as a function of mass loss (After Khodachenko et al. 2012)

a reasonable to ignore the open field lines material flux and put $\gamma = 1$. This yields a equation for the Alfvénic surface equatorial radius r_A:

$$r_A{}^5 = \frac{4\pi\delta\theta\; B_{d0}^2 r_p}{\mu_0\, \omega_p\, \left(dM_p^{th}/dt\right)} \tag{10.11}$$

Figure 10.7 displays the solution of Eq. 10.11 as a function of the planetary thermal mass loss, assuming $V_{esc} \sim V_{cor} = \omega_p r_A r_p$ is the speed of corotation at the Alfvénic surface. This assumption corresponds to the case of a dominating centrifugally driven nature of material outflow outside the Alfvénic surface taken for the definiteness sake. This is of course an idealizing assumption, which in view of many other uncertainties regarding plasma and magnetic field parameters of hot Jupiters, may nevertheless be taken as a suitable simplification. Additional arguments in support of this approach may be taken from the analogy with the Solar System Jupiter, where spacecraft measurements indicate about an outflow of a sub-corotating material which takes place beyond the Jovian Alfvénic radius $R_{AJ} \sim 20R_J$. In particular, according to Galileo measurements, the module of plasma velocity for $r > R_{AJ}$ sometimes varies strongly, but in average remains to be close to 200 km/s (the corotation velocity at 20 R_J) in vicinity of R_{AJ}. By this, the azimuthal component of plasma velocity decreases with distance as $1/r^2$, which is typical for the case of a free material outflow with conservation of angular momentum. Similar result is provided also at large distances by the Hill model (Hill 1979). As it can be seen in Fig. 10.7, the values of R_A corresponding to the mass loss rates typical for the Jupiter-type test planet at orbital distances between 0.045 and 0.3 AU fall in the interval from 3.3 r_p to 7.3 r_p and are high enough to neglect by the factor γ dependence on R_A in Eq. 10.11. In other words, the plasma lost along the open filed lines does not produce strong influence on the overall value of the mass load to the magnetodisk in the considered region of parameters. The effect of neutral fraction in the escaping material is also not too strong. Thus, without making a huge mistake in estimation of R_A one may take $\gamma = 0.9\left(1 - \frac{1}{2\,r_A}\right) \sim 0.9 \sim 1$, which may be used for quick estimations of the R_A value for hot Jupiters. The difference between the precise result followed from the equation Eq. 10.11 and the estimations given by Eq. 10.12 do not exceed several percents of the last (∼3–6 %) the planetary magnetic dipole moments to certain power k of ω_p one can write that $\frac{B_{d0}}{B_{d0J}} = \frac{\omega_p^k}{\omega_J^k}$, where B_{d0J} is the Jovian magnetic dipole field at the surface of the planet in the equatorial plane. Then, using the estimative Eq. 10.12 and assuming a similarity of the magnetodisk geometric parameters (e.g., $\delta\theta$) for the considered hypothetic hot Jupiter and the Solar System Jupiter, we obtain:

$$\frac{R_A}{R_{AJ}} = \left(\frac{\omega_J}{\omega_p}\right)^{(2k-1)/5} \left(\frac{dM_J/dt}{dM_p^{(th)}/dt}\right)^{1/5} \tag{10.12}$$

Note that in the case of $k = 1/2$, i.e. with the planetary magnetic dipole scaling model of Stevenson (1983), the ratio R_A/R_{AJ}, according to Eq. 10.12 it is controlled only by the mass loss rate and does not depend on the planetary angular velocity.

Conclusion

It is shown that the general paraboloid magnetospheric model that was constructed on the base of the terrestrial one and other magnetic planets in the Solar System (Alexeev 1986) can be adopted also to exoplanets (Khodachenko et al. 2012). We show that one has to took into account changes of the magnetopause shape by introducing a new parameter s, that characterizes the flaring. The high solar wind plasma conductivity prevents the penetration of the magnetic field caused by the internal magnetospheric sources into the magnetosheath. We have demonstrated the method to calculate more accurately the planetary dipole screening magnetopause current field, which allows us to use the presented model more efficiently.

Based on the paraboloid magnetosphere model concept, we introduced in the present study a more complete view of an exoplanetary magnetosphere. This is applied to the case of a close orbit hot Jupiter with a continuously expanding and outward flowing hydrogen atmosphere. The advantage of the paraboloid magnetosphere model consists in the consequent account of the whole variety of magnetospheric key current systems and magnetic field sources. Of special importance, in the context of exoplanetary physics, is the flexibility of the paraboloid magnetosphere model with respect to the

(continued)

modelled object, and the possibility to apply it for reconstruction of magnetospheres of different types of exoplanets. The paraboloid magnetosphere model was successfully applied in the magnetosphere study of solar system planets and gave good agreement with the spacecraft in-situ measurements. However, the present work is the first time when the model is applied to exoplanets.

The key element and major specifics of the hot Jupiter magnetosphere considered in the paper consists in the presence of the magnetodisk, which origin is connected with an expanding and escaping atmospheric material of a close orbit giant exoplanet, i.e., its significant mass loss. The escaping atmospheric gas is ionized by the stellar radiation and contributes the build up of a disk around the planet. The rotation (even slow) of the planetary magnetic dipole plays an important role in the process of disk formation.

The considered magnetodisk is assumed to be located outside the Alfvénic surface, at which the equality of energy of the planetary dipole magnetic field and of the co-rotating plasma kinetic energy is achieved. Beyond this surface the rotating magnetic field of a planet can not drive equatorial plasma in rigid co-rotation, and the outflowing sub-corotating plasma changes the topology of the magnetic field by creating a thin equatorial current sheet of the disk. Taking the thermal mass loss of a hot Jupiter as the material source for the equatorial plasma disk requires that the particle escape height is less then the inner radius of the disk. Such an assumption seems to be a realistic one in view of the recent estimates of the upper thermosphere boundary for the close-in hot Jupiter's HD209458b (Koskinen et al. 2010), which yield the values of about 3 planetary radii. Above these heights the atmosphere of HD 209458b is mostly ionized. Altogether, according to the picture adopted in the paper, a hot Jupiter's thermal mass loss process plays an important role (as the major material source) in the build-up of magnetodisk and shaping of the magnetosphere. Mass loss processes, due to ion pick-up by the stellar wind (see Chap. 7 (Kislaykova et al. 2014)), as well as via the production of extended coronas, suprathermal atoms and energetic neutral hydrogen atoms (ENAs) (Chap. 4 (Fossati et al. 2014), Chap. 6 (Shematovich et al. 2014), and Chap. 11 (Grießmeier 2014)) are responsible for the mass loss of close-in hot Jupiters. The information on a realistic size and shape of the exoplanetary magnetosphere is of significant importance for the efficiency of these loss processes.

Stellar wind interaction with hot Jupiter's have also been simulated numerically, using resistive MHD, by Ip et al. (2004) (extremely close-in case, no shock) and by Preusse et al. (2007). The formation of induced magnetospheres near moderately close-in unmagnetized terrestrial type exoplanets has been modelled by Lipatov et al. (2005) on the basis of a hybrid code and in the drift-kinetic approximation, as well as by Johansson et al. (2009), using a hybrid

(continued)

code. However, in all these numerical studies the planetary obstacle has been implemented just as a spherical boundary at which zero particle velocity and constant density were imposed together with the prescribed particle removal and production mechanisms to model surface absorption and atmosphere expansion. Therefore the effects of rotating planetary intrinsic magnetic fields and corresponding dynamics of the surrounding plasma, including the formation of planetary magnetodisks were not incorporated into these models.

In Fig. 10.8 the magnetospheric structure, which have been modelled by substellar point pressure balance condition are shown. Inside the magnetopause the pressure of the total magnetic field, jointly produced by the screened magnetic dipole, magnetodisk and magnetotail, as well as of the disk plasma pressure pmp are taken into account. The PMM provides a self-consistent approach to the calculation of a 3D magnetic field structure in a planetary magnetosphere, and determines R_{ss}, based on the consequent evaluation of the total magnetic field in the substellar point which incorporates the contribution of all mentioned above components (current systems) of the model (i.e., magnetodisk, magnetotail and magnetopause currents). The view of magnetic field lines in the *ZX*-plane of a hot Jupiter's magnetosphere for a Jupiter-type planet orbiting at different distances around a solar-analogue star is shown in Fig. 10.8.

The performed modelling clearly indicates that the presence of a magnetodisk distinctively changes the character of the magnetosphere of a close orbit giant exoplanet which appears to be a magnetodisk-dominated one, in contrast to the dipole-dominated magnetospheres of the Solar System planets (except of Jupiter). In that sense the magnetodisk-dominated magnetospheres, typical for close orbit hot Jupiter's with a strong mass loss, appear to be a new type of planetary magnetospheres unknown in the Solar System, which require further investigations. A more realistic structure of a hot Jupiter magnetosphere predicted by the paraboloid magnetosphere model may be up to 40 $\sim$ 70 % larger size, as compared to a simple dipole-type case (Khodachenko et al. 2012).

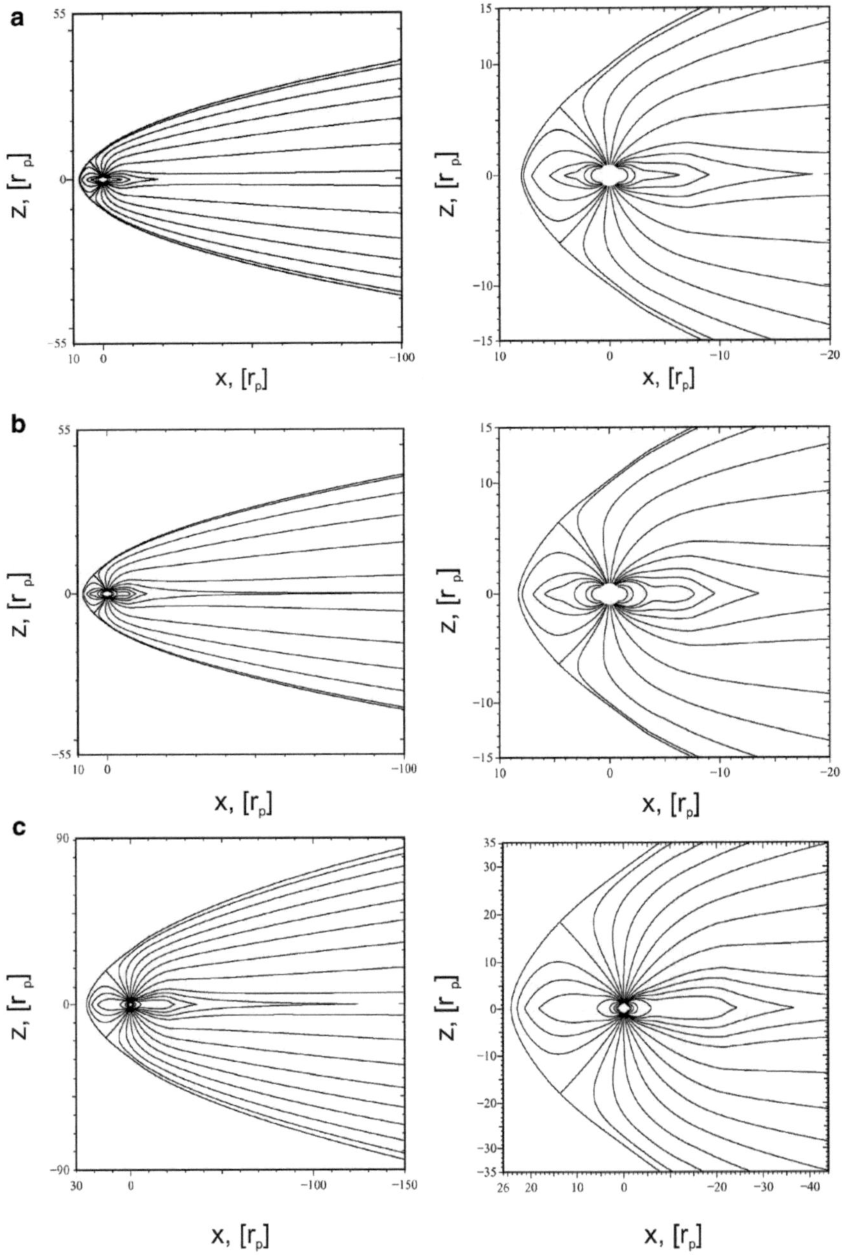

Fig. 10.8 Magnetic field lines (*left panel* shows a zoomed view) in the *ZX*-plane ($y = 0$ plane) of a hot Jupiter magnetospheres for the Jupiter-type planet ($M_p = M_J$ and $r_p = R_J$), orbiting at different distances around a solar-analogue *G*-type star. (a) 0.045 AU; (b) 0.1 AU; (c) 0.3 AU. The atmospheric mass losses, $\dot{M}$ [kt/s] are 10, 000; 1, 800; 184.. The magnetopause stand-off distances at substellar point, R_{ss} [r_p] are 8.0; 8.27; 24.2 and the tailward size of the magnetospheres, L_n [r_p] are 71.8; 74.2; 217.0 (Khodachenko et al. 2012)

Acknowledgements The authors acknowledge the support by the International Space Science Institute (ISSI) in Bern, Switzerland and the ISSI team *Characterizing stellar- and exoplanetary environments*. The authors acknowledge the support by the Austrian Research Foundation FWF NFN project S116 'Pathways to Habitability: From Disks to Active Stars, Planets and Life', and the related FWF NFN subproject, S116 606-N16 'Magnetospheric Electrodynamics of Exoplanets'. This work was also supported by the FWF project P21197-N16. Finally, the authors are thankful to the European FP7 project IMPEx (No. 262863).

References

Alexeev, I. I., & Shabansky, V. P. (1972). *Planetary and Space Science, 20*, 117.

Alexeev, I. I. (1978). *Gemagnetizm i Aeronomiya, 18* (in Russian); *Geomagnetism and Aeronomy, 18*, 447 (in English).

Alexeev, I. I., Kropotkin, A. P., & Veselovsky, I. S. (1982). *Solar Physics, 79*, 385.

Alexeev, I. I. (1986). *Journal of Geomagnetism and Geoelectricity, 38*, 1199.

Alexeev, I. I., & Kalegaev, V. V. (1995). *Journal of Geophysical Research, 100*, 267.

Alexeev, I. I., & Bobrovnikov, S. Y. (1997). *Geomagnetizm i Aeronomiya, 37*(6), 24.

Alexeev, I. I., Belenkaya, E. S., & Sibeck, D. G. (1998), *Geomagnetizm i Aeronomiya, 38*(1), 9.

Alexeev, I. I., Sibeck, D. G., & Bobrovnikov, S. Y. (1998). *Journal of Geophysical Research, 103*(A4), 6675.

Alexeev, I. I., Belenkaya, E. S., Bobrovnikov, S. Y., & Kalegaev, V. V. (2003). *Space Science Reviews, 107*, 7.

Alexeev, I. I., & Belenkaya, E. S. (2005). *Annals of Geophysics, 23*, 809.

Alexeev, I. I., Kalegaev, V. V., Belenkaya, E. S., Bobrovnikov, S. Y., Bunce, E. J., Cowley, S. W. H., & Nichols J. D. (2006). *Geophysical Research Letters, 33*, L08101.

Alexeev, I. I., Belenkaya, E. S., Slavin, J. A., Korth, H., Erson, B. J., Baker, D. N., Boardsen, S. A., Johnson, C. L., Purucker, M. E., Sarantos, M., & Solomon, S. C. (2010). *Icarus, 209*, 23.

Alexeev, I. I., Belenkaya, E. S., Bobrovnikov, S. Y., Slavin, & J. A., Sarantos, M. (2008). *Journal of Geophysical Research , 113*, A12210.

Arridge, C. S., Achilleos, N., Dougherty, M. K., Khurana, K. K., Russell, C. T. (2006). *Journal of Geophysical Research, 111*, A11227.

Belenkaya, E. S. (2004). *Planetary and Space Science, 52*, 499.

Belenkaya, E. S., Bobrovnikov, S. Y., Alexeev, I. I., Kalegaev, V. V., & Cowley, S. W. H. (2005). *Planetary and Space Science, 53*, 863.

Belenkaya, E. S., Alexeev, I. I., Kalegaev, V. V., & Blokhina, M. S. (2006). *Annals of Geophysics, 24*, 1145.

Belenkaya, E. S., Cowley, S. W. H., & Alexeev, I. I. (2006). *Annals of Geophysics, 24*, 1649.

Chapman, S., & Ferraro, V. C. A. (1931). *Journal of Geophysical Research , 36*, 77.

Coroniti, F. V., & Kennel, C. F. (1977). *Geophysical Research Letters, 4*, 211.

Cowley, S. W. H., & Bunce, E. J. (2003). *Planetary and Space Science, 51*, 57.

Fairfield, D. H. (1971). *Journal of Geophysical Research, 76*, 6700.

Fossati, L., Haswell, C. A., Linsky, J. L., & Kislyakova, K. G. (2014). H. Lammer & M. L. Khodachenko (Eds.), *Characterizing stellar exoplanetary environments* (pp. 59). Heidelberg/New York: Springer.

García Muñoz, A. (2007). *Planetary and Space Science, 55*, 1426.

Greene, J. M., & Miller, R. L. (1994). *Planetary and Space Science, 42*, 895.

Grießmeier, J.-M. (2014). H. Lammer & M. L. Khodachenko (Eds.), *Characterizing stellar exoplanetary environments* (pp. 213). Heidelberg/New York: Springer.

Herbert, F. (2009). *Journal of Geophysical Research, 114*, A11206.

Hill, T. W. (1979). *Journal of Geophysical Research, 84*, 6554.

Huddleston, D. E., Russel, C. T., Kivelson, M. G., Khurana, K. K., & Bennet, L. (1998). *Journal of Geophysical Research, 103*, 20075.
Ip, W.-H., Kopp, A., & Hu, J.-H. (2004). *Astrophysical Journal, 602*, L53.
Johansson, E. P. G., Bagdonat, T., & Motschmann, U. (2009). *Astronomy and Astrophysics, 496*, 869.
Joy, S. P., Kivelson, M. G., Walker, R. J., Khurana, K. K., Russell, C. T., & Ogino, T. (2002). *Journal of Geophysical Research, 107*, 1309.
Johnson, C. L., Purucker, M. E., Korth, H., Erson, B. J., Winslow, R. M., Al Asad, M. M. H., Slavin, J. A., Alexeev, I. I., Phillips, R. J., Zuber, M. T., & Solomon, S. C. (2012). *Journal of Geophysical Research, 117*, E00L14.
Kanani, S. J., Arridge, C. S., Jones, G. H., Fazakerley, A. N., Mcrews, H. J., Sergis, N., Krimigis, S. M., Dougherty, M. K., Coates, A. J., Young, D. T., Hansen, K. C., & Krupp, N. (2010). *Journal of Geophysical Research, 115*, A06207.
Kislyakova, K. G., Holmström, M., Lammer, H.,& Erkaev, N. V. (2014). H. Lammer, & M. L. Khodachenko (Eds.), *Characterizing stellar exoplanetary environments* (pp. 137). Heidelberg/ New York: Springer.
Kivelson, M. G., & Southwood, D. J. (2003). *Planetary and Space Science, 51*, 891.
Khodachenko, M. L., Alexeev, I., Belenkaya, E., Lammer, H., Grießmeier, J.-M., Leitzinger, M., Odert, P., Zaqarashvili, T., & Rucker, H. O. (2012). *Astrophysical Journal, 744*(70), 16pp
Landau, L. D., & Lifshitz E. M. (1966). Fluid Mechanics (Volume 6 of Course of Theoretical Physics, 3rd rev. English ed., Chapter 13, paragraph 114, p 458, Eq. (114.1)). Oxford: Pergamon Press.
Landau, L. D. & Lifshitz E. M. (1966). Fluid Mechanics, Third revised English edition (Volume 6 of Course of Theoretical Physics) Pergamon Press.
Lipatov, A. S., Motschmann, U., Bagdonat, T., & Grießmeier, J.-M. (2005). *Planetary and Space Science, 53*, 423.
Mestel, L. (1968). *Monthly Notices of the Royal Astronomical Society, 138*, 359.
Preusse, S., Kopp, A., Büchner, J., & Motschmann, U. (2007). *Planetary and Space Science, 55*, 589.
Russell, C. T. (1977). *Geophysical Research Letters, 4*, 387.
Shematovich, V. I., Bisikalo, D. V., & Dmitry, E. I. (2014). H. Lammer & M. L. Khodachenko (Eds.), *Characterizing stellar exoplanetary environments* (pp. 105). Heidelberg/New York: Springer.
Shue, J.-H., Song, P., Russell, C. T., Steinberg, J. T., Chao, J. K., Zastenker, G., Vaisberg, O. L., Kokubun, S., Singer, H. J., Detman, T. R., & Kawano, H. (1998). *Journal of Geophysical Research, 103*, 17961.
Slavin, J. A., & Holzer, R. E. (1979). *Journal of Geophysical Research, 84*, 2076.
Slavin, J. A., Smith, E. J., Spreiter, J. R., & Stahara, S. S. (1985). *Journal of Geophysical Research, 90*, 6275.
Slavin, J. A., Acuña, M. H., Erson, B. J., Baker, D. N., Benna, M., Boardsen, S. A., Gloeckler, G., Gold, R. E., Ho, G. C., Korth, H., Krimigis, S. M., McNutt, R. L., Raines, J. M., Sarantos, M., Schriver, D., Solomon, S. C., Trávníček, P., & Zurbuchen, T. H. (2009). *Science, 324*, 606.
Spreiter, J. R., Summers, A. L., & Alksne, A. Y. (1966). Hydromagnetic flow around the magnetosphere. Planetary and Space Science, 14, 223.
Stevenson, D. J. (1983). *Reports on Progress in Physics, 46*, 555.
Vasyliunas, V. M. (1983). A. J. Dessler (Ed.) *Physics of the jovian magnetosphere* (p. 395). Cambridge: Cambridge University Press.
Yelle, R. V. (2004). *Icarus, 170*, 167.

Chapter 11
Detection Methods and Relevance of Exoplanetary Magnetic Fields

Jean-Mathias Grießmeier

Abstract In analogy to the planets of the Solar System, most extrasolar planets are expected to have an intrinsic, internally generated magnetic field. These magnetic fields are believed to influence a number of physical processes, so that planets with and without fields may not behave and evolve the same way, and implications for the planet are manyfold. Clear observational evidence for such fields is however difficult to find, and no unambiguous detection has yet been achieved. Over the past few years, a number of methods have been suggested with which an exoplanetary magnetic field could be detected remotely. Some of these methods could even be used to characterize the planetary magnetic field strength quantitatively. The present work describes the different ways in which a planetary magnetic field may modify the planetary evolution and reviews the different methods that have been suggested to detect these fields. These methods are compared and we evaluate which techniques have the highest potential for future detection of exoplanetary magnetic fields.

11.1 Introduction: Planetary Magnetic Fields

With the possible exception of Venus, all planets in the Solar System either currently have or had in their past internally generated magnetic fields (e.g. Stevenson 2003, see also Chap. 10 by Alexeev et al. 2014). For extrasolar planets, one might naively expect the situation to be similar, but our experience with exoplanets has taught us that analogies with the Solar System have to be used with care, and that nature maybe surprisingly diverse. Theories for estimations of exoplanetary magnetic fields abound, sometimes with surprising results and conclusions. If, however, we want to test these theories, we need observations. Such observations may indeed be possible, as magnetic fields are believed to influence the planetary state and evolution in a

J.-M. Grießmeier (✉)
LPC2E – Université d'Orléans/CNRS, France and Station de Radioastronomie de Nançay, Observatoire de Paris – CNRS/INSU, USR 704 – University Orléans, OSUC, route de Souesmes, 18330 Nançay, France
e-mail: jean-mathias.griessmeier@cnrs-orleans.fr

H. Lammer, M. Khodachenko (eds.), *Characterizing Stellar and Exoplanetary Environments*, Astrophysics and Space Science Library 411,
DOI 10.1007/978-3-319-09749-7_11

number of ways. Some of these effects are subtle and indirect, but others may be observable more or less directly. Currently, there is no single measurement which unambiguously proves the existence of an exoplanetary magnetic field, but a number of methods have been suggested with which such a field could be detected, and in some cases even be quantified, remotely.

This works studies two questions related to exoplanetary magnetic fields: Why do we care?, and How could we observe? To answer these questions the aim of this work is to discuss the ways in which a planetary magnetic field may modify the planet and its evolution. For each interaction channel, the following questions have to addressed:

- Does the interaction lead to a noticeable effect, or will it be masked by other effects? Only in the case of a noticeable effect the interaction pathway can be used to study exoplanetary magnetic fields.
- Is the same interaction possible without a planetary magnetic field? Only if false positives can be excluded, the interaction pathway can be used to study exoplanetary magnetic fields qualitatively.
- Is it possible that a magnetic field exists even if the interaction is absent? Only if false negatives can be excluded, the interaction pathway can be used to qualify a planet as non-magnetized.

As one will see, not all effects that have detectable signatures are suited as a detection method. This does not necessarily mean that for such effects the magnetic influence is weaker or less interesting – it just means that it cannot be exploited directly.

We will proceed as follows: Sect. 11.2 studies the ways in which a planetary magnetic field may be important for gas giants. In Sect. 11.3, we will look at terrestrial planets. All interaction pathways are compared in "Conclusion".

11.2 Effects of Magnetic Fields on Gas Giants

In this section, we will go through the effects through which a planetary magnetic field can influence the evolution of a gas giant.

11.2.1 Gas Giants: Superflares

Some very large flares (up to 10^7 times more energetic than the largest solar flare) were observed for normal F and G stars on, or very close to, the main sequence (Schaefer et al. 2000). It has been suggested by Rubenstein and Schaefer (2000) that these superflares may be caused by magnetic interaction between the star and a (yet undetected) magnetized hot Jupiter. They suggest that the tangled magnetic fields of the host star and the planet can lead to massive reconnection events, which may trigger a superflare. A first quantitative estimate of this interaction was presented by Cuntz et al. (2000), finding that the strength of the interaction is proportional to the planetary magnetic field strength. Thus, it was soon suggested that the detection of such magnetic interaction would confirm the existence of exosolar planetary magnetospheres and dynamo activity (Cuntz and Shkolnik 2000). Numerical simulations and estimations confirmed that the reconnection could provide an important energy source and trigger flares (Ip et al. 2004).

Lanza (2009) presents a more detailed model in which large stellar flares may be triggered by close-in planets. The author argues that magnetic reconnection between the stellar coronal field and the planetary field is not sufficient to account for the flare power. Instead, a different interaction mechanism is suggested. In this model, a hot Jupiter contributes to the magnetic helicity budget by increasing the helicity dissipation that triggers an additional magnetic energy release in the stellar corona. Ideally, the planet should be magnetized (a planetary magnetic field value of 5 G is tentatively assumed). However, even if the planetary magnetic field is negligible, some dissipation by the currents induced in the planetary conductive interior are expected.

Pillitteri et al. (2010) observed a large flare in the HD189733 system that seems to come from active regions displaced by $\sim$75–78° with respect to the sub-planetary point. The size of the flaring region was estimated to be of the order of the stellar radius. The authors find good agreement between their observation and a simple MHD model (assuming a planetary field strength of half of that of Jupiter). However, this is based on a single observation in 2009, which does not allow firm conclusions.

The huge amount of light curves collected by the Kepler satellite allowed to extend the observational database. In total, 1547 superflares were detected on 279 G dwarf stars (Maehara et al. 2012; Shibayama et al. 2013). As with the study of Schaefer et al. (2000), none of the stars with observed superflares has a hot Jupiter planet, indicating that superflares related to exoplanets are rare (Maehara et al. 2012; Shibayama et al. 2013) (see also Sect. 9.2.5 by Guenther and Geier (2014) for a more detailed review).

To conclude: Recent Kepler data indicate that superflares are unlikely to be related to magnetized exoplanets.

11.2.2 Gas Giants: Planetary migration

The magnetic field of a giant planet has been suggested to have an important effect on the migration of newly formed planets.

Using a Weber Davis stellar wind model (Weber and Davis 1967), Lovelace et al. (2008) study the influence of the azimuthal ram pressure of the magnetized wind on the planet. Depending on the planet's orbital velocity compared to the stellar rotation speed, this ram pressure can either increase of decrease the planetary angular momentum. In other words, if the planet orbits beyond a certain critical distance, the wind will move it outward; if it is within the critical distance, it will move inward. The migration timescale is inversely proportional to the planetary effective cross section (the planetary radius if unmagnetized, otherwise the magnetopause radius). The magnetopause radius, in turn, depends on the planetary magnetic field. The migration timescales connected to this effect are of the order of 2–20 Myr for a planet with a surface field of 100 G (compared to a maximum surface field strength of ∼16 G for Jupiter (Connerney et al. 1998)).

Vidotto et al. (2009) and Vidotto et al. (2010) study the same effect, but for weak-lined T Tauri stars. Vidotto et al. (2009) examine the case of a stellar magnetic dipole aligned with the stellar rotation axis. Using a stellar wind model with a smaller wind magnetic field strength, they find larger timescales than Lovelace et al. (2008). This analysis is extended by Vidotto et al. (2010), who study the case of a tilted stellar magnetosphere. A tilt of the magnetic field by 30° reduces the migration timescale by a factor of ∼2. Still, even for a planet with a polar surface field of 100 G, these timescales seem to be larger than those resulting from other processes (e.g. the interaction between the protoplanet and the disk), so that the effect of the planetary magnetic field on its migration history is not likely to be the dominating effect.

To conclude: Other effects are likely to dominate over the influence of the planetary magnetic field on the planetary migration history.

11.2.3 Gas Giants: H_3^+ *Emission*

Infrared emission from H_3^+ molecules is the dominant cooling mechanism in Jupiter's thermosphere. For hot Jupiters, the extra heating by the nearby star leads to stronger cooling, and the H_3^+ emission is expected to be orders of magnitude stronger than on Jupiter. Observations have been performed at NASA's Infrared Telescope Facility, but no emission was detected (Shkolnik et al. 2006).

It has been suggested that this emission might be modified in the presence of a planetary magnetic field. The field would direct precipitating electrons towards the poles. This enhancement could in principle be quite large (Shkolnik et al. (2006) mention orders of magnitude). However, as the observation would not be able to resolve the planetary disk, this distribution seems difficult to discern. It is not clear whether (and how much) the disk-integrated H_3^+ signal would be affected by a planetary magnetic field. Also, atmospheric composition and structuring will have a strong influence on the strength of the signal, probably allowing false positives and false negatives when identifying H_3^+ emission with a magnetic field.

To conclude: The infrared emission by H_3^+ molecules can be modified by a planetary magnetic field. The spatial distribution is certainly modified, but this will be difficult to measure. In principle, the amplitude of the emission could be modified, but careful theoretical studies would be required to avoid confusion with other effects.

11.2.4 Gas Giants: Planetary Mass Loss

Hot Jupiters are characterized by the proximity to their host star. Consequently, their atmospheres are strongly heated, which can lead to the atmosphere expanding by up to several planetary radii (Lammer et al. 2003, 2009). This can lead to strong atmospheric loss of either neutral atoms or ionized material. In some cases, the expanded planetary atmosphere can even reach up to the Roche lobe. This extreme atmospheric expansion is even more true in the case of hot Jupiters in orbit around young host stars. In that case, the planetary atmosphere is more extended due to the intense heating by the high stellar X-ray and EUV flux (Lammer et al. 2003; Grießmeier et al. 2004; Ribas et al. 2004) (see also Chap. 1 (Linsky and Güdel 2014)). At the same time, the stellar wind of a young star is denser and faster (Wood et al. 2002; Wood 2007; Grießmeier et al. 2004; Wood et al. 2005a,b; Wood 2006; Grießmeier et al. 2007a,b; Holtzwarth and Jardine 2007; Wood and Linsky 2010), (see also Chap. 2 (Wood et al. 2014)). Also, the star is likely to have a stronger CME activity (Grießmeier et al. 2007a; Khodachenko et al. 2007a). Combined, these factors indicate that hot Jupiters may be subject to strong erosion, especially when the stellar wind or stellar CMEs can come close to the extended atmosphere. This enhanced particle loss can be reduced if the planet is shielded by a strong planetary magnetic field.

The effect of planetary magnetic fields on the planetary mass loss was first studied by (Grießmeier et al. 2004). They found that the stellar wind conditions and XUV photons flux provided by a young star could lead to huge mass loss rates though hydrodynamic expansion when the stellar winds compresses the magnetosphere close to the (extended) upper atmosphere. Additional loss could occur via ion pick-up (Erkaev et al. 2005). Atmospheric erosion by stellar CMEs was first discussed by Khodachenko et al. (2007a).

With a more detailed, energy-limited mass loss calculation and a realistic heating efficiency coefficient, Lammer et al. (2009) found that non-thermal stellar plasma-induced H^+ pick-up erosion of a weakly magnetized hot Jupiter is most likely negligible over the evolutionary timescale, provided that the planet interacts with the ordinary stellar wind or average CMEs. Integrated over the planet's lifetime, the mass loss was found to be $\leq 12.5\,\%$ of its total mass. Fast CMEs, however, could lead to high non-thermal escape rates.

The planetary magnetic field not only keeps the stellar wind further away from the planetary atmosphere, but it can also change the topology of the planetary outflow and guide the flow. According to Adams (2011), magnetic fields will dominate the flow provided that the field strength near the planet is greater than $\sim$1 G. In that case, the geometry of the flow pattern is set by the field structure, and the mass loss rate can be strongly reduced (Adams 2011). For typical cases, they find that only 10 % of the planetary surface can support outflow, and the mass loss can be reduced by a factor $\sim$3.

It was noted by Lammer et al. (2009) that a giant planet, even without an intrinsic magnetic field, will be partially protected by an induced field, unless it is situated extremely close to its host star. More importantly, in the case of weak magnetic shielding, the enhanced particle loss can give rise to an extended magnetodisk, which in turn could provide better protection against erosion by stellar wind and CMEs (Khodachenko et al. 2012a). This is supported by the observations of a large number of hot Jupiters, showing that those planets survive even under extreme conditions (Khodachenko et al. 2012b). It seems these planets may be better protected than previously assumed. Whether this indeed is due to protective magnetodisks or due to a stronger intrinsic magnetic field remains to be seen. In either case, the assumption of hot Jupiters with virtually no magnetic protection may not be as realistic as previously assumed.

To conclude: Hot Jupiters without magnetic protection are less common than previously thought. The effects of a small to moderate magnetic field on the planetary mass loss rate remains to be studied in more detail. Under most conditions, the mass loss is, however, likely to be much smaller than the planetary mass, even when integrated over the planetary lifetime. This may be different when frequent, strong CMEs are considered.

11.2.5 Gas Giants: Chromospheric Emission

Besides the possibility of superflares triggered by the magnetic interaction between a planet and its host star, Cuntz and Shkolnik (2000) also suggested that magnetic interaction could lead to increased non-radiative energy production providing additional chromospheric and coronal heating, which could lead to detectable chromospheric emission.

First studies of observational data used one line of the Ca II infrared triplet at 8662 Å. Seven planet-hosting stars were observed, but no identification of a planetary signature in the stellar emission was possible (Saar and Cuntz 2001). Subsequent observations in the optical resonance lines of Ca II H and K at 3933 and 3968 Å were more successful and showed signs of an increase of the chromospheric variability (not emission, as is sometimes understood) of about 1–2 % caused by planets around the stars HD 179949 (Shkolnik et al. 2003, 2004), υ And (Shkolnik et al. 2005), and possibly τ Bootes (Walker et al. 2008; Shkolnik et al. 2008) and HD 189733 (Shkolnik et al. 2008). These observations indicated one maximum per planetary orbit, a hot spot in the stellar chromosphere which precedes the calculated planetary passage (the subplanetary point) by 0.17 and 0.47 in phase, corresponding to a lead angle of 60° and 169°, respectively. The existence of a non-zero lead angle seems reasonable, because the field lines between the star and the planet are not be straight, but bent. Later observations of HD 179949 and υ And found the previously detected emission to be absent, which was interpreted as a sign that the emission may be sporadic and may not occur at all times (Shkolnik et al. 2008). Miller et al. (2012) found no sign of chromospheric emission from WASP-18. This is remarkable, as this is one of the most extreme objects known (a planet with 10 times the mass of Jupiter and an orbital period less than 1 day). However, the lack of observed emission can be interpreted in different ways: Miller et al. (2012) reminds that the emission may be transient, Shkonik (2013) explains the lack of emission by a possible small stellar magnetic field, whereas for Guenther and Geier (2014), the missing activity from this planet casts doubts on previous claimed detections. A more detailed review of different observations and their statistical analysis is given in Chap. 9 by Guenther and Geier (2014).

The exact mechanism responsible for the observed chromospheric heating is not determined. Two main models were suggested, none of which can currently be ruled out. In the first model, both the star and the planet are magnetized. Depending on the orientation of the magnetic fields, reconnection might occur between stellar and planetary fieldlines (Ip et al. 2004), thereby creating hot plasma. The plasma then travels along the magnetic field lines down to the footpoints, constituting an additional source of heat in the stellar chromosphere. It was quickly suggested that, if this scenario is true, these observations could constitute an observational indication for extrasolar magnetospheres (Cuntz and Shkolnik 2000).

However, as was first indicated by Saar et al. (2004), an alternative explanation is possible. In this second model, a non-magnetized planet could act as a unipolar inductor like Io around Jupiter, and still cause reconnection. While the observations seem to indicate that the source of the emission is close to the stellar surface, this would also be possible for a unipolar inductor (Saar et al. 2004). The authors argue that the observed phase shift between the chromospheric hot-spot and the sub-planetary point are difficult to explain for a unipolar inductor model. This difficulty was removed in later studies: In a more detailed version of this model, the unipolar inductor is replaced by an analytical Alfvén-wing model (Preusse et al. 2006), which allows to explain the lead angles observed by Shkolnik et al. (2003) and Shkolnik et al. (2005) using realistic stellar wind parameters. This was subsequently confirmed by numerical simulations (Kopp et al. 2011). Thus, the observation of a magnetic interaction should not be be interpreted as an indication for the presence of a planetary dipole field.

To conclude: Chromospheric emission can be induced by a close-in planet, but that planet does not have to be magnetized. For this reason, the tentative observations of the modification of chromospheric emission should not be taken as a proof of an exoplanetary magnetic field.

11.2.6 Gas Giants: Early Transit Ingress and Bow Shock Modelling

HST/COS observations of the transiting exoplanet WASP-12b show an early ingress in NUV at a wavelength corresponding to MgII resonance line cores (see also Chap. 4 (Fossati et al. 2014) and 8 (Vidotto et al. 2014)). The transit egresses, however, is compatible with the optical transit time, showing there is extra absorption by MgII during transit ingress (Fossati et al. 2010).

Almost immediately, different explanations for this asymmetry between ingress and egress have been suggested. In one of those explanations, the asymmetry was attributed to the presence of a bow shock (Vidotto et al. 2010), which arises due to the relative velocity between the stellar wind and the planet (taking into account the planetary orbital motion). Vidotto et al. (2010) assume the shock is co-located with the magnetopause, for which they determine the position using a pressure balance argument, retaining only the magnetic pressure of the star (based on a stellar dipole field) and the magnetic pressure of the planet. They then calculate the magnetopause position from the depth of the extra absorption (assuming the volume is filled by MgII), and use the measured upper limit for the stellar magnetic field to calculate the stellar magnetic pressure. From this, they estimate the planetary magnetic field. Qualitatively, the hypothesis that a bow shock could explain the observed early ingress has been validated by Monte Carlo radiative transfer simulations (Llama et al. 2011). Quantitative estimations, however, rely on the detailed plasma conditions (composition, density, temperature). The analysis if further complicated by the fact that the expected signature is time variable on the timescales of stellar magnetic field and stellar wind variations (Llama et al. 2013): Both the transit duration and the ingress time can vary when compared to optical light curves. Thus, consecutive near-UV transit light curves may vary significantly. One the one hand, the signature can therefore provide an insight into the structure and evolution of the stellar wind; on the other hand, this makes a quantitative treatment of this multi-parameter problem challenging at the least.

More problematic, however, is the fact that other explanations (without a magnetized planet) are possible, so that the observation of planetary magnetic fields using this method cannot rule out false positives. For example, Fossati et al. (2010) speculate the planet could be surrounded by an absorbing cloud which overfills the Roche lobe. The asymmetry could be due to compressed matter in front of the planet. The asymmetry is however better explained by the model of Lai et al. (2010), who suggest that the early ingress can be caused by mass transfer through a Lagrangian point towards the star.

The validity of this scenario is confirmed by numerical simulations, which have shown that a pile-up of material can generate a signature identical to the one observed (Bisikalo et al. 2013a,b). Using 3D gas dynamic simulations, they show that the overfilling of the planetary Roche lobe leads to outflow from the upper atmosphere in the direction of the L1 and L2 points. This leads to a stationary, non-spherical shape of the planetary envelope (see also Chap. 5 (Bisikalo et al. 2014) and 8 (Vidotto et al. 2014)).

Heating of the planet envelope by the bow shock can also account for the transit depth. It is unclear how such a scenario without a magnetized planet could be distinguished from the bow shock scenario described above.

To conclude: The early ingress of a transiting planet in NUV has been suggested to be the signature of a planetary magnetic field (see Chap. 8 by Vidotto et al. (2014)). Quantitative estimations, however, rely on the detailed plasma conditions (composition, density, temperature). More importantly, alternative explanations are possible, so that false positives cannot be ruled out.

11.2.7 *Gas Giants: Transit Profile and Ly-α Absorption Modelling*

Observations of the transiting exoplanet HD 209458b have shown a broad absorption signature of the stellar Lyman-α emission during transit, suggesting the presence of large amounts of atomic hydrogen. The observations have also revealed the presence of high-velocity atomic hydrogen at great distances from the planet. Initially, this has been interpreted as hydrogen atoms escaping from the planet's exosphere, possibly undergoing hydrodynamic blow-off, and being accelerated by stellar radiation pressure. Later, it has been shown by Holmström et al. (2008) that these atoms are probably formed by charge exchange, which leads to the creation of Energetic Neutral Atoms (ENAs).

As the stellar wind protons are the source of the observed energetic neutral atoms, this provides a way of probing stellar wind conditions. Also, parametric numerical simulations allow to study the influence exospheric parameters and of the location of the planetary magnetopause on the line shape. In this way Ekenbäck et al. (2010) and Lammer et al. (2011) find that the Lyman-α observations of HD 209458b before and during transit are best explained by a magnetic obstacle (magnetopause) generated by a planetary magnetic dipole moment of $\sim$40 % of Jupiter's value. While a number of different parameters influence the absorption, this method does not rely on a single data point; rather, frequency-resolved observations contain information on the behavior of particles of different velocity, allowing to break the degeneracy.

To conclude: Energetic neutral atoms might provide the explanation for Lyman-α absorption observed at HD 209458b. The amount of absorption depends on the position of the obstacle, which is determined by the stellar wind and the planetary magnetic field strength. If applied carefully, this method offers great potential. A number of parameters determine the absorption, but velocity-resolved observations may allow to break this degeneracy.

11.2.8 *Gas Giants: Radio Emission*

All magnetized planets of the Solar System emit radio waves via the cyclotron maser instability (Zarka 1998), the brightest emission being that of Jupiter. This particular mode of emission leads to radio waves with a frequency close to the local gyrofrequency. Thus, the maximum emission frequency $f^{\max}$ is given by

$$f^{\max} = \frac{eB_{\mathrm{P}}^{\max}}{2\pi m_{\mathrm{e}}}, \tag{11.1}$$

where m_{e} and e are the electron mass and charge, and $B_{\mathrm{P}}^{\max}$ is the maximum magnetic field strength close to the polar cloud tops (Farrell et al. 1998). Because the other planets of the Solar System have magnetic field strengths much lower than Jupiter, their maximum emission frequency is below the terrestrial ionospheric cutoff (~10 MHz), which makes their emission unobservable from the ground.

The same cyclotron maser instability is believed to operate at extrasolar planets. However, due to their distance (10^5 times more distant than for Solar System planets, leading to a flux lower by a factor of 10^{10}), the flux of an exo-Jupiter would be undetectable, unless a mechanism is found which can enhance the emission considerably.

Theoretical studies (see below) indicate that this might indeed be the case. As a second condition, either the planet or the star has to have a sufficiently strong magnetic field. As seen from Eq. (11.1), the maximum emission frequency can then be above the terrestrial ionospheric cutoff. Both these conditions, the radio flux and the cutoff frequency, are the subject of a number of theoretical studies, as detailed below.

11.2.8.1 Theoretical studies

In the last few years, a number of theoretical studies of exoplanetary radio emission have been undertaken. A few important results should be mentioned here.

- At least in most cases, exoplanetary radio emission is expected to exceed the emission of the planetary host star (Zarka et al. 1997; Grießmeier et al. 2005a).
- Even a purely planetary signal will be partially modulated by the stellar rotation period (Fares et al. 2010). This will complicate the discrimination between a stellar and a planetary radio signal.
- For a certain number of planets, the plasma frequency in the stellar wind is expected to be of the same order of magnitude as the maximum emission frequency. In these cases, escape of the radio emission from its source towards the observer may not be possible (Grießmeier et al. 2007b; Hess and Zarka 2011).
- For the emission to be detectable from the ground, either a magnetized planet is required, or a strongly magnetized star (Zarka et al. 2001; Zarka 2006a; Grießmeier et al. 2007b).
- The planetary magnetic moment is an ill-constrained, yet important quantity for estimating exoplanetary radio flux. Different theoretical arguments have led to two main approaches: Farrell et al. (1998) and Grießmeier et al. (2004) assume the planetary magnetic moment can be calculated by a force balance, and find a planetary magnetic field which depends on the planetary rotation rate. On the other hand, Reiners and Christensen (2010) assume the planetary magnetic moment to be primarily driven by the energy flux from the planetary core. Thus, they find no dependence on the planetary rotation rate; however, they obtain stronger magnetic fields and thus more favorable observing conditions for young planets. Planetary radio observations may be one way to discriminate between these two models.

In those cases where strong emission is possible, the expected radio flux depends on the source of available energy. Different energy sources have been suggested (a more detailed comparison of the first four can be found in Grießmeier et al. (2007b)):

- The kinetic energy flux of the solar wind protons impacting on the magnetopause (Zarka et al. 1997; Farrell et al. 1998; Zarka et al. 2001; Lazio et al. 2004; Stevens 2005; Grießmeier et al. 2005a, 2007c,b), which had already been suggested to be the source of Solar System planetary radio emission (Desch and Kaiser 1984).
- The magnetic energy flux or electromagnetic Poynting flux of the interplanetary magnetic field (Zarka et al. 2001; Zarka 2006b, 2007; Grießmeier et al. 2007b; Jardine and Cameron 2008). From the data obtained in the Solar System, it is not possible to distinguish which of these models is more appropriate, so that both models have to be considered.
- Unipolar interaction, where the star-planet system can be seen as a giant analog to the Jupiter-Io system. This emission is unlikely to be detectable, except for stars with an extremely strong magnetic field (Zarka et al. 2001; Zarka 2006b, 2007; Grießmeier et al. 2007b).
- The kinetic energy of stellar CMEs impacting the planetary magnetosphere. During periods of such CME-driven radio activity, considerably higher radio flux levels can be achieved than during quiet stellar conditions (Grießmeier et al. 2006, 2007c,b).
- The rapid rotation of a planet with strong internal plasma sources (Nichols 2011, 2012).

For all of these models, the radio emission from a number of planets is expected to be intense enough and in the right frequency range to allow for detection by modern radio telescopes (see Fig. 11.1 for an example).

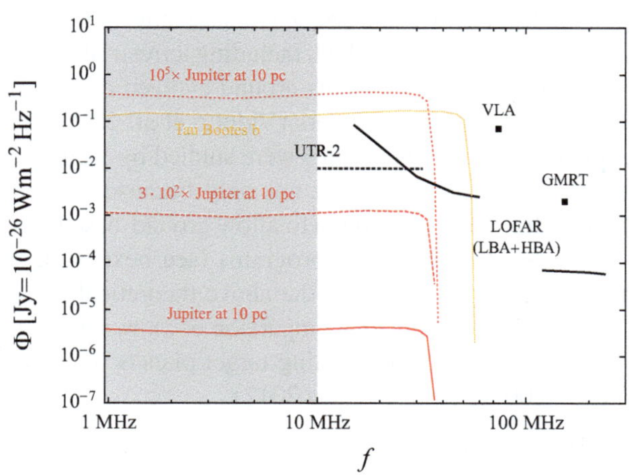

Fig. 11.1 Example for predicted radio flux, compared to the approximate detection limits of modern low-frequency radio telescopes (based on Grießmeier et al. 2011). The emission is detectable by a telescope when it is above the instrument's detection limit

Further results include:

- Most models favor close-in planets, especially hot Jupiters (Zarka et al. 2001; Zarka 2007; Grießmeier et al. 2007b,c, 2011). However, rapidly rotating planets with strong internal plasma sources can also produce radio emission at detectable levels at orbital distances of several AU from their host star (Nichols 2011, 2012).
- Calculating the acceleration of electrons, Jardine and Cameron (2008) find that for planets inside the closed magnetosphere of the host star the output power may saturate rather than increase with decreasing distance.
- Because the stellar wind parameters strongly depend on the stellar age, the expected radio flux is a function of the age of the exoplanetary host star (Stevens 2005; Grießmeier et al. 2005a). The radio flux of a planet around a young star may be orders of magnitude higher than for a planet in an older system.
- For the same reason, the uncertainty on the estimated radio flux at Earth is dominated by the uncertainty in the stellar age (Grießmeier et al. 2007c), which is usually considerable.
- Because the radio emission depends on the stellar wind parameters, it is important to take into account the fact that the stellar wind velocity varies with orbital distance (Grießmeier et al. 2007c).
- Terrestrial planets have been studied by Driscoll and Olson (2011). They conclude that an anomalously strong field (a factor of 3 larger than the most optimistic prediction) is required for emission above the Earth's ionospheric cutoff; furthermore, the expected flux levels are very low.
- Hess and Zarka (2011) recently performed simulations to study how physical information on the star-planet system can be extracted from radio observations. In particular, they show that the interaction mode (i.e. exoplanet-induced stellar emission vs. planetary radio emission) and the orbital inclination can be obtained through repeated radio observations.
- Not only planets hosted by main sequence stars are interesting targets. More exotic environments have been studied, including terrestrial planets around white dwarfs (Willes and Wu 2005), planets around evolved cool stars (Ignace et al. 2010), and planets around T Tauri stars (Vidotto et al. 2010). Interstellar rogue planets (i.e. planets not bound to a star) were studied by Vanhamäki (2011).
- The theoretical studies are important not only because they indicate that the anticipated radio flux is strong enough to allow ground-based detection, but they also serve to guide the observation programs (see next paragraph) and select the most promising targets. Some of the above theoretical considerations have served as the basis for a systematic comparison of known exoplanetary systems, which led to lists of the most interesting target planets (e.g. Lazio et al. 2004; Grießmeier et al. 2007b, 2011; Nichols 2012).

To conclude: Theoretical studies agree on the fact that the expected radio flux is close to the detection limit of current radio telescopes. The cutoff frequency would indicate the planetary magnetic field strength, allowing quantitative studies. While false negative detections cannot be ruled out, false positives are not possible, which leads to a strong motivation for observations.

11.2.8.2 Observation Attempts

In addition to those theoretical studies, a number of observation attempts were carried out. Maybe surprisingly, the first attempts at observation of exoplanetary radio emission go back at least to 1977 (Yantis et al. 1986). So far, three different search strategies have been employed:

- At the beginning, exoplanetary radio observations were necessarily unguided ones, as exoplanets had not yet been discovered, or (a few years later) not discovered in great numbers. For this reason, early observations were *targeted observations of stars without known planets*. Such observations have been performed at Clark Lake (Yantis et al. 1986), at the VLA (Winglee et al. 1986), and at UTR-2 (Zarka et al. 1997).
- For later observation campaigns, a larger number of exoplanetary systems were known, allowing *targeted observations of stars with known planets*. Such observations have been performed at UTR-2 (Zarka et al. 1997; Rayabov et al. 2004; Zarka 2011), at the VLA (Bastian et al. 2000; Farrell et al. 2003, 2004a; Lazio and Farrel 2007; Lazio et al. 2010a), with the Effelsberg radio telescope by Guenther (see Grießmeier et al. 2005b), with the Mizusawa telescope (Shiratori et al. 2006), with GMRT (Winterhalter et al. 2006; Majid et al. 2005; George and Stevens 2007; Lecavelier des Etangs et al. 2009, 2011, 2013; Hallinan et al. 2013), and with the GBT (Smith et al. 2009).
- In a few cases, exoplanets have also been searched for by *analysis of archival survey data*. Such observations have used data from the VLA Lazio et al. (2004, 2010b) and from GMRT Sirothia et al. (2014).

In a few cases, highly tentative and uncertain detections have been reported (Smith et al. 2009; Lecavelier des Etangs et al. 2013; Sirothia et al. 2014), but so far, no unambiguous detection has been achieved. The current non-detections can be explained in a number of ways (Bastian et al. 2000; Farrell et al. 2004b). For this reason, a non-detection is not easily convertible into meaningful upper limits. At the same time, the total observing time spent in the appropriate frequency range is rather limited, motivating the planned and ongoing observations such as those with the Very Large Array (VLA), Giant Metrewave Radio Telescope (GMRT), with the Ukrainian T-shaped Radio telescope-2 (UTR-2) and with LOFAR (Zarka 2011; Grießmeier et al. 2011).

To conclude: No unambiguous detection has yet been achieved. Observations are ongoing at a number of radio telescope, searching for the radio signature of an exoplanetary magnetic field.

11.3 Effects of Magnetic Fields on Terrestrial Planets

In this section, we will go through the effects though which a planetary magnetic field can influence the evolution of a terrestrial planet.

11.3.1 Terrestrial Planets: Atmospheric Escape

Similarly to the situation for gas giants (see Sect. 11.2.4), the lack of a strong magnetic moment can pose a severe threat to the atmospheres of terrestrial planets. In the case of strong atmospheric erosion, this can render otherwise perfectly habitable worlds non-habitable (see also Chap. 7 by Kislyakova et al. (2014)).

11.3.1.1 Situation in the Present-Day Solar System

A planet's magnetic field shields its atmosphere by diverting the incoming solar wind along the magnetic field lines (see also Chap. 10 by Alexeev et al. 2014). In the simplest picture, these particles are fully diverted and do not interact with the planetary atmosphere at all. In this case, a strong magnetic field strongly reduces the atmospheric loss. In a more realistic approach, the particles are diverted towards the polar regions, where part of them interact with the planetary atmosphere. The Solar System provides us with observations of Venus, Earth and Mars which allow to test how far the simple intuition holds. The comparison of such observations shows that the current oxygen loss rate from Earth is about one order of magnitude smaller than that from Venus or Mars (Seki et al. 2001), favoring indeed a magnetic shielding effect even under present-day conditions. This is at least partially due to particle recycling: The O^+ loss measured in the magnetosphere is smaller by an order of magnitude than the O^+ outflow from the polar ionosphere (Seki et al. 2001). It seems that most of the ions escaping the atmosphere are effectively recycled and returned to the system, contributing to atmospheric shielding.

The importance of a planetary magnetic field even for present-day Solar System conditions is confirmed by a recent study by Driscoll and Bercovici (2013). They introduce a new limit for escape processes. This magnetic limited escape acts in combination and competition with the well-known limits of energy-limited (or hydrodynamic) escape and diffusion-limited escape. Using this concept, Driscoll and Bercovici (2013), performed numerical simulations for a Venus-like model planet. They calculate the ion loss via Kelvin-Helmholtz instabilities operating close to the magnetopause, comparing cases with and without a strong magnetic field. The case without strong magnetic field leads to rapid loss of Venus' total water inventory (even without stellar evolution), and thus is compatible with observations. An analogous planet with an Earth-like magnetic field would have been able to retain most of its atmospheric water over timescales of billions of years.

11.3.1.2 Stellar Wind

If one looks at the early Solar System, or planets around a young star, the situation is very different when compared to the present-day Solar System, mainly for two reasons.

First, the young star has a much higher activity in the X-ray and EUV than the present-day Sun (Ribas et al. 2004), which leads to an strong atmospheric expansion (Lammer et al. 2003; Grießmeier et al. 2004; Kulikov et al. 2007). The exobase height rises, and can reach altitudes close to – or even beyond – the planetary radius (see also Chap. 7 (Kislyakova et al. 2014)).

Second, the stellar wind of a young star is much denser and faster, compressing the magnetosphere (Grießmeier et al. 2004, 2007c, 2010). Both effects combined lead to exosphere altitudes which are much closer to the magnetopause than for present-day Earth. If, as a third ingredient, the planetary magnetic field is weak, the magnetosphere may be close enough to the atmosphere to allow strong atmospheric erosion.

Grießmeier et al. (2010) explore the region in parameter space where the stellar wind and CMEs can compress the magnetosphere down to a certain altitude level. They examine in which cases the stellar winds can compress the magnetosphere down to levels of approximately 1.15 times the Earth radius, i.e. an altitude of 1,000 km above the planetary surface. This limit corresponds to the exobase altitude at a time when the solar XUV flux was 70 times higher than today, and can be considered to be a typical limit for strong atmospheric erosion (Khodachenko et al. 2007b; Lammer et al. 2007). Under these conditions, Grießmeier et al. (2010) found strong atmospheric erosion for weakly magnetized planets, especially for stellar ages ≤ 0.7 Gyr.

Using atmospheric model simulations, Kulikov et al. (2007) look at the evolution of the planets Earth, Mars and Venus during the early phases of the Solar System. They calculate the extent of atmospheric expansion, and find exobase levels close to those expected to magnetic standoff distances for weakly magnetized planets, showing that a magnetic field indeed plays an important role in protecting the atmosphere.

They also found that the atmospheric CO_2 mixing ratio plays an important role. Atmospheric CO_2 acts as a highly efficient cooling agent. During the phase of strong solar XUV heating, it reduces the exobase temperature, thus limiting the atmospheric expansion. When the exobase is close to the planetary magnetopause, this effect can thus dramatically reduce the atmospheric erosion. According to Lichtenegger et al. (2010), the terrestrial atmosphere must have benefitted from such CO_2 cooling during its early stages, as a nitrogen-rich atmosphere as the one we know today would have been quickly eroded.

To conclude: Even without stellar evolution, the presence of a magnetic field would have modified the atmospheric composition of Venus. In addition, during the early stage of the host star's evolution, a strong magnetic field seems important to protect the atmosphere of a terrestrial planet against strong erosion by the stellar wind. If the atmosphere is not protected, complete atmospheric erosion may happen, rendering the planet non-habitable.

11.3.1.3 Stellar CMEs

Similar to the effect of the stellar wind, stellar CMEs can strongly compress the planetary magnetosphere and bring it close to the expanded upper atmosphere that is expected due to strong stellar XUV heating. Stellar CMEs should be frequent during the early stages of stellar evolution, and a close-in planet such as a habitable planet around an M star may be subject to quasi-constant CME bombardment: At distances < 0.1 AU, all CMEs launched in the direction of the planet will affect it, while at 1 AU, only 20 % of the CMEs are still strong enough to have a noticable effect (Khodachenko et al. 2007b).

Using ion pick-up simulations, Lammer et al. (2007) calculate the atmospheric loss as a function of the position of the obstacle that deflects the stellar wind. In this, the obstacle can be e.g. the magnetopause of a magnetized planet or the ionopause of a non-magnetized planet (their case I). They find that, in the case of strong CMEs, weakly magnetized Earth-like exoplanets can experience high non-thermal atmospheric loss rates of the order of tens, hundreds, or even thousand of bars. As in the case of stellar wind erosion, atmospheric CO_2 mixing ratio plays an important role as a highly efficient cooling agent.

Khodachenko et al. (2007b) and Grießmeier et al. (2010) explore the region in parameter space where stellar CMEs can compress the magnetosphere down to a certain altitude level. They examine in which cases the stellar winds can compress the magnetosphere down to levels of approximately 1.15 times the Earth radius, i.e. an altitude of 1,000 km above the planetary surface. This limit can be considered to be a typical limit for strong atmospheric erosion (Khodachenko et al. 2007b; Lammer et al. 2007). For strong CMEs, Grießmeier et al. (2010) found strong atmospheric erosion for all weakly magnetized planets within the habitable zone of M stars.

To conclude: A strong magnetic also field seems important to protect the atmosphere of close-in terrestrial planets against erosion by strong stellar CMEs, which are expected to be frequent for young stars. Without magnetic protection, complete atmospheric erosion is rather likely to happen, rendering the planet non-habitable.

11.3.2 Terrestrial Planets: Protection Against Cosmic Rays

The universe is full of high energy particles, called cosmic rays. Fortunately, these particles do not reach the surface of a terrestrial planet unhindered. In total, three major shielding layers contribute to the reduction of the particle flux: The astrosphere (determined by the stellar wind particle flux and interplanetary magnetic field), the planetary magnetosphere (determined by the planetary magnetic field strength), and the planetary atmosphere (mainly determined by the atmospheric depth). For the purpose of this review, we will concentrate on the effect of the

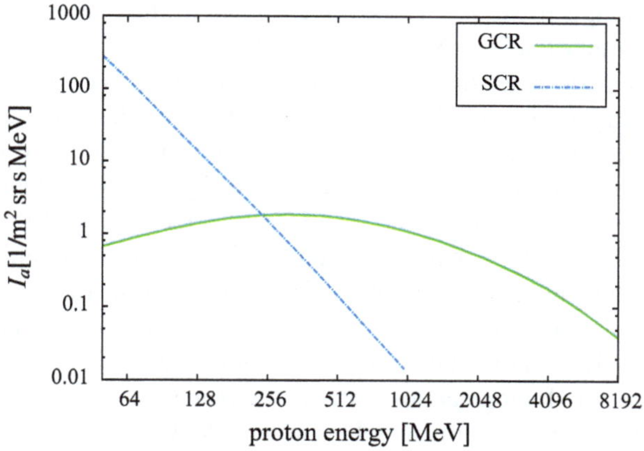

Fig. 11.2 Cosmic ray particle populations: galactic cosmic rays (*solid green line*) and average solar cosmic rays during solar maximum solar maximum (*dash-dotted blue line*)

planetary magnetic field. In particular, a weak magnetic field may result in a high particle flux to the top of the planetary atmosphere, with various implications for the planetary environment. In the following, we will look at two different sources of cosmic rays.

11.3.2.1 Galactic Cosmic Rays

Galactic cosmic rays (GCRs) are high-energy particles that are accelerated in astrophysical sources; the main contribution is believed to originate from supernova remnants (Aharonian et al. 2004; Ackermann et al. 2013). The GCRs in the Solar System represent an integrated effect from all galactic sources, and can be considered as isotropic. After passing through the astrosphere, galactic cosmic rays have a rather broad intensity peak around 0.5–1 GeV (Fig. 11.2). The propagation of galactic cosmic ray particles through exoplanetary magnetospheres has been numerically analyzed. Grießmeier et al. (2005c) have studied the dependence of magnetospheric shielding against cosmic rays on the stellar age, which is a defining parameter for the stellar wind parameters. The influence of the orbital distance has been studied by Grießmeier et al. (2009), and the influence of tidal locking via its influence on the planetary magnetic field by Grießmeier et al. (2005c) and Grießmeier et al. (2009). The effect of the planetary size and type, via the estimated magnetic field, was tentatively discussed by (Grießmeier et al. 2009).

More recently, Grießmeier et al. (2014a,b) take a slightly different approach, and systematically study the influence of the planetary magnetic field. Instead of applying a model to estimate the planetary magnetic moment, they show how magnetic protection varies as a function of the planetary magnetic dipole moment. They evaluate the efficiency of magnetospheric shielding as a function of the particle energy (in the range 16 MeV $\leq$ E $\leq$ 500 GeV) and of the planetary magnetic field strength (in the range 0 $M_\oplus \leq$ M $\leq$ 10 $M_\oplus$). These studies show that in the absence of a strong magnetic field, the flux of GCRs to the planetary atmosphere can be greatly enhanced (up to two orders of magnitude for particles energies <256 MeV).

The modification of atmospheric chemistry and the associated destruction of atmospheric biosignature or biomarker molecules by such an enhanced GCR flux was discussed by Grenfell et al. (2007) and Grießmeier et al. (2014b). In particular, they find that ozone may be depleted by up to 20 % above 40 km altitude, which is probably not sufficient to be detectable in spectroscopic observations. In Grießmeier et al. (2005c) qualitatively potential biological implications have been discussed. A more quantitative treatment of this is given in Atri et al. (2013) and Grießmeier et al. (2014b), where biological dose rates at the planetary surface are calculated as a function of the cosmic ray flux and the surface UV flux is evaluated.

They find that the surface biological dose rate may increase by a factor of two for a vanishingly small planetary magnetic field. In comparison, the shielding by planetary atmosphere was found to be considerably more efficient (the biological dose rate increases by a factor of several hundred for a thin atmosphere).

To conclude: For galactic cosmic rays, ozone is partially destroyed ($\sim$20 %), which is probably not sufficient to be detectable in spectroscopic observations. Concerning the biological dose rate, the shielding of the planetary surface is dominated by the planetary atmosphere rather than its magnetosphere.

11.3.2.2 Stellar Cosmic Rays

The second source of energetic particles is the planetary host star. Stellar cosmic rays are associated to stellar flares and CMEs, and the flux at the planet is highly anisotropic. Compared to galactic cosmic rays, they have a much steeper energy spectrum (Fig. 11.2). Thus, they dominate for energies below $\sim$200 MeV; for higher particle energies, they are usually negligible compared to the galactic cosmic ray contribution. The effect of stellar cosmic rays on extrasolar planets has been studied for the case of an unmagnetized planet. The modification of atmospheric chemistry and the destruction of atmospheric biosignature or biomarker molecules are evaluated by Segura et al. (2010) and Grenfell et al. (2012). Both find strong removal of atmospheric ozone (>90 %) in the case of a strong stellar flare. A more recent study, taking into account updated ion pair production rates and HO_x chemistry is in preparation, showing that the effect may be partially compensated by HO_x chemistry (Tabataba-Vakili et al. 2014).

To conclude: Due to their larger flux, stellar cosmic rays have a potentially much stronger impact than galactic cosmic rays. A strong stellar flare is believed to strongly modify the planetary atmosphere and to change the planetary spectrum, without however totally removing the ozone lines. The extent of this modification is currently under study.

11.3.3 Terrestrial Planets: Comet-Like Exosphere

Transit observations of close-in super-Earths have been suggested to provide information about a planetary magnetosphere (Mura et al. 2011). The idea is that neutral atoms such as Na will be blown away from the planet by the radiation pressure. Ionized atoms, however, might accumulate in the planetary magnetosphere, which has a different shape: Its direction is defined by the stellar wind velocity relative to the planet (taking into account the orbital velocity).

Based on this idea, Mura et al. (2011) use a numerical model to simulate exospheric and magnetospheric distributions of different particle populations, including neutral sodium, ionized calcium and ionized magnesium for the planet CoRoT-7b. Orbiting at just 0.017 AU, CoRoT-7b is the first super-Earth planet that was detected, and still is one of the most extreme objects currently known. Due to its close proximity and its extreme temperature, the planet should have an extended exosphere, making it a prime target for such observations. A sodium neutral atom tail is likely to form behind the planet resulting from the enormous radiation pressure acceleration. Such a tail extends from the planet in the anti-sunward direction. Exospheric Ca atoms are ionized very rapidly by stellar photons; the resulting CaII-ions are directed by the stellar wind (MgII is also a good candidate, as its abundance at the surface is likely to be high). Given the high transverse velocity of the planet the ion tail is expected to be inclined by about 45° with respect to the star-planet line or the line of sight during the transit. Thus, the combined observation of NaI and CaII would give information about the magnetosphere, and thus the magnetic field, of CoRoT-7b.

This prediction was quickly followed-up by observations: Using the UVES instrument on the VLT, Guenther et al. (2011) observed CoRoT-7b in- and out-of-transit. They searched for emission and absorption lines originating in the planetary exosphere, whith a focus on Ca I, Ca II, and Na. Although their measurements were highly sensitive, they did not find spectral lines originating from the planet. This non-detection raises the question whether our understanding of CoRoT-7b is as complete as was assumed. In any case, this remains a promising method, provided that the signal is strong enough to be detected.

To conclude:An ion-filled magnetosphere would change the transit shape at specific wavelengths. Theoretical calculations indicate that this might be a good way to detect exoplanetary magnetospheres, but the detection of the weak signal remains challenging.

Conclusion
The different methods that have been suggested for the detection of exoplanetary magnetic fields are summarized in Table 11.1. One notices that a number of methods allow for false negatives, i.e. the effect may be absent even in

(continued)

Table 11.1 Comparison of different methods that have been suggested for the diction of exoplanetary magnetic fields. Column 1: observation method. Column 2: section in which this method is described. Column 3: according to theoretical estimations, do we expect a magnetic field to generate a noticable effect? Column 4: are there alternative explanations (i.e. without a magnetic field) for a signal? Column 5: is it possible that a magnetic field exists even in the absence of a signal? Column 6: is this method suitable for the search for exoplanetary magnetic fields?

Observation	Section	Expected effect	False positives?	False negatives?	Suitable
Superflares	11.2.1	Weak or none	Yes	Yes	No
Planetary migration	11.2.2	Weak	Yes	Yes?	No
H_3^+ emission	11.2.3	Yes?	Yes	Yes	No
Gas giant mass loss	11.2.4	Yes	Yes	Yes	No
Chrom. emission	11.2.5	Yes	Yes	Yes	No
Early ingress	11.2.6	Yes	Yes	No?	No
Transit profiles (ENAs)	11.2.7	Yes	No?	No?	?
Radio emission	11.2.8	Yes	No	Yes	Yes
Atmospheric loss	11.3.1	Yes	Yes	Yes	No
Cosmic rays	11.3.2	Yes	Yes?	Yes?	No
Comet-like exosphere	11.3.3	Yes	No?	Yes	?

the presence of a planetary magnetic fields. More importantly, most methods allow for false positives, i.e. signals with could erroneously be interpreted as the signature of a planetary magnetic field. The notable exception to this is the observation of exoplanetary radio emission, for which false positives can be excluded. As an added benefit, this method goes beyond the qualitative detection, and would allow to quantify the magnetic field value. It seems that this method is still our best hope for the observation and study of exoplanetary magnetic fields.

Acknowledgements J.-M. Grießmeier acknowledges the support by the International Space Science Institute (ISSI) in Bern, Switzerland and the ISSI team *Characterizing stellar- and exoplanetary environments* and discussions with A. Reiners from the Institut für Astrophysik in Göttingen, Germany.

References

Ackermann, M. et al. (2013). *Science, 339*, 807.

Adams, F. C. (2011). *Astrophysical Journal, 730*, 27.

Aharonian, F. A. et al. (2004). *Nature, 432*, 75

Alexeev, I. I., Grygoryan, M. S., Belenkaya, E. S., Kalegaev, V. V., & Khodachenko, M. L. (2014). In H. Lammer & M. L. Khodachenko (Eds.), *Characterizing stellar and exoplanetary environments* (pp. 189). Heidelberg/New York: Springer.

Atri, D., Hariharan, B., & Grießmeier, J.-M. (2013). *Astrobiology, 13*, 910.

Bastian, T. S., Dulk, G. A., & Leblanc, Y. (2000). *Astrophysical Journal, 545*, 1058.

Bisikalo, D. V., Kaigorodov, P. V., Ionov, D. E., & Shematovich, V. I. (2013a). *Astronomy Reports, 90*(10), 779.

Bisikalo, D. V., Kaygorodov, P. V., Ionov, D. E., Shematovich, V. I., Lammer, H., & Fossati, L. (2013b). *Astrophysical Journal, 764*, 19.

Connerney, J. E. P., Acuña, M. H., Ness, N. F., & Satoh, T. (1998). *Journal of Geophysical Research, 103*(A6), 11929.

Cuntz, M., & Shkolnik, E. (2000). *Astronomische Nachrichten, 323*, 387.

Cuntz, M., Saar, S. H., & Musielak, Z. E. (2000). *Astrophysical Journal, 533*, L151.

Desch, M. D., & Kaiser, M. L. (1984). *Nature, 310*, 755.

Driscoll, P., & Olson, P. (2011). *Icarus, 213*, 12.

Driscoll, P., & Bercovici, D. (2013). *Icarus, 226*, 1447.

Ekenbäck, A., Holmström, M., Wurz, P., Grießmeier, J.-M., Lammer, H., Selsis, F., & Penz, T. (2010). *Astrophysical Journal, 709*, 670.

Erkaev, N. V., Penz, T., Lammer, H., Lichtenegger, H. I. M., Biernat, H. K., Wurz, P., Grießmeier, J.-M., & Weiss, W. W. (2005). *Astrophysical Journal Supplement, 157*, 396.

Fares, R., Donati, J. F., Moutou, C., Jardine, M. M., Grießmeier, J.-M., Zarka, P., Shkolnik, E. L., Bohlender, D., Catala, C., & Cameron, A. C. (2010). *Monthly Notices of the Royal Astronomical Society, 406*, 409.

Farrell, W. M., Desch, M. D., & Zarka, P. (1998). *Journal of Geophysical Research, 104*(E6), 14025.

Farrell, W. M., Desch, M. D., Lazio, T. J., Bastian, T., & Zarka, P. (2003). In D. Deming & S. Seager (Eds.), *Scientific frontiers in research of extrasolar planets* (ASP, Vol. 294, pp. 151). San Francisco: Astronomical Society of the Pacific.

Farrell, W. M., Lazio, T. J. W., Desch, M. D., Bastian, T. S., & Zarka, P. (2004a). In R. Norris & F. Stootman (Eds.), *Bioastronomy 2002: life among the stars* (IAU Symposium, Vol. 213, pp. 73). San Francisco: Astronomical Society of the Pacific.

Farrell, W. M., Lazio, T. J. W., Zarka, P., Bastian, T. J., Desch, M. D., & Ryabov, B. P. (2004b). *Planetary and Space Science, 52*(15), 1469.

Fossati, L., Haswell, C. A., Froning, C. S., Hebb, L., Holmes, S., Kolb, U., Helling, C., Carter, A., Wheatley, P., Cameron, A. C., Loeillet, B., Pollacco, D., Street, R., Stempels, H. C., Simpson, E., Udry, S., Joshi, Y. C., West, R. G., Skillen, I., & Wilson, D. (2010). *Astrophysical Journal, 714*, L222.

Fossati, L., Haswell, C. A., Linsky, J. L., & Kislyakova, K. G. (2014). In H. Lammer & M. L. Khodachenko (Eds.), *Characterizing stellar and exoplanetary environments* (pp. 59). Heidelberg/New York: Springer.

George, S. J., & Stevens, I. R. (2007). *Monthly Notices of the Royal Astronomical Society, 382*, 455.

Grenfell, J. L., Grießmeier, J.-M., Patzer, B., Rauer, H., Segura, A., Stadelmann, A., Stracke, B., Titz, R., & von Paris, P. (2007). *Astrobiology, 7*, 208.

Grenfell, J. L., Grießmeier, J.-M., von Paris, P., Patzer, A. B. C., Lammer, H., Stracke, B., Gebauer, S., Schreier, F., & Rauer, H. (2012). *Astrobiology, 12*, 1109.

Grießmeier, J. M., Stadelmann, A., Penz, T., H. Lammer, H., Selsis, F., Ribas, I., Guinan, E. F., Motschmann, U., Biernat, H. K., & Weiss, W. W. (2004). *Astronomy and Astrophysics, 425*, 753.

Grießmeier, J.-M., Motschmann, U., Mann, G., & Rucker, H. O. (2005a). *Astronomy and Astrophysics, 437*, 717.

Grießmeier, J.-M., Motschmann, U., Glassmeier, K. H., Mann, G., & Rucker, H. O. (2005b). In L. Arnold, F. Bouchy, & C. Moutou (Eds.), *Tenth anniversary of 51 Peg-b: Status of and prospects for hot Jupiter studies* (Vol. 259). Paris: Platypus Press.

Grießmeier, J.-M., Stadelmann, A., Motschmann, U., Belisheva, N. K., Lammer, H., & Biernat, H. K. (2005c). *Astrobiology, 5*, 587.

Grießmeier, J.-M., Motschmann, U., Khodachenko, M. L., & Rucker, H. O. (2006). In H.O. Rucker, W.S. Kurth, & G. Mann (Eds.) *Planetary radio emissions VI* (Vol. 571). Austrian Academy of Sciences Press, Vienna.

Grießmeier, J.-M., Preusse, S., Khodachenko, M. L., Motschmann, U., Mann, G., & Rucker, H. O. (2007a). *Planetary and Space Science, 55*, 618.

Grießmeier, J.-M., Zarka, P., & Spreeuw, H. (2007b). *Astronomy and Astrophysics, 475*, 359.

Grießmeier, J.-M., Preusse, S., Khodachenko, M. L., Motschmann, U., Mann, G., & Rucker, H. O. (2007c). *Planetary and Space Science, 55*, 618.

Grießmeier, J.-M., Stadelmann, A., Grenfell, J. L., Lammer, H., & Motschmann, U. (2009). *Icarus, 199*, 526.

Grießmeier, J.-M., Khodachenko, M. L., Lammer, H., Grenfell, J. L., Stadelmann, A., & Motschmann, U. (2010). In A. G. Kosovichev, A. H. Andrei, & J. P. Rozelot (Eds.), *IAU Symposium 264: Solar and stellar variability: Impact on Earth and Planets*, Rio de Janeiro p. 385–394. Cambridge: Cambridge University Press.

Grießmeier, J.-M., Zarka, P., & Girard, J. N. (2011). *Radio Science, 46*, RS0F09.

Grießmeier, J.-M., Tabataba-Vakili, F., Stadelmann, A., Grenfell, J. L., & Atri, D. (2014a). Astronomy & Astrophysics, to be submitted

Grießmeier, J.-M., Tabataba-Vakili, F., Stadelmann, A., Grenfell, J. L., & Atri, D. (2014b). Astronomy & Astrophysics, to be submitted

Guenther, E. W., Geier, S. (2014). In *characterizing stellar and exoplanetary environments* (pp. 169). Heidelberg/New York: Springer.

Guenther, E. W., Cabrera, J., Erikson, A., Fridlund, M., Lammer, H., Mura, A., Rauer, H., Schneider, J., Tulej, M., von Paris, P., & Wurz, P. (2011). *Astronomy and Astrophysics, 525*, A24.

Hallinan, G., Gopal-Krishna, S.K. Sirothia, Antonova, A., Ishwara-Chandra, C. H., Bourke, S., Doyle, J. G., Hartman, J., & Golden, A. (2013). *Astrophysical Journal, 762*, 34.

Hess, S. L. G., & Zarka, P. (2011). *Astronomy and Astrophysics, 531*, A29.

Holmström, M., Ekenbäck, A., Selsis, F., Penz, T., Lammer, H., & Wurz, P. (2008). *Nature, 451*, 970.

Holzwarth, V., & Jardine, M. (2007). *Astronomy and Astrophysics, 463*, 11.

Ignace, R., Giroux, M. L., & Luttermoser, D. G. (2010). *Monthly Notices of the Royal Astronomical Society, 402*, 2609.

Ip, W. H., Kopp, A., & Hu, J. H. (2004). *Astrophysical Journal, 602*, L53.

Jardine, M., & Cameron, A. C. (2008). *Astronomy and Astrophysics, 490*, 843.

Khodachenko, M. L., Lammer, H., Lichtenegger, H. I. M., Langmayr, D., Erkaev, N. V., Grießmeier, J.-M., Leitner, M., Penz, T., Biernat, H. K., Motschmann, U., & Rucker, H. O. (2007a). *Planetary and Space Science, 55*, 631.

Khodachenko, M. L., Ribas, I., Lammer, H., Grießmeier, J.-M., Leitner, M., Selsis, F., Eiroa, C., Hanslmeier, A., Biernat, H. K., Farrugia, C. J., & Rucker, H. O. (2007b). *Astrobiology, 7*(1), 167.

Khodachenko, M. L., Alexeev, I., Belenkaya, E., Lammer, H., Grießmeier, J.-M., Leitzinger, M., Odert, P., Zaqarashvili, T., & Rucker, H. O. (2012a). *Astrophysical Journal, 744*, 70.

Khodachenko, M. L., Alexeev, I. I., Belenkaya, E. S., & Lammer, H. (2012b). In C. Stehlé C. Joblin, & L. d'Hendecourt (Eds.) *European conference on laboratory astrophysics – ECLA*, Paris (EAS publications series, Vol. 58, pp. 233).

Kislyakova, K. G., Holmström, M., Lammer, H., & Erkaev, N. V. (2014). In H. Lammer & M. L. Khodachenko (Eds.), *Characterizing stellar and exoplanetary environments* (pp. 137). Heidelberg/New York: Springer.

Kopp, A., Schilp, S., & Preusse, S. (2011). *Astrophysical Journal, 729*, 116.

Kulikov, Yu. N., Lammer, H., Lichtenegger, H. I. M., Penz, T., Breuer, D., Spohn, T., Lundin, R., & Biernat, H. K. (2007). *Space Science Reviews, 129*, 207.

Lai, D., Helling, C., & van den Heuvel, E. P. J. (2010). *Astrophysical Journal, 721*, 923.

Lammer, H., Selsis, F., Ribas, I., Guinan, E. F., Bauer, S. J., & Weiss, W. W. (2003). *Astrophysical Journal, 598*, L121.

Lammer, H., Lichtenegger, H. I. M., Kulikov, Yu. N., Grießmeier, J.-M., Terada, N., Erkaev, N. V., Biernat, H. K., Khodachenko, M. L., Ribas, I., Penz, T., & Selsis, F. (2007). *Astrobiology, 7*, 185.

Lammer, H., Odert, P., Leitzinger, M., Khodachenko, M. L., Panchenko, M., Kulikov, Yu. N., Zhang, T. Z., Lichtenegger, H. I. M., Erkaev, N. V., Wuchterl, G., Micela, G., Penz, T., Biernat, H. K., Weingrill, J., Steller, M., Ottacher, H., Hasiba, J., & Hanslmeier, A. (2009). *Astronomy and Astrophysics, 506*, 399.

Lammer, H., Kislyakova, K. G., Holmström, M., Khodachenko, M. L., Grießmeier, J.-M., Wurz, P., Selsis, F., & Hanslmeier, A. (2011). In H.O. Rucker, W.S. Kurth, P. Louarn, & G. Fischer (Eds.), *Planetary radio emissions VII* (Vol. 303). Austrian Academy of Sciences Press, Vienna.

Lanza, A. F. (2009). *Astronomy and Astrophysics, 505*, 339.

Lazio, T. J. W., Farrell, W. M., Dietrick, J., Greenless, E., Hogan, E., Jones, C., & Hennig, L. A. (2004). *Astrophysical Journal, 612*, 511.

Lazio, T. J. W., & Farrell, W. M. (2007). *Astrophysical Journal, 668*, 1182.

Lazio, T. J. W., Shankland, P. D., Farrell, W. M., & Blank, D. L. (2010a). *Astronomical Journal, 140*, 1929.

Lazio, T. J. W., Carmichael, S., Clark, J., Elkins, E., Gudmundsen, P., Mott, Z., Szwajkowski, M., & Hennig, L. A. (2010b). *Astronomical Journal, 139*, 96.

Lecavelier des Etangs, A., Gopal-Krishna, S. K. Sirothia, & Zarka, P. (2009). *Astronomy and Astrophysics, 500*, L51.

Lecavelier des Etangs, A., Gopal-Krishna, S. K. Sirothia, & Zarka, P. (2011). *Astronomy and Astrophysics, 533*, A50.

Lecavelier des Etangs, A., Gopal-Krishna, S.K. Sirothia, Zarka, P. (2013). *Astronomy and Astrophysics, 552*, A65.

Lichtenegger, H. I. M., Lammer, H., Grießmeier, J.-M., Kulikov, Yu. N., von Paris, P., Hausleitner, W. Krauss, S., & Rauer, H. (2010). *Icarus, 210*, 1.

Linsky, J. L., & Güdel, M. (2014). In H. Lammer & M. L. Khodachenko (Eds.), *Characterizing stellar and exoplanetary environments* (pp. 3). Heidelberg/New York: Springer.

Llama, J., Wood, K., Jardine, M., Vidotto, A. A., Helling, C., Fossati, L., & Haswell C. A. (2011). *Monthly Notices of the Royal Astronomical Society, 416*, L41.

Llama, J., Vidotto, A. A., Jardine, M., Wood, K., Fares, R., & Gombosi, T. I. (2013). *Monthly Notices of the Royal Astronomical Society, 436*, 2179.

Lovelace, R. V. E., Romanova, M. M., & Barnard, A. W. (2008). *Monthly Notices of the Royal Astronomical Society, 389*, 1233.

Maehara, H., Shibayama, T., Notsu, S., Notsu, Y., Nagao, T., Kusaba, S., Honda, S., Nogami, D., & Shibata, K. (2012). *Nature, 485*, 478.

Majid, W., Winterhalter, D., Chandra, I., Kuiper, T., Lazio, J., Naudet, C., & Zarka, P. (2005). In H.O. Rucker, W.S. Kurth, G. Mann (Eds.), *Planetary radio emissions VI* (Vol. 589). Austrian Academy of Sciences Press, Vienna.

Miller, B. P., Gallo, E., Wright, J. T., & Dupree, A. K. (2012). *Astrophysical Journal, 754*, 137.

Mura, A., Wurz, P., Schneider, J., Lammer, H., Grießmeier, J.-M., Khodachenko, M. L., Weingrill, J., Guenther, E., Cabrera, J., Erikson, A., Fridlund, M., Milillo, A., Rauer, H., & von Paris, P. (2011). *Icarus, 211*, 1.

Nichols, J. D. (2011). *Monthly Notices of the Royal Astronomical Society, 414*, 2125.

Nichols, J. D. (2012). *Monthly Notices of the Royal Astronomical Society, 427*, L75.

Pillitteri, I., Wolk, S. J., Cohen, O., Kashyap, V., Knutson, H., Lisse, C. M., & Henry, G. W. (2010). *Astrophysical Journal, 722*, 1216.

Preusse, S., Kopp, A., Büchner, J., & Motschmann, U. (2006). *Astronomy and Astrophysics, 460*, 317.

Reiners, A., & Christensen, U. R. (2010). *Astronomy and Astrophysics, 522*, A13.

Ribas, I., Guinan, E. F., Güdel, M., & Audard, M. (2004). *Astrophysical Journal, 622*, 680.

Rubenstein, E. P., & Schaefer, B. E. (2000). *Astrophysical Journal, 529*, 1031.

Ryabov, V. B., Zarka, P., & Ryabov, B. P. (2004). *Planetary and Space Science, 52*(15), 1479.

Saar, S. H., & Cuntz, M. (2001). *Monthly Notices of the Royal Astronomical Society, 325*, 55.

Saar, S. H., Cuntz, M., & Shkolnik, E. (2004). In A. K. Dupree & A. O. Benz (Eds.), *Stars as suns: Activity, evolution, and planets* (IAU Symposium, Vol. 219, pp. 355).
Schaefer, B. E., King, J. R., & Deliyannis, C. P. (2000). *Astrophysical Journal, 529*, 1026.
Segura, A., Walkowicz, L. M., Meadows, V., Kasting, J., & Hawley, S. (2010). *Astrobiology, 10*, 751.
Seki, K., Elphic, R. C., Hirahara, M., Terasawa, T., & Mukai, T. (2001). *Science, 291*, 1939.
Shibayama, T., Maehara, H., Notsu, S., Notsu, Y., Nagao, T., Honda, S., Ishii, T. T., Nogami, D., & Shibata, K. (2013). *Astrophysical Journal Suppliment Series, 209*, 5.
Shiratori, Y., Yokoo, H., Sasao, T., Kameya, O., Iwadate, K., & Asari, K. (2006). In L. Arnold, F. Bouchy, & C. Moutou (Eds.), *Tenth anniversary of 51 Peg-b : Status of and prospects for hot Jupiter studies* (Vol. 290). Platypus Press: Paris.
Shkolnik, E., Walker, G. A. H., & Bohlender, D. A. (2003). *Astrophysical Journal, 597*, 1092.
Shkolnik, E., Walker, G. A. H., & Bohlender, D. A. (2004). *Astrophysical Journal, 609*, 1197.
Shkolnik, E., Walker, G. A. H., Bohlender, D. A., Gu, P. G., & Kürster, M. (2005). *Astrophysical Journal, 622*, 1075.
Shkolnik, E., Gaidos, E., & Moskovitz, N. (2006). *Astronomical Journal, 132*, 1267.
Shkolnik, E., Bohlender, D., Walker, G. A. H., & Collier Cameron A. (2008). *Astrophysical Journal, 676*, 628.
Shkolnik, E. L. (2013). *Astrophysical Journal, 766*, 9.
Sirothia, S. K., Lecavelier des Etangs, A., Gopal-Krishna, S. K. Sirothia, Kantharia, N. G., & Ishwar-Chandra, C. H. (2014). *Astronomy and Astrophysics, 562*, A108.
Smith, A. M. S., Collier Cameron, A., Greaves, J., Jardine, M., Langston, G., & Backer, D. (2009). *Monthly Notices of the Royal Astronomical Society, 395*, 335.
Stevens, I. R. (2005). *Monthly Notices of the Royal Astronomical Society, 356*, 1053.
Stevenson, D. J. (2003). *Earth and Planetary Science Letters, 208*, 1.
Tabataba-Vakili, F. Grenfell, J. L. Grießmeier, J.-M. & Rauer, H. (2014). Astronomy & Astrophysics, to be submitted
Vanhamäki, H. (2011). *Planetary and Space Science, 59*, 862.
Vidotto, A. A., Opher, M., Jatenco-Pereira, V., & Gombosi, T. I. (2009). *Astrophysical Journal, 703*, 1734.
Vidotto, A. A., Opher, M., Jatenco-Pereira, V., & Gombosi, T. I. (2010). *Astrophysical Journal, 720*, 1262.
Vidotto, A. A., Bisikalo, D. V., Fossati, L., & Llama, J. (2014). In H. Lammer & M. L. Khodachenko (Eds.), *Characterizing stellar and exoplanetary environments* (pp. 153). Heidelberg/ New York: Springer.
Walker, G. A. H., B. Croll, B., Matthews, J. M., Kuschnig, R., Huber, D., Weiss, W. W., Shkolnik, E. Rucinski, S. M., Guenther, D. B., Moffat, A. F. J., & Sasselov, D. (2008). *Astronomy and Astrophysics, 482*, 691.
Weber, E. J., & Davis, Jr. L. (1967). *Astrophysical Journal, 148*, 217.
Willes, A. J., & Wu, K. (2005). *Astronomy and Astrophysics, 432*, 1091.
Winglee, R. M., Dulk, G. A., & Bastian, T. S. (1986). *Astrophysical Journal, 209*, L59.
Winterhalter, D., Kuiper, T., Majid, W., Chandra, I., Lazio, J., Zarka, P., Naudet, C., Bryden, G., Gonzales, W., & Treumann, R. (2006). In H. O. Rucker, W. S. Kurth, & G. Mann (Eds.), *Planetary radio emissions VI* (Vol. 595). Austrian Academy of Sciences Press, Vienna.
Wood, B. E. (2006). *Space Science Reviews, 126*, 3.
Wood, B. E. (2007). *Living Review Solar Physics, 1*, 2.
Wood, B. E., & Linsky, J. L. (2010). *Astrophysical Journal, 717*, 1279.
Wood, B. E., Müller, H. R., Zank, G. P., & Linsky, J. L. (2002). *Astrophysical Journal, 574*, 412.
Wood, B. E., Redfield, S., Linsky, J. L., Müller, H. R., & Zank, G. P. (2005a). *Astrophysical Journal Supplement, 159*, 118.
Wood, B. E., Müller, H. R., Zank, G. P., Linsky, J. L., & Redfield, S. (2005b). *Astrophysical Journal, 628*, L143.
Yantis, W. F., Sullivan III., W. T., & Erickson, W. C. (1977). *Bulletin of the American Astronomical Society, 9*, 453.

Zarka, P. (1998). *Journal of Geophysical Research, 103*(E9), 20159.
Zarka, P. (2006a). *Planetary and Space Science, 55*, 598.
Zarka, P. (2006b). In H.O. Rucker, W. S. Kurth, & G. Mann (Eds.), *Planetary radio emissions VI* (Vol. 543). Austrian Academy of Sciences Press, Vienna.
Zarka, P. (2007). *Planetary and Space Science, 55*, 598.
Zarka, P. (2011). In H.O. Rucker, W.S. Kurth, P. Louarn, & G. Fischer (Eds.), *Planetary radio emissions VII* (Vol. 287). Austrian Academy of Sciences Press, Vienna.
Zarka, P., Queinnec, J., Ryabov, B. P., Ryabov, V. B., Shevchenko, V. A., Arkhipov, A. V., Rucker, H. O., Denis, L., Gerbault, A., Dierich, P., & Rosolen, C. (1997). In H.O. Rucker, S.J. Bauer & A. Lecacheux (Eds.), *Planetary radio emissions IV* (Vol. 101). Austrian Academy of Sciences Press, Vienna.
Zarka, P., Treumann, R. A., Ryabov, B. P., & Ryabov, V. B. (2001). *Astrophysics and Space Science, 277*, 293.

Chapter 12
Alfvén Radius: A Key Parameter for Astrophysical Magnetospheres

Elena S. Belenkaya, Maxim L. Khodachenko, and Igor I. Alexeev

Abstract The Alfvén radius is a distance where the magnetic energy density is equal to the kinetic energy density, or bulk velocity equals the Alfvén velocity. In this paper we discuss a role of Alfvén radius for different types of magnetospheres and magnetosphere-disk systems. Among the astrophysical disks considered here are the magnetic disks surrounding the outer planets in the Solar System (Jupiter and Saturn) and exoplanets, the heliospheric current sheet, accretion disks of neutron stars, pulsars, millisecond X-ray pulsars, white dwarfs and black holes, disks in the X-ray binaries, disks of young stars and active galactic nuclei (AGNs). We note that mainly in the magnetosphere-disk system, the inner edge of astrophysical disk (independently of its origin, direction of motion and material in it) in the presence of a strong magnetic field is located close to the Alfvén radius. For magnetized planets a concept of Alfvén radius is important as for the interaction with the solar/stellar wind, either for the inter-magnetospheric processes.

12.1 Introduction: Alfvén Radius and Astrophysical Magnetic Environments

The astrophysical disks are rather complex and quite diverse. They are associated with a large variety of objects, such as the outer planets in the Solar System (Jupiter and Saturn), exoplanets, Sun and young stars, pulsars, millisecond X-ray pulsars and X-ray binaries, white dwarfs, neutron stars, black holes and active galaxy nuclei. Here we pay attention to one common feature of all these disks: in the presence of

E.S. Belenkaya (✉) • I.I. Alexeev
Skobeltsyn Institute of Nuclear Physics (MSU SINP), Lomonosov Moscow State University, 1(2), Leninskie gory, GSP-1, Moscow 119234, Russian Federation
e-mail: elena@dec1.sinp.msu.ru; alexeev@dec1.sinp.msu.ru

M.L. Khodachenko
Skobeltsyn Institute of Nuclear Physics (MSU SINP), Lomonosov Moscow State University, 1(2), Leninskie gory, GSP-1, Moscow 119234, Russian Federation

Austrian Academy of Sciences, Space Research Institute, Graz, A-8042, Austria
e-mail: maxim.khodachenko@oeaw.ac.at

H. Lammer, M. Khodachenko (eds.), *Characterizing Stellar and Exoplanetary Environments*, Astrophysics and Space Science Library 411,
DOI 10.1007/978-3-319-09749-7_12

a strong magnetic field the location of the inner edge of the disks is close to the Alfvén radius (Belenkaya et al. 2011, 2012; Belenkaya and Khodachenko 2012).

The location of the inner edge of astrophysical disk plays a significant role for a lot of physical processes. The accretion disk luminosity depends on it; it controls the jets position and determines the maximum temperature in the disk. This parameter influences on the location and the wave diapason of emissions. Field-aligned currents in the heliospheric current sheet and in the Jovian magnetodisk are concentrated near the inner boundaries of these current-carrying disks. It is suggested that the inner edge of a disk surrounding black hole is close to the last stable orbit (Shakura and Sunyaev 1973). That's why a determination of the disk's inner edge position is very significant for a large range of tasks (Belenkaya et al. 2011, 2012; Belenkaya and Khodachenko 2012).

Planets and compact objects are surrounded by magnetospheres (Vasyliunas 1979; Bagenal 1992; Belenkaya 2009; Beskin 2010). If an object possesses intrinsic magnetic field, the magnetosphere is formed around it (see also Chap. 11, Grießmeier (2014)), and its size is determined by the pressure balance (Vasyliunas 1979). It was noted by Vasyliunas (1979), that inside the magnetosphere of a magnetized compact object plasma flows reverse to a stellar wind. Istomin and Komberg (2002) wrote that when the shock encounter the magnetic field of a white dwarf or a neutron star, a magnetotail can be formed stretched along the shock propagation. The authors showed that in the magnetospheres of a compact object the Alfvén radius (the distance where magnetic and plasma pressures are equal) determines the main characteristic parameters of magnetosphere.

If a compact object is surrounded by a disk, the disk is connected with the magnetospheric boundary being out of the magnetosphere (see e.g., Fu and Lai 2012). If a planet possesses disk, the disk is placed inside the planetary magnetosphere. In this paper we consider some common properties of the magnetospheres in the presence of a strong magnetic field and emphasize a role of Alfvén radius and Alfvén velocity.

12.2 "Non-local Alfvén Radius" in Magnetized Planetary Magnetospheres

Alfvén velocity, V_A, may be interpreted as the magnetic energy density per particle (Burton et al. 1970). $V_A = B(\mu_0\rho)^{-1/2}$, where B is magnetic field strength, ρ – plasma density, and $\mu_0 = 4\pi \cdot 10^{-7}$ H m^{-1} is the permeability of the vacuum. Alfvén velocity is the velocity of propagation of Alfvén waves along magnetic field. Depending on local parameters Alfvén velocity significantly varies inside the planet magnetospheres. Thus, Burton et al. (1970) noted that the measured Alfvén velocity in the Earth's magnetosphere ranges from 4,800 km s^{-1} outside the plasmapause to 490 km s^{-1} at the equatorial region inside the plasmasphere.

The terrestrial magnetosphere is formed in the solar wind flow, and its boundary (magnetopause) separates the outer and inner parts of the space preventing penetration of fields and matter (however, partially such penetration takes place). At the Earth's magnetosphere, the characteristic scale determining the size of the magnetosphere is the distance to the subsolar magnetopause point, R_1. In a rough approximation in this point the dynamic pressure of the solar wind, p_{sw}, equals to the magnetic pressure of the terrestrial magnetosphere, ${B_m}^2/2\mu_0$.

$$p_{sw} = B_m^2/2\mu_0, \tag{12.1}$$

where B_m is z component of the magnetospheric magnetic field just near the nose magnetopause. This field includes mainly the Earth's dipole field and the field of its shielding currents at the magnetopause (Chapman-Ferraro currents). The pressure $p_{sw} = k\rho_{sw}{V_{sw}}^2$, where ρ_{sw} and V_{sw} are the solar wind density and velocity, respectively; the coefficient k is 0.88 for a gasdynamic model. It should be mentioned that for $k = 2$, $p_{sw} = 2\rho_{sw}{V_{sw}}^2$ presents the pressure for the perfectly elastic collisions at the magnetopause, while $k = 1$ corresponds to the particles sticking to the magnetopause. It was noted by Shue et al. (2011) that the "reduction factor" k of the dynamic pressure, determines the real pressure sustaining the magnetospheric pressure at the noon terrestrial magnetopause. For the average values of the solar wind parameters (5 cm^{-3} and 400 km s^{-1}), the solar wind dynamic pressure $\rho_{sw}{V_{sw}}^2$ is ~1.3 nPa, and the subsolar magnetopause distance $R_1 \sim 11\ R_E$ (where R_E is the Earth's radius).

From the pressure balance conditions we obtain that $V_{sw} = (B_m^2/k\mu_0\rho_{sw})^{1/2}$, or it is

$$V_{sw} = k^{-1/2}B_m(\mu_0\rho_{sw})^{-1/2} \approx 1.1B_m(\mu_0\rho_{sw})^{-1/2} \tag{12.2}$$

for $k = 0.88$. The solar wind velocity estimated with Eq. (12.2) almost coincides (within the accuracy of $k^{-1/2}$) with formula for Alfvén speed in which magnetic field is measured just inside of the noon magnetopause, but plasma density and velocity are taken for the undisturbed solar wind in the vicinity of the Earth. We can call velocity determined in such way as a non-local Alfvén velocity (V_{nl-A}):

$$V_{nl-A} = B_m(\mu_0\rho_{sw})^{-1/2}. \tag{12.3}$$

The same is true for the Mercury's magnetosphere. Slavin et al. (2009) use $k = 0.88$ in the subsolar pressure equilibrium equation at the noon magnetopause. The outer pressure was considered by the authors to be equal to a dynamic solar wind pressure, while the inner pressure was magnetic pressure for a dipolar magnetosphere (indeed the Chapman-Ferraro currents also give contribution to the magnetospheric field). The distance from the planetary center to the subsolar magnetopause point at Mercury is ~1.4 R_M, where R_M is Mercury's radius. Thus for Mercury $V_{sw} = 1.1V_{nl-A}$, where V_{nl-A} according Eq. (12.3), connects the plasma density in the undisturbed solar wind at the Mercury orbit with the magnetospheric magnetic field at the dayside magnetopause (see Eq. (12.3)).

For the magnetospheres of Jupiter and Saturn the magnetopause subsolar point where the solar wind flow stops, is also located at the place where the internal magnetospheric pressure equals the external pressure of the interplanetary medium. As in the cases of Earth and Mercury, the solar wind pressure is mainly equal to its dynamic pressure ($p_{sw} = k\rho_{sw}{V_{sw}}^2$) at the outer planet orbits, however, the magnetospheric pressure just near the noon magnetopause includes not only the magnetic pressure ${B_m}^2/2\mu_0$ (note that for Jupiter and Saturn the magnetodisk and ring current give significant contribution in it), but also the pressure of the magnetospheric plasma (p_{pl}). This is especially significant for Jupiter where both of these pressures are of the same order of magnitude (see Slavin et al. 1985, and references therein). For this reason, the pressure balance at the noon magnetopause could be rewritten as

$$p_{sw} = p_{pl} + B_m^2/2\mu_0. \tag{12.4}$$

If we introduce $\beta = p_{pl}/({B_m}^2/2\mu_0)$, Eq. (12.4) will look as:

$$p_{sw} = (\beta + 1)B_m^2/2\mu_0.. \tag{12.5}$$

Krimigis et al. (1979) based on the Voyager 2 data reported that the outer magnetosphere of Jupiter up to $\sim 30\ R_J$ (where R_J is the Jupiter's radius), is populated by hot plasma ions of hydrogen, oxygen, and sulfur. Magnetic field of the Jovian magnetosphere at the noon magnetopause includes the planetary dipole field, the field of Chapman-Ferraro currents, the field of the magnetodisk surrounding Jupiter and of its shielding currents at the magnetopause (Belenkaya 2004; Alexeev and Belenkaya 2005). The magnetic moment of the magnetodisk field exceeds the dipole moment of Jupiter by ~ 2.6 (as estimated by comparison of their contributions at the subsolar magnetopause) (Belenkaya 2004, 2009). Thus, the magnetospheric magnetic field for Jupiter is determined mainly by its magnetodisk created due to the rapid rotation of the planet and the internal magnetospheric sources of plasma (it is supposed that Io gives the main contribution to plasma population). Based on the Pioneer and Voyager data Slavin et al. (1985) found the mean magnetopause nose distance for Jupiter $\sim$68 R_J.

As at the noon Jovian magnetopause $\beta \sim 1$, it follows from Eq. (12.5) that $p_{sw} = B_m^2/\mu_0$, or $k\rho_{sw}V_{sw}^2 = B_m^2/\mu_0$, and $V_{sw} = B_m/(k\rho_{sw}\mu_0)^{1/2}$. For Jupiter and Saturn Slavin et al. (1985) found that $k = 1.16$. Thus, from Eq. (12.3) it follows that at the subsolar Jovian magnetopause $V_{sw} \approx 0.9\ V_{nl\text{-}A}$. This means that for the Jupiter the solar wind velocity almost equals to the "non-local Alfvén velocity".

At Saturn, there are also the internal sources of the magnetospheric plasma. The main one is Enceladus, but the atmospheres of Saturn and Titan, as well as icy moons and rings also give their contributions. However plasma β just inside of the subsolar magnetopause is $\sim$0.1 or less (Slavin et al. 1985, and references therein). Therefore, these authors assumed that the magnetospheric pressure at the noon magnetopause is mainly determined by the magnetic field and that the mean value of the stand-off magnetopause distance for Saturn is $\sim$19 R_S, where R_S is the Saturn's radius (so, the largest satellite Titan in the Saturn system located at the distance ~20.3 R_S sometimes leaves the Saturn's magnetosphere). For Saturn, the pressure balance at the noon magnetopause can be written in the form of Eq. (12.4), which using the estimation $\beta < 10^{-1}$ turns to Eq. (12.1), where $p_{sw} = k\rho_{sw}V_{sw}^2$ with $k = 1.16$. Thus, $1.16\,\rho_{sw}V_{sw}^2 = B_m^2/2\mu_0$ or $V_{sw} = B_m/(2.32\mu_0\rho_{sw})^{1/2} = 0.7V_{nl\text{-}A}$.

Venus does not have magnetic field and the intrinsic field of Mars is very small, so these planets possess induced magnetospheres. In the flow passing by the unmagnetized body, electric currents are closed in the induced magnetosphere through the ionosphere, or the body's surface. At the noon boundary of the induced magnetosphere, a pressure balance is between the dynamic pressure outside and the thermal plasma pressure inside. Venus has a bow shock upstream the magnetopause, but the magnetospheric magnetic field exists only in a form of disturbances of the solar wind magnetic field. Mars also has induced magnetosphere with a bow shock before it and the solar wind interacts with Mars ionosphere as well.

Thus, at the magnetopause subsolar point, within the accuracy of the coefficient of the order of unity, the magnetospheric magnetic field is connected with the velocity and density measured in the undisturbed solar wind flow upstream the planetary orbit by Eq. (12.3) for non-local Alfvén velocity. Therefore according Eq. (12.2), at this point the solar wind velocity is about the "non-local Alfvén velocity", and by analogy with the Alfvén radius the stand-off distance of the planetary magnetosphere can be named the "non-local Alfvén radius".

The planets exist not only around Sun, but also around other stars (exoplanets). The magnetized exoplanets possess magnetospheres. At the substellar magnetopause the pressure balance takes place between the stellar wind pressure and the exoplanet's magnetospheric pressure, both of which include the magnetic and plasma pressures (see Khodachenko et al. (2012), and references therein). The magnetospheric magnetic pressure is $B_m{}^2/2\mu_0$, and the magnetospheric plasma pressure is $\rho_{msph}V^2{}_{msphth}$, where ρ_{msph} is the magnetospheric plasma density, $V_{msphth} = (k_B T_{msph}/m_{msph})^{1/2}$ is the thermal magnetospheric velocity, $k_B = 1.38 \cdot 10^{-23}$ kg m^2 s^{-2} is the Boltzmann constant, T_{msph} and m_{msph} are the temperature and average mass per particle in the magnetosphere. The stellar wind pressure includes the dynamic pressure $k\rho_{stw}V_{stw}{}^2$, where ρ_{stw} is the stellar wind density and V_{stw} is the relative velocity between stellar wind and exoplanet which includes the orbital velocity of a planet also (Khodachenko et al. 2012). The magnetic pressure of the interplanetary magnetic field (IMF) is $B_{stw}{}^2/2\mu_0$, where B_{stw} is the stellar wind magnetic field. Thus, the pressure balance condition at the substellar point looks as the following:

$$k\rho_{stw}V_{stw}^2 + B_{stw}^2/2\mu_0 = B_m^2/2\mu_0 + \rho_{msph}V_{msphth}^2/2. \tag{12.6}$$

So, in a case when we ignore the IMF and magnetospheric plasma pressure for exoplanet, we receive: $k\rho_{stw}V_{stw}{}^2 = B_m{}^2/2\mu_0$, or again $V_{stw} = (B_m{}^2/2\mu_0 k\rho_{stw})^{1/2} = (1/2k)^{1/2}V_{nl-A}$ (see Eq. (12.3)). Therefore, in this approximation the substellar distance of the exoplanet magnetopause can be considered as a "non-local Alfvén radius", where the relative velocity and plasma density at the planetary orbit in the undisturbed stellar wind are described by formula for Alfvén velocity with the magnetic field just inside the stand-off magnetopause.

12.3 Alfvén Radius in the Magnetized Planet Magnetospheres Including Disks

Belenkaya et al. (2012) and Belenkaya and Khodachenko (2012) described the current-carrying disks location in the magnetospheres of Jupiter and Saturn in the terms of Alfvén radius. These disks are formed due to the action of fast planetary rotation and the presence of internal sources of the magnetospheric plasma. As the main motion in the inner parts of these disks is in the azimuthal direction (co-rotation effect), the azimuthal magnetospheric plasma speed was taken for calculation of the bulk velocity in these regions. It was noted by Hill et al. (1974) and Hill (1979) that magnetic field is stretched in a disk-like geometry beyond the Alfvén radius.

It was shown that inside the magnetospheres of gas-giant planets the velocity of azimuthal flow is equal to the Alfvén in the inner edges of disks (in the case of Jupiter, disk is called a magnetodisk, while in the case of Saturn, it is called a ring-current). Comparison with observations and calculations in the paraboloid magnetospheric magnetic field model constructed for these planets (Alexeev and Belenkaya 2005; Alexeev et al. 2006) gave the distances to the inner disk edges 18.4 R_J and 6.5 R_S, for Jupiter and Saturn, respectively. It was also observed in-situ by spacecraft that the Jovian magnetic field lines are stretched in the typical for a disk form beyond $\sim$20 R_J, where plasma β exceeds unity. For Saturn the plasma outflow was registered by Pioneer 11 at the distances from the planetary centre $\sim$6–8 R_S (Frank et al. 1980), where plasma β approaches unity. Magnetic disk causes there the reconfiguration of the field lines structure. So, beyond the Alfvén radius determined by the azimuthal velocity, the current-carrying disks arise in the Jovian and Kronian magnetospheres.

The same was assumed by Khodachenko et al. (2012) for the magnetized exoplanets located at distances <0.5 AU from their host stars and having the masses of the order of the Jupiter's one. In this case the magnetospheric plasma source was assumed to be connected with the thermal escape of the upper planetary atmosphere material heated and ionized by stellar radiation. The inserting of the current-carrying disk in the exoplanet's magnetosphere allowed the authors to explain the larger size of hot Jupiter magnetospheres (see also Chap. 10, Alexeev et al. (2014)), stronger magnetospheric magnetic field, and better protection of the exoplanetary atmosphere from the errosion by the stellar wind. At the Alfvén radius the azimuthal magnetospheric plasma velocity approaches the local Alfvén velocity, and field lines become to be stretched in a disk-like form.

12.4 Alfvén Radius in the Magnetospheres of Magnetized Stars

The region around a magnetized star with the dominated presence of stellar magnetic field could be considered as its magnetosphere. A rapidly rotating magnetosphere model of helium-strong stars in which the stellar wind plasma on the nearly rigid field lines accumulates at the places with minimum gravitational plus centrifugal potential was described by Townsend (2008). As a result, the magnetosphere with a warped disk is developed. A formation of a disk around hot stars (young white dwarfs or young neutron stars) due to the high light pressure dominated over the gravitation has been considered by Bespalov and Zheleznyakov (1990). They called such disks radiative diskons. Similar to the description of Townsend (2008), plasma accumulates in the mostly distant parts of magnetic field lines, creating an equatorial disk. In radiative diskons plasma moves slowly to the disk periphery.

Young stars are surrounded by protoplanetary disks. Slow rotating classical T Tauri stars usually are considered to have accretion disks (protoplanetary disks). The boundary between the magnetosphere of T Tauri star and surrounding accretion disk coincides with the Alfvén radius. At this place the high-speed jets arise (Pudritz et al. 2006). It was also noted by Abubekerov and Lipunov (2003) that the magnetosphere of the T Tauri star is bounded by the Alfvén radius.

The heliospheric current sheet with the total current about $3 \cdot 10^9$ A is located in the heliosphere. Alfvén considered the Sun as a unipolar generator producing this current. Current in the distant disk is closed by the field-aligning currents coming to the Sun's high-latitude atmosphere, while in the inner edge it is closed by the field-aligned currents going from the auroral solar regions to the disk. It was stated by Zhao and Hoeksema (2010) that the inner edge of the heliosphere is the boundary between the sub-Alfvénic coronal expansion and the super-Alfvénic solar wind. Based on the STEREO A and B data in 2008 they found that this boundary was at 14 R_{Sun}, where R_{Sun} is the solar radius. Beyond the Alfvén radius the radial component prevails in the ambient magnetic field (which is typical for the disk), while below it the magnetic field has a multipolar structure. Thus, the coronal multipolar magnetic field exists up to the Alfvén radius bounded the solar magnetosphere, beyond which the current-carrying heliospheric sheet is located.

12.5 Alfvén Radius in the Magnetospheres of Compact Objects in the Presence of a Strong Magnetic Field

Astrophysical compact objects are white dwarfs, neutron stars, black holes, and supermassive black holes (the most compact objects known). Compact objects can be isolated or existing in binary systems. They are extremely dense remnants of star evolution. In their vicinity the high-energy X-ray and γ-ray radiation, high-frequency oscillations, and relativistic jets arise. Near black holes the Einstein's General Theory of Relativity should be applied.

After explosion of a supernova, the envelope is thrown away, while a core of the star collapses to a compact object which could be a neutron star, if its mass is between 1.4 and 3 M_{Sun}, where $M_{Sun} = 2 \cdot 10^{30}$ kg is a solar mass. Some of neutron stars have accretion disks, while the others are accreting directly from a stellar wind. A disk can be formed if its specific angular momentum l is bigger than the Keplerian one at the neutron star surface, $l_K = (GM^* R^*)^{1/2}$, where M^* is the mass of the star and R^* is its radius, G is Newton's gravitational constant ($G = 6.673 \cdot 10^{-11}\,\mathrm{m^3\,kg^{-1}\,s^{-2}}$).

The location of the disk's inner edge depends, in particular, on the presence of magnetic field in the neutron star. For a non-magnetized neutron star, the disk can reach its surface. For the neutron star with a strong magnetic field, Ghosh and Lamb (1979a,b) found that the inner boundary of the disk is close to the place where the integrated magnetic stress acting on the disk is comparable to the integrated material stress associated with the inward radial drift and orbital motion of plasma. Thus Ghosh and Lamb (1979a,b), considered the interaction of the disk's plasma with the dipole magnetic field of a neutron star, found that the inner edge of disk roughly coincides with the Alfvén radius. This statement was supported by Cheng et al. (1993) who studied the neutron star–accretion disk system. The location of the inner edge, R_0, of an accretion disk surrounding a magnetized object, as a rule, is expressed as $R_0 = \xi R_A$, where R_A is the Alfvén radius for the spherical accretion and parameter ξ depends on the fraction of the stellar magnetic flux threading the disk; usually ξ is taken to be 0.5 (Frank et al. 1992). The location of the inner edge of accretion disk depends also on the accretion rate: when it increases, the disk approaches the star.

The mode of interaction of neutron star with surrounding disk is determined by the location of the inner edge of a disk relative to other main characteristic radii: the corotation radius, R_c, and the light cylinder radius, R_L. The corotation radius is defined as $R_c = (GM^*/\Omega^{*2})^{1/3}$, where the angular velocities of the disk and the star (Ω^*) are the same. The light cylinder radius is defined as $R_L = c/\Omega^*$, where c is the speed of light. There are three basic modes of interaction of a neutron star with the surrounding disk: accretor, propeller, and ejector (Shvartsman 1970a,b; Illarionov and Sunyaev 1975). If the inner radius of the disk is beyond the corotation radius, but smaller than the light cylinder radius, the system is in the propeller stage. The ejector stage takes place when the inner radius is beyond both the corotation radius and the light cylinder radius. For the accretor mode, the inner radius of the disk is smaller than the light cylinder radius, and accretion to the central object occurs when the corotation radius becomes larger than Alfvén radius, it stops where the object's magnetic pressure becomes equal to the kinetic flow pressure (Shvartsman 1970a,b) or at the Alfvén radius.

In propeller state the falling material penetrates down to the Alfvén radius, due to the action of the centrifugal barrier. For ejectors the relativistic outflowing momentum flux is larger than the ram pressure of the surrounding material. Possenti et al. (1998) analyzing the change of state of a neutron star from ejector to accretor, found that the plasma flowing to neutron star is blown away by the radiation pressure of the central object until the surrounding plasma's ram pressure starts exceeding the neutron star momentum flux, after that plasma flows to it, as to accretor. This process is accompanied by the spinning down of the central object. It was concluded by Shvartsman (1970a,b) that a fast-rotating neutron star at first appears as an ejector (radio pulsar), then in a course of slowing down it evolves to the propeller stage, and finally, as a slowly rotating neutron star, it becomes accretor. Accretion is connected with a loss of the angular momentum of plasma (Beskin 2010).

Pulsar is a small, extremely dense, and rapidly rotating (the typical period 0.1–5 s) neutron star. It emits narrow beams of radiation along its magnetic dipole axis which does not coincide with the spin axis. Pulsar has a very strong dipole-type magnetic field (10^{11}–10^{15} G). In particular, the effect of X-ray pulsar may be obtained during accretion to a rapidly spinning neutron star with the magnetic dipole axis inclined relative the rotation axis. Radio-pulsars (mainly, single neutron stars) are the mostly numerous class of pulsars. Millisecond pulsars (a half of which are in binaries) have the spin period 1–10 ms and magnetic field 10^8–10^9 G. It was mentioned by Zhang and Dai (2010) that the magnetized disk is thermally unstable before the Alfvén radius (where the ram pressure of the accreted matter equals the magnetic pressure), and becomes stable beyond it. The size of the magnetosphere of the accreting star is of the order of the Alfvén radius (Abubekerov and Lipunov 2003).

Black hole (with mass $>3\ M_{\mathrm{Sun}}$) is formed as a result of collapse of a core of massive stars. A rapidly rotating disk is formed around black hole. In spite of a fact that black hole has no own magnetic field, the surrounding plasma (including accretion disk) can generate it. For Schwarzschild black holes Shakura and Sunyaev (1973) assumed that the inner edge of the disk is at 3 R_{G}, since no stable circular orbits are possible near the black hole at $R < 3\ R_{\mathrm{G}}$. Here $R_{\mathrm{G}} = 2\mathrm{GM}^*/c^2$ is the gravitation Schwarzchild radius. It was suggested by Fendt and Greiner (2001) and Pudritz et al. (2006) that at the distance equal to the Alfvén radius the disk is destroyed and plasma acceleration takes place with jet ejection. For bright AGNs the inner edges of disks are also close to the Alfvén radii.

Conclusion

If the relative speed between a planet and a wind from the host star is sub-magnetosonic and sub-Alfvénic, an Alfvénic-wing-type magnetosphere arises. For the super-Alfvénic relative speed, a magnetosphere with a bow shock, a magnetopause, and a tail is formed. The interaction of the solar or stellar wind with the magnetized planet has specific features dependent on the concrete planet, but here we summarize the mutual property of such coupling. It occurs that the characteristic scale of magnetosphere (the stand-off magnetopause distance) could be considered as a length of the order of

(continued)

"non-local Alfvén radius", at which the magnetospheric magnetic field close to the noon magnetopause is connected with the solar/stellar wind velocity and density in the undisturbed upstream flow at the planetary orbit by the expression for Alfvén velocity.

If the magnetized planet possesses a disk inside the magnetosphere, its inner edge is located at the Alfvén radius determined by the azimuthal (co-rotation) velocity of the magnetospheric plasma.

For the magnetized neutron stars and pulsars the magnetospheric boundary coincides with the Alfvén radius. If they are surrounded by accretion disks, the disk's inner edge is located at this distance also. The inner edges of disks associated with the X-ray and radio pulsars, white dwarfs, and black holes in the presence of a strong magnetic field are placed close to their Alfvén radii. This effect is caused by the fact that the strong magnetic field near the central body guides the plasma in the inner regions of the disk independently of the material of which the disk is built and the direction of material motion in the disk. Thus, at the place where magnetic field controls plasma, disk disappears. When the magnetic energy density becomes equal to the kinetic energy density, the disk formes or disrupt, depending on the prevailing material motion. This happens at the Alfvén radius.

Acknowledgements The authors acknowledge the support by the International Space Science Institute (ISSI) in Bern, Switzerland and the ISSI team *Characterizing stellar- and exoplanetary environments*. This work was carried out at the Institute of Nuclear Physics, Moscow State University and was supported by the RFBR Grants 11-05-00894-a, 12-05-00219-a, and 12-02-92600-KO-a, by the EU FP7 projects EUROPLANET/JRA3 and IMPEX (no. 262863). The authors acknowledge also the support by the Austrian Research Foundation FWF NFN project S116 'Pathways to Habitability: From Disks to Active Stars, Planets and Life', and the related FWF NFN subproject, S116 606-N16 'Magnetospheric Electrodynamics of Exoplanets'. This work was also supported by the FWF project P21197-N16.

References

Abubekerov, M. K., & Lipunov, V. M. (2003). *Astronomy Reports, 47*, 679.

Alexeev, I. I, & Belenkaya, E. S. (2005). *Annales Geophysicae, 23/3*, 809.

Alexeev, I. I., Kalegaev, V. V., Belenkaya, E. S., Bobrovnikov, S. Z., Bunce, E.J., Cowley, S. W. H., & Nichols, J. D. (2006). *Geophysical Research Letters, 33*, L08101.

Alexeev, I. I., Grygoryan, M. S., Belenkaya, E. S., Kalegaev, V. V., & Khodachenko, M. L. (2014). In H. Lammer & M. L. Khodachenko (Eds.), *Characterizing stellar and exoplanetary environments* (pp. 189). Heidelberg/New York: Springer.

Bagenal, F. (1992). *Annual Review of Earth and Planetary Sciences, 20*, 289.

Belenkaya, E. S. (2004). *Planetary and Space Science, 52*, 499.

Belenkaya, E. S. (2009). *Physics Uspekhi, 52/8*, 765.

Belenkaya, E. S., Alexeev, I. I., & Khodachenko, M. L. (2011). EPSC-DPS Joint Meeting 2011 6, EPSC-DPS2011-4-1

Belenkaya, E. S., Alexeev, I. I., & Khodachenko, M. L. (2012). *Astrophysics and Space Science Proceedings, 33*, 217.

Belenkaya, E. S., & Khodachenko, M. L. (2012). *International Journal of Astronomy and Astrophysics, 2*, 81.

Beskin, V. S. (2010). Astron. Astrophys. Library (Springer, Berlin/Heidelberg) doi:10.1007/978-3-642-01290-7

Bespalov, P. A., & Zhelyaznyakov, V. V. (1990). *Pis'ma v Astronomicheskii Zhurnal, 16*, 1030.

Burton, R. K., Russell, C. R., & Chappell, C. R. (1970). *Journal of Geophysical Research, 75/28*, 5582.

Cheng, K. S., Yu, K. N., & Ding, K. Y. (1993). *Astronomy and Astrophysics, 275*, 53.

Fendt, C., & Greiner, J. (2001). *Astronomy and Astrophysics, 369*, 308.

Frank, L. A., Burek, B. G., Ackerson, K. L., Wolfe, J. H., & Mihalov, J. D. (1980). *Journal of Geophysical Research, 85*, 5695.

Frank, J., King, A. R., & Raine, D. (1992). Accretion Power in Astrophysics, 2nd edition. Cambridge University Press, Cambridge.

Fu, W., & Lai, D. (2012). *Monthly Notices of the Royal Astronomical Society, 423*, 831.

Ghosh, P., & Lamb, F. K. (1979a). *Astrophysics Journal, 232*, 259.

Ghosh, P., & Lamb, F. K. (1979b). *Astrophysics Journal, 234*, 296.

Grießmeier, J.-M. (2014). In H. Lammer & M. L. Khodachenko (Eds.), *Characterizing stellar and exoplanetary environments* (pp. 213). Heidelberg/New York: Springer.

Illarionov, A. F., & Sunyaev, R. A. (1975). *Astronomy and Astrophysics, 39*, 185.

Istomin, Ya. N., & Komberg, B. V. (2002). *Astronomy Reports, 46*, 908.

Hill, T. W. (1979). *Journal of Geophysical Research, 84*, 6554.

Hill, T. W., Dessler, A. J., & Michel, F. C. (1974). *Geophysical Research Letters, 1*, 3.

Khodachenko, M. L., Alexeev, I. I., Belenkaya, E. S., Lammer, H., Grießmeier, J.-M., Leitzinger, M., Odert, P., Zaqarashvili, T. V., & Rucker, H. O. (2012). *Astrophysics Journal, 744*, 70.

Krimigis, S. M., Armstrong, T. P., Axford, W. I., Bostrom, C. O., Fan, C. Y., Gloeckler, G., Lanzerotti, L. J., Keath, E. P., Zwickl, R. D., Carbary, J. F., & Hamilton, D. C. (1979). *Science, 206*, 977.

Possenti, A., Colpi, M., D'Amico, N., & Burderi, L. (1998). *Astrophysics Journal, 497*, L97.

Pudritz, R. E., Ouyed, R., Fendt C., & Brenburg, A. (2006). Protostars Planets V, University of Arizona Press, Tucson, 277

Shakura, N. I., & Sunyaev, R. A. (1973). *Astronomy and Astrophysics, 24*, 337.

Shue, J.-H., Chen, Y.-S., Hsieh, W.-C., Nowada, M., Lee, B. S., Song, P., Russell, C. T., Angelopoulos, V., Glassmeier, K. H., McFadden, J. P., & Larson, D. (2011). *Journal of Geophysical Research, 116*, A02203.

Shvartsman, V. F. (1970a). Radiofizika. Izvestiya Vyzshich Uchebnych Zavedeniy, 12, 1852.

Shvartsman, V. F. (1970b). *Astronomicheskii Zhurnal, 47/3*, 660.

Slavin, J. A., Smith, E. J., Spreiter, J. R., & Stahara, S. S. (1985). *Journal of Geophysical Research, 90*, 6275.

Slavin, J. A., Erson, B. J., Zurbuchen, T. H., Baker, D. N., Krimigis, S. M., Acuna, M. H., Benna, M., Boardsen, S. A., Gloeckler, G., Gold, R. E., Ho, G. C., Korth, H., McNutt Jr., R. L., Raines, J. M., Sarantos, M., Schriver, D., Solomon, S. C., & Travnicek, P. (2009). *Geophysical Research Letters, 36*, L02101.

Townsend, R. H. D. (2008). *Monthly Notices of the Royal Astronomical Society, 389*, 559.

Vasyliunas, V. M. (1979). *Space Science Reviews, 24*, 603.

Zhang, D., & Dai, Z. G. (2010). *Astrophysics Journal, 718*, 841.

Zhao, X. P., & Hoeksema, J. T. (2010). *Solar Physics, 266*, 379.

Part IV
Space and Ground-Based Exoplanet Observation and Characterization Tools

Various space agencies recognize that the question with regard to life beyond Earth in general, and the associated issue of the existence and study of exoplanets in particular, of one of the greatest importance. The characterization of exoplanetary systems, their evolution, their atmospheres, their magnetospheric and plasma environment and finally the detection of Earth-like planets orbiting nearby stars, and their ability to host life forms will be one of the great challenges for astrophysics in the coming decades. In the following articles the authors describe the most relevant space- and ground-based projects/missions that are currently planed and build during the next 10 years. It is expected that the space and ground-based telescopes which are discussed in the following chapter, will enhance our knowledge how numerous planets such as Earth in the Galactic neighborhood are. Because observations in the UV spectrum are of crucial relevance for the characterization of upper planetary atmospheres, their plasma and magnetic environment as well as non-hydrostatic behavior and related atmospheric escape an article takes a closer look to UV astronomy's HST-successor, the WSO-UV space observatory.

Chapter 13
Living with Stars: Future Space-Based Exoplanet Search and Characterization Missions

Malcolm Fridlund, Heike Rauer, and Anders Erikson

Abstract The study of exoplanets have now been ongoing for more than 20 years. After a first phase of essentially discovery, a remarkable diversity has been found among the more than 1,700 objects that have been detected. Planetary bodies orbiting other stars have begun to be characterized physically, but the next stage of exploration, given an increase in sensitivities and resolutions will provide for data that can be used for comparative planetology in the real sense of the term, i.e. a direct comparison, both between objects outside the Solar System, as well as with objects within our home system. In this chapter we describe briefly all the space assets that are either in orbit, intended to be launched within the near future and also those that are decided for the intermediate time perspective.

13.1 Introduction and Background

Carl Sagan once presented a snapshot of how he saw the process that begins with the Big bang and ends with the present status, where mankind have evolved on a small rocky planet in the outskirts of the Milky Way Galaxy and begun to ask questions about itself. He expressed it as "This is what Hydrogen atoms do – given 15 billion years of evolution!"

This process contains indeed some of the most enduring questions and issues pursued by mankind. This can be formulated as

- How do we understand ourselves?
- Where do we come from?
- Where are we going?

M. Fridlund (✉)
Institute for Planetary Research, DLR, Rutherfordstrasse, 124 89, Berlin, Germany

Leiden Observatory, Leiden, The Netherlands
e-mail: fridlund@strw.leidenuniv.nl

H. Rauer • A. Erikson
Institute for Planetary Research, DLR, Rutherfordstrasse, 124 89, Berlin, Germany
e-mail: heike.rauer@dlr.de; anders.erikson@dlr.de

H. Lammer, M. Khodachenko (eds.), *Characterizing Stellar and Exoplanetary Environments*, Astrophysics and Space Science Library 411,
DOI 10.1007/978-3-319-09749-7_13

- What are the conditions under which life arises, evolves and interact within the Universe?

In order to be able to address these philosophical issues and make any progress in understanding ourselves we have to formulate research questions that can be answered by experiments. Such a set of questions could be

- Is the Earth unique or has life also developed elsewhere?
- What makes a planet habitable?
- How do planetary systems form and evolve?

Here we have of course assumed that life only can form and evolve on or near planetary surfaces. The great English scientist Fred Hoyle wrote a novel[1] in the 1960s where he assumed that life also could form in molecular interstellar clouds and that actually this was the most common form of life in the Universe while life on the Earth was the exception. We chose, however, to disregard such ideas here and assume that the surfaces of rocky planets, or their immediate surroundings (atmospheres, oceanic depths, etc) are the abodes of life.

So our list above can be made more detailed by asking another set of questions

- Did the Earth form in a special place and/or under extraordinary circumstances?
- How diverse are planets and planetary systems?
- What are the general characteristics of terrestrial planets in the habitable zones of different types of stars?
- How do stars and planets evolve as systems? What is the impact of the star on the planet and vice versa.

To understand the lofty questions formulated above, this is of course not enough. As Carl Sagan did in the 1980s we would really need to go back to the Big Bang, and formulate an understanding of the fundamental natural laws of the Universe. We would need to understand the way our own Solar System works, since it is the only system we can study in detail and in situ. And, to begin with, we need to understand how gas and dust form stars and planets, how common planets are in the Universe, and how life arose and evolved in the Solar System. All of these issues are being pursued by modern astrophysics and space research, but one of the most dramatic breakthroughs that have been made since the days of Sagan is the actual discovery of exoplanets in the last 20 years. As this is being written, the announcement that the number of worlds known to orbit stars other than the Sun is rapidly reaching 2,000,[2] (Marcy et al. 2014).

To continue to develop our understanding, and to place our own Solar System in context we need to carry out *Comparative Planetology* across interstellar distances. This will require us to define observable parameters that can be used for such a comparison, as well as observe a large enough sample to be statistically significant.

[1]The Black Cloud, William Heinemann Ltd, 1957, ISBN 0-451-11432-9.

[2]http://exoplanets.eu

This can then be used to make suitable comparisons, both with exo-systems and with relevant Solar System objects where we in addition to observations also can utilize in-situ measurements.

But we can not be satisfied with only learning about planets. The example of the Solar System teaches us that we are indeed living with a star, as expressed by the American space administration, NASA, some years ago. The understanding of the star – planet connection is extremely important in a number of areas, e.g.

- How do the formation process impact on both star and planet(s)?
- How does the stellar evolution impact the planetary evolution – and vice versa?
- How does the star and its evolution impact the habitability of a planet, as well as life itself?

In order to carry out the observations and analysis mentioned above, we need to find and study many exoplanets of all kinds there are. Particularly we are now looking for small, Earth- or Super-Earth size objects. For all such planets we need to determine as many physical parameters as possible but as a minimum, we need the masses, $M_{\rm p}$, the radii, $R_{\rm p}$ and the ages of the individual exoplanets, with a high precision, and, as said, for a large sample. It is going to be demonstrated in the next section, that in order to do so we are also going to need the same parameters with equally high precision, for the host stars. We therefore need to sample a large number of different types of stars of different ages.

13.1.1 What Do We Currently Know About the Physics of Exoplanets?

There are several methods through which we can learn about exoplanets and they will be briefly described in this section with emphasis on their application on space based platforms.

After being the object of the speculation of scientists and philosophers for millennia, the first confirmed detection of a planetary body outside our own Solar System came in 1992. At this time, it was however, not so surprising that a detection was made, as much as where it was made. It was not made by studying nearby solar-type stars like e.g. α Centauri. Indeed, the first several terrestrial-mass planets (confirmed) have turned out to orbit the Pulsar PSR B1257+12. They are therefore more likely to be the result of the debris of a stellar supernova explosion accreting into planetary bodies, and not the product of the actual star formation process (where material left over from the formation of the star itself out of the interstellar medium form into planets). These detections were achieved through timing measurements of the pulsar itself, detecting the minuscule changes introduced into the timing of the radio pulses, by the gravitational influence of the planet on the neutron star – something possible because of the inherent stability of the pulsar signals (Wolszczan and Frail 1992).

It was not so surprising that a discovery was made, even though it took place around objects where no planets were expected, and using methods that had not been suggested in the context of exoplanets. This was because the search for exoplanets had been ramping up over the last half dozen decades or so. Already during the nineteenth century, detections of exoplanets were claimed and during the middle of the twentieth century the number of reported candidates were increasing. Essentially all of these detections were made, using the astrometric technique, where the gravitational deflection caused by an orbiting planet on a star, as it moves across the sky, was measured. Although none of these detections were ever confirmed (instead being due to systematics not being fully understood), it was clear that the availability of techniques, and the precision of these were approaching the level of sophistication where detections of extra-solar systems were a clear possibility – as long as such systems existed in significant numbers.

Measuring the astrometric deflections is difficult, demanding μarcsecond precision and stability in the observed data. The most obvious of planet detection techniques two decades ago, were instead the Radial Velocity deflection (RV, the stellar velocity component along the line-of-sight) of an exoplanet, detectable because the planet and its host star both orbit around the center-of-mass of the system (in the Solar System, the deflection of the center-of-mass from the geometrical center of the Sun is caused mainly by Jupiter, and depending on the orbital position of the other planets, the center-of-mass is located either inside the Sun or just outside). This movement around the center-of-mass cause a change in the RV of the star, that is detectable by spectrographs on Earth through the Doppler effect, and proportional to the stellar AND planetary masses. This detection of exoplanets was first proposed by Struve in 1952, when he also suggested (based on the observations of close binary stars) that the first planets to be detected would be Jupiter-mass (or larger) objects orbiting very close to their host stars. (At the time of Struve's prediction in 1952, the change in RV caused in the Sun by Jupiter, 12 m/s, could not be detected, but more massive planets orbiting very close to the host star would indeed be detectable).

In 1988 the Canadian astronomers Campbell, Walker, and Yang published a paper reporting the radial velocity observations of a few dozens of solar type stars and claiming the indication of Jupiter mass companions in 7 of these stars (Campbell et al. 1988). Most astronomers, however, ignored this "indication" since all the detections were marginal. Of the seven candidates, one, the planet orbiting γ Cephei, has been confirmed in 2003.

It did not make a large impact either, when Dave Latham and co-workers published a paper entitled "The unseen companion of HD 114762 – A probable brown dwarf" in May 1989 (Latham et al. 1989). An object with 11 Jupiter masses were orbiting a solar type star every 84 days. In the abstract, Latham writes: "This leads to the suggestion that the companion is probably a brown dwarf star, and may even be a giant planet".

At the time, astronomers made a distinction purely based on theoretical considerations that an object with a mass larger than 13 solar masses is a star of the type Brown Dwarf since it presumably generates energy from some kind of nuclear reactions. A smaller object would be a planet, since it would not generate any significant energy internally. We note here that the radial velocity method does only provide a minimum mass of any orbiting body, since by itself it does not tell us what the inclination of the system is and therefore how much of the true space velocity of the planet is along the line of sight. It turns out that Latham's object HD 114762b really has a small inclination and thus a mass of 11 $M_{Jupiter}$. HD 114762b, can thus, together with γ Cephei b, be considered the first exoplanets detected.

The first discovery of a bona fide planet, recognized immediately as such, was reported in the fall of 1995 (Mayor and Queloz 1995). It concerned the planet 51 Pegasi b. It turned out that 51 Peg b, was a Jupiter-mass object orbiting with a period of only a few days and thus causing a deflection easily detectable in 1995. The next objects, 70 Vir b and 47 Uma b (Marcy and Butler 1996; Butler and Marcy 1996) were of the same type. At the time it was somewhat surprising since it was mostly supposed that exo-systems would all resemble our own, based mainly on the philosophical principle that there should not be anything special with our system. It is however true that the RV-method is heavily biased towards detecting massive planets orbiting very close to the host star, and it was assumed at the time that eventually the systems resembling our own system would be picked up eventually.

The RV method has been very successful in detecting more and more planets of lower and lower mass since 1995. The original 1 σ detection limit of about 50–200 m/s, 50 years ago, is now <0.7 m/s. Literally hundreds of smaller planets have been detected and the lower limit to the mass registered is of order one or a few Earth-masses.

In that same little paper in 1952 (Struve 1952), noted that there would be eclipses or rather transits. If the orbital plane happens to intersect the line-of-sight to the star, the planet will occasionally transit (part of) the stellar surface. The random orientation of the orbit of a close in planet would be such that in more than 5 % of the cases, the planet would regularly transit its star as seen from the Earth resulting in a drop of the stellar light intensity of about 2 % – something easily detectable with the new (in 1952) photoelectric detectors.

Today, after discovering the first transiting objects in 2000 (Brown et al. 2001) literally hundreds and maybe thousands (see section below about the CoRoT and Kepler space missions) of these objects are known. We now know that the probability of a chance alignment varies between about 0.5 % for a $1R_{\oplus}$ planet at 1 AU from a solar type star to several tens of percent for gaseous giant planets that are orbiting very close to red dwarf stars. By observing the light output of the star when the planet is transiting the stellar surface, the shape and depth of the light curve can tell us a lot about the size and other physical parameters of the planet (and of the star). Specifically, the transit constrains the inclination of the exoplanetary orbit so well that the true mass of the planet can be found from the radial velocity curve. By observing both the occultation of part of the light of the star by the planet, as well as the associated radial velocity curve, exact measurements of the planetary mass and diameter allow a determination of the body's average density and thus its mineralogy. Since this method – known most often as the transit method – requires a chance alignment to occur, any systematic investigation calls for the simultaneous observations of a large number of stars during an as long as possible time with an as high as possible duty cycle of the observations.

Struve thus, in 1952 (Struve 1952), in principle, set out the modern methods of searching for Exo-planets and today when reading his short paper one wonders why it would take until the years 1995 (RV detection) and 2000 (transit detection) respectively, before the first results according to these suggestions were reported? The difficulty with the detectors (mainly that the observations had to be detected on photographic plates in the case of RV or that the transit observations had very little chance of a detection when the stars are observed one and one, together with the emerging belief that all Solar Systems more or less *HAD* to look like our own, is probably the answer.

Today we know almost two thousand planets (and many more candidates) orbiting stars other than our own Sun. Many methods have been used to find them and a large fraction have been discovered from space. Almost all of these latter have been found using the NASA's Kepler space telescope (the majority) and the French space agency-European CoRoT mission (see below) – both of which utilize the transit method. The drawback with this method is that only a very small percentage of all objects transits in front of their host stars as seen from the Earth (or its immediate vicinity). The larger fraction of exoplanets is, however, still found with the radial velocity method which, when brought to bear on bright enough host stars also can find planets that are not transiting. This method is currently exclusively used from the ground because of the large, stable and precise spectrographs required.

While the transit method provides the radius of the exoplanet, and the radial velocity method produces the mass of the planet, both parameters are expressed in terms of those of the host star. And this is the crux of the matter. The precision of the key physical parameters required in order to begin to study exoplanets in a detail that would be enough to draw conclusions on both their sizes, their geology, and their evolutionary status is governed by the precision we can achieve with the host star parameters. And stellar physics have problems in this respect. Masses and radii are generally not known with the required precision. In order to determine stellar radii and masses we are clearly not achieving the required precision ($\approx \pm 1\,\%$ in radii, and $\approx \pm 5\,\%$ in mass – see Figs. 13.2 and 13.3)

So what have we learnt from the discovery and observations of the close to 2,000 exoplanetary objects? The most fundamental fact discerned from the ground- and space-based observations made hitherto, is the enormous diversity of planets. This is especially so since before the results of the observations began to appear in earnest in 1995, it was expected more or less that all exoplanetary systems would resemble the Solar System closely. This turned out not to be so, and we can now say that albeit there are individual similarities between either individual objects or systems (but hardly both), so far, no system resembling our own has been detected. The most similar system so far may be Kepler 90 which contains 7 planets, with 2 Earth-size planets in 7 and 9 day orbits, 3 super-Earths ($\approx$2–3 R_{Earth}) in 60, 90 and 125 day orbits, a Saturn-size planet in a 210 day orbit and a Jupiter at 1 AU in a 331 day orbit (Cabrera et al. 2014).

What we have also found is that there are planets that appear to be broadly resembling our own Earth, although for most objects for which we have determined either a reliable diameter, or a reliable mass, we are lacking the other parameter.

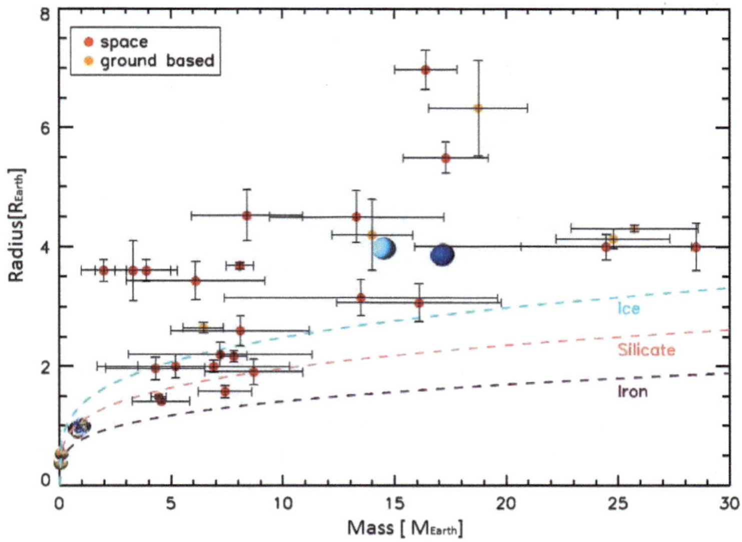

Fig. 13.1 Mass-radius diagram for planets with different bulk compositions. Water ice (*blue line*), silicate rock (*orange line*), iron (*purple line*) (see Wagner et al. 2011, for details) are compared to known low-mass planets (with 1 sigma error bars) (After Rauer et al. 2013)

And even in the best of cases, we are hampered by the lack of precision in the stellar parameters who give us the individual planetary ones. This becomes very clear by studying Fig. 13.1. This figure displays the current knowledge about planets smaller than about 8 R_{Earth} and 30 M_{Earth} for which both diameter and mass have been measured. Note that the error bars plotted in this figure are 1 σ.

And as far as evolutionary studies are concerned the situation turns real bad. For the main sequence solar type stars we are mostly concerned with in these cases, the errors in the evolutionary status of the host star may be as large as one to several giga years.

So why do we need both masses radii and ages with high precision in order to expand on hitherto achieved results? It can be seen from the Figs. 13.2 and 13.3, that in order to discern the composition and internal structure of rocky planets we require a much larger precision than what we have so far been able to achieve. Such an increase in precision will also be required in order to properly interpret eventual spectroscopic observations. As an example, one can mention that without knowing the structure of the planet (including its average density) it will be impossible to interpret a detection of O_2 or O_3 as biomarkers, since it will not be possible to discern where on the planet the oxygen molecules are located – at the surface, or at the top of the atmosphere where it could be abiotically produced. As will be seen below, this calls for a completely new technique, namely the one selected for ESA's new PLATO 2.0 mission (see Sect. 13.3). An increase in the precision of

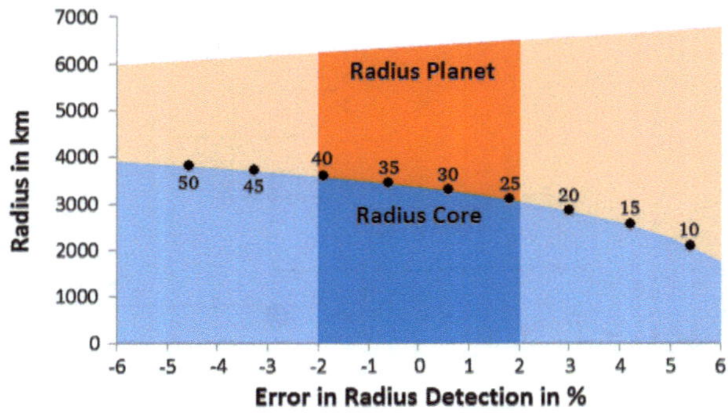

Fig. 13.2 Radius of planet and its core depending on the uncertainty in radius R_{planet} We assume a planet of 1 M_{Earth} and vary the radius (0 corresponds to 1 R_{Earth} within current uncertainties. The *shaded region* demonstrate the expected improvement with the PLATO 2.0 mission (±2 % in radius at the fainter end of the host star magnitude range representing a worst case scenario) (After Rauer et al. 2013)

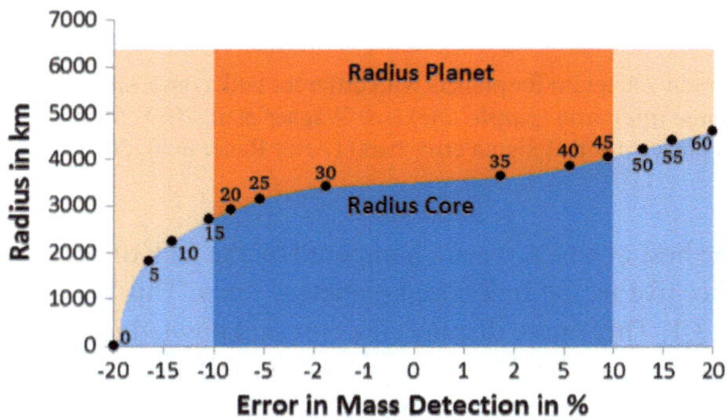

Fig. 13.3 Radius of planet and its core depending on the uncertainty in planetary mass M_{planet}, but keeping the radius fixed at 1 R_{Earth} and varying the planet mass within current uncertainties (±20 %). Numbers at *black dots* provide the core mass fraction as percentage of total mass. The *dark shaded* regions illustrate the expected PLATO 2.0 accuracy (±2 and ±10 % in radius and mass, respectively – again a worst case scenario with the faintest host stars) (After Rauer et al. 2013)

the observed radii, masses and age of exoplanetary host stars would, however, also benefit stellar evolution theory.

13.2 Current and Near Future Observations from Space

In this section, we describe recent, and future space missions with significant exoplanetary objectives.

13.2.1 *The Hubble Space Telescope*

The 2.6 m orbital Hubble Space Telescope (HST) of NASA has carried out a number of important investigations regarding exoplanets over the years, including spectroscopy. To the HST goes the credit for making the first detection of an exoplanet from space. Today, with ≈2,000 known exoplanets one hardly remembers that for the first few years after the detection of 51 Peg b and 47 Uma b, there remained some doubt that these planets existed. It was suggested by authors like Gray and Hatzes (1997) that the radial velocity detection was caused by starspots and other activity on the stellar surface and one was essentially observing the stellar rotation or a harmonic thereof. It was the detection of Brown et al. (2001) and Charbonneau et al. (2000) with the HST and ground based equipment, of the transit of the RV planet HD 209458b, that once and for all dispelled the notion that astronomers were observing something other than exoplanets in these cases. Albeit in configurations very different from our own Solar System.

HST has carried out many observations of the transmission spectra of transiting giant exoplanets, reporting the detection of e.g. Magnesium, Sodium and water (Nikolov et al. 2014; Vidal-Madjar et al. 2013; Huitson et al. 2013). A very exciting application is the use of the fine guidance sensors (FGS) of the HST in order to map out the astrometric deflection caused by exoplanets such as e.g. υ And c and d (McArthur et al. 2010) and in other systems (including upper limits), demonstrating also the possibility for later astrometric missions such as Gaia and possibly the Nearby Earth Astrometric Telescope (NEAT) (ESA).

Hubble's high resolution and spectroscopical capability has also been valuable concerning the study of gas and dust disks and jets from young stellar objects.

13.2.1.1 MOST

The Microvariability and Oscillations of STars telescope, (MOST), launched by Canada in 2003 is a small spacecraft (53 kg) equipped with a 150 mm aperture telescope that feeds two frame-transfer CCDs, one for tracking and the other for science. MOST was launched on 2003 June 30 into a low-Earth, Sun-synchronous, polar orbit allowing stars between $-19°$ and $+36°$ declination to be viewed continuously for up to 60 days. It is the first spacecraft dedicated to the study of asteroseismology, but also intended to search for transits of known exoplanets. As such MOST have made some crucial observations confirming the detection made previously with radial velocity measurements but is challenged by small planets of the same type as CoRoT-7b (see below), i.e. rocky planets with short orbital periods, and with masses in the super Earth category. Results have, however, been reported for such objects orbiting the very bright stars 55 CnC ($m_v \approx 5$) (Dragomir et al. 2013a) and HD 97658 ($m_v \approx 5$) (Dragomir et al. 2013b).

13.2.1.2 SPITZER

After 10 years in space, NASA's Spitzer Space Telescope[3] which was designed for 'general purpose' IR observations of galactic and extra-galactic sources and especially for the study of star forming and debris disks around Young Stellar Objects, has changed into a magnificent tool for the studying of exoplanets. When the observatory was designed back in the 1990s the first exoplanets had just been detected when the telescope design was frozen. Thanks to its extraordinary and somewhat unexpected thermal stability, as well as some changes implemented after launch, Spitzer now has capabilities not envisaged originally. Further, after the on-board coolant was exhausted, the spacecraft is passively held to a temperature of about −240 C, the so-called warm SPITZER mission, which allows the infrared cameras to operate and transforms the telescope into an excellent observatory for studying exoplanetary transits, as well as to detect some of the infrared light emitted by exoplanetary atmospheres and surfaces. The list of exoplanetary Spitzer results is long and distinguished. Among the results can be mentioned the first secondary eclipse measurements from 2005 (Deming et al. 2005). Spitzer has continued with producing a long list of excellent exoplanet results. One can mention thermal phase curves for atmospheric dynamics and heat transport constraints (Knutson et al. 2012) inference of clouds (Demory et al. 2013), as well as a simultaneous discovery of the transits of 55 Cnc e (Sect. 13.2.1.1), (Demory et al. 2011). The telescope is currently expected to be operational for several more years.

13.2.2 CoRoT: The First Dedicated Space Mission Related to Exoplanets

The first space telescope, specifically designed in order to discover and study transiting extra solar planets, and launched into a 900 km polar orbit, was the CoRoT (Convection, ROtation and planetary Transits) mission of the French Space Agency (CNES), with the Europan Space Agency (ESA) and other international partners. It was launched in Dec. 2006 by a Soyuz 2.1b rocket (Baglin and Fridlund 2006; Baglin et al. 2009). It began to take data on 2 February 2007 and continued doing so until the 2 November 2012 when a power failure made the last Data Processing Unit fail. The mission's two scientific objectives were to search for extrasolar planets with short orbital periods, particularly those of large terrestrial (super-Earth) size, and also to perform asteroseismology by measuring solar-like oscillations in stars. Basically a 27 cm off-axis focal telescope with a detector package consisting of 4 CCD's, 2 each of which were dedicated to the two scientific objectives. CoRoT could observe fields of 4 deg^2 in two 10 deg wide regions of the sky – one in the general direction of the galactic center and one in the constellation of Monoceros

[3]e.g. http://www.spitzer.caltech.edu

in the opposite region of the sky. Each region could be followed for a maximum of 180 days after it was necessary to do a 180 deg change in pointing and follow the other region. This was due to Earth's orbital motion which after this period allowed scattered solar light to enter the baffle of the telescope, disturbing the photometry. Between 10,000 and 11,000 objects could be followed each observing run. During its mission, around 150,000 light curves of individual objects (monitored with a cadence around 95 % during 30 days and up to 180 days) were acquired.

CoRoT discovered it's first planet, CoRoT-1b in 2007, and since then it has recorded more than 35 planets, and orbiting brown dwarfs, that have been (or are in the process of being) confirmed through several procedures including the detection of a radial velocity signature from the star, providing an estimate of the planets mass. Among the confirmed objects are the first transiting terrestrial planet, CoRoT-7b, and the CoRoT team succeeded in measuring its basic parameters, (Léger et al. 2009; Queloz et al. 2009; Hatzes et al. 2010). The exquisite photometric precision of this space-based facilities have enabled the first detections of the secondary transit and phase curve at optical wavelengths (Snellen et al. 2009; Alonso et al. 2009; Borucki et al. 2009).

Around 250 candidates remain to be followed up, but most of these are fainter than 16th magnitude and there are no spectrographs today that can provide follow-up – at least as what concerns smaller planets. These observations have to await the implementation of more sophisticated spectrographs on 8 m class telescopes such as the Very Large telescope of ESO, equipped with the ESPRESSO spectrograph (first light expected in 2016).

13.2.3 Kepler

CoRoT was followed by NASA's Kepler mission (e.g. http://kepler.nasa.gov) – with very similar objectives to CoRoT and which was launched in March 2009 (Borucki et al. 2009). It operated successfully until mid 2013. Kepler differs from CoRoT in that it is larger (95 cm aperture), and have a larger field of view (100 deg^2 as compared to 4 deg^2). This translate into more and brighter target stars, as well as the possibility of detecting smaller transiting planets. Also the method was different in that it observed the same field (located in the constellation of Cygnus) during the whole period. This field had about 150,000 target stars (which were 1–2 mag brighter than the targets of CoRoT). From these targets, this second space mission, has to date confirmed 250 planets (recently 750 more planets were claimed, using statistical methods (Rowe et al. 2014; Lissauer et al. 2014)), as well as registering 3,000 candidates. It has detected systems with 5–7 planets, including both gas giants, hot Neptune, and smaller planets (Borucki et al. 2010; Fridlund et al. 2014; Cabrera et al. 2014; Borucki et al. 2013).

CoRoT and Kepler have been the pioneering space missions that, to a large extent, have changed exoplanetology into its current form. Both missions have carried out asteroseismology as well as exoplanetology and discovered the p-mode oscillations in red giant stars, enabling elements of gyrochronology to be carried out for the first time in our galaxy, as well as many other discoveries in this nascent field. Kepler has also managed to measure the asteroseismological signature of the first planet-hosting star (HAT-P7-b, (Christensen-Dalsgaard et al. 2010)). Kepler thus become a precursor of the PLATO mission (see Sect. 13.3). The two missions have thus set the scene for both ESA's next medium class mission PLATO and NASA's explorer class mission TESS (see below).

13.2.4 Gaia: Bulk Observations

Gaia is ESA's next generation astrometric satellite, recently successfully launched in December 2013 and as this is written in the commissioning phase at its permanent L2 orbit. Gaia is based on the Hipparcos mission, carried out by ESA 1989–1993. Gaia's fundamental challenge: to create an extraordinarily precise three-dimensional map including the movements of about one billion stars throughout our Galaxy and beyond.

The scientific measurements of the Gaia mission cover three principal areas: astrometry, photometry and spectroscopy, the latter good enough for the acquisition of radial velocities and astrophysical parameters. Among problems related to the origin, structure, and evolutionary history of our Galaxy the highly accurate data will allow the detection of tens of thousands of exoplanets in sizes Neptune and larger, both from the astrometric wobble, and from transit observations detected photometrically. As such, Gaia will likely be the first space mission to detect exoplanets astrometrically, in bulk. HST (see Sect. 13.2.1) has previously carried out observations of a few suitable target systems using its Fine Guidance Sensors to perform relative astrometry.

But Gaia's highly precise stellar distances of essentially all of the future PLATO's (see Sect. 13.3) target stars will also enormously help that later mission by improving the asteroseismic data that PLATO will acquire through the removal of some of the properties of the exoplanetary host stars as a free parameter. As important examples one can mention that by measuring the distance to the target very accurately, both the luminosity (and the effective temperature) and the radii of the object can be held fixed in the retrieval of the astroseismic parameters, thus further increasing the precision in the planets physical parameters.

13.2.5 The Immediate Future: TESS, CHEOPS and the James Webb Space Telescope

13.2.5.1 The Transiting Exoplanet Survey Satellite: TESS

The Transiting Exoplanet Survey Satellite (TESS) has been selected for launch as a NASA Explorer program space mission in 2017. It consists of a platform equipped with 4 wide angle 10 cm telescopes with CCD cameras with a total size of 67 megapixels. The spacecraft can observe up to 10,000 stars per field and the data is transmitted to the Earth every 2 weeks.

Interestingly enough, for the first time, NASA will utilize a so-called lunar resonant orbit (or P/2). It is a highly elliptical 100,000 × 375,000 km orbit which is stable as long as the apogee of the orbit is timed to be away from the moon. The advantage of this orbit, as compared to e.g. orbits like the one of CoRoT or MOST is that one is located outside the Van Allen belts and their damaging radiation. The data is downloaded when the spacecraft is at perigee when a high data rate can be maintained.

The mission will observe essentially the whole sky, although most of it for only 1–2 months per field, concentrating on G- and K-type stars as well as about 1,000 of the brightest red dwarfs. It can detect stars brighter than magnitude 12 and it should thus be possible to detect a large number of planets that are Neptune, Super-Earth or even Earth-sized (for the absolutely brightest stars) with orbital periods that could be as long as about 1 month. Such objects will be prime targets for the JWST mission (see below) to attempt a first characterization of the transit atmospheric spectra of planets in the Neptune range.

13.2.5.2 CHEOPS

CHEOPS is ESA's first mission in the 'small mission class'. The small mission program is supposed to originate in one or more of the ESA member states and ESA's involvement is limited to (currently) 50 MEuro. Such missions should have a development time not exceeding 3.5–4 years. CHEOPS itself is dedicated to the photometric follow-up of exoplanets discovered through radial velocity observations of bright – and thus nearby – stars, where the parameters discerned from the RV observations make it likely that a transit will occur. The measurement of a highly accurate transit light curve will allow for the determination of precise planetary parameters. The goal is to achieve a 10 % accuracy for the radii of planets in the super-Earth to Neptune size range. The targets will be bright, and thus the precision will be the best before the launch of PLATO 2.0 (see Sect. 13.3) in 2024. The spacecraft is expected to be launched in 2017.

The main science objective of the CHEOPS mission will be the study of exoplanets, with sizes ranging between one and six Earth radii, specifically:

- The mass-radius relation in a planetary mass range for which only a handful of data exist and with a precision not previously achieved;
- The identification of planets with significant atmospheres over a range of planetary masses,
- Distances from the host star and stellar parameters from Gaia;
- Place constraints on possible planet migration paths followed during the formation and evolution of planets;
- Probe the atmospheres of known hot Jupiters;
- Provide unique targets for future ground-based (e.g. E-ELT) and space-based (e.g. JWST) observatories with spectroscopic capabilities.

CHEOPS will be a small tri-axially stabilized spacecraft with a total launch mass of about 250 kg and a pointing accuracy of 8 arc s. The baseline is to use a standard small satellite platform with some modifications. The spacecraft design will be consolidated in the forthcoming phase A/B1 study, which will also lead to selection of the platform. The CHEOPS mission will be equipped with a single medium-size telescope of 0.3 m aperture. The telescope is thus similar to the one of CoRoT, but the targets will be 6–12 mag, while those of CoRoT were 11–16 mag, and the telescope will therefore have a 1–2 order-of-magnitudes higher sensitivity (CoRoT achieved 1–20 ppm photometry for targets in the 6–9 magnitude range but this was for astroseismic targets and no transiting exoplanet host star was observed with this precision).

All platform requirements are aimed at supporting the functionality of the telescope and its high photometric precision. The main implications for the platform are related to pointing capabilities and the thermal environment for the payload. The reference scenario calls for a shared Vega or Soyuz launch. The baseline orbit is Sun-synchronous, with an altitude of 650–800 km and a mean local time of the ascending node of 06:00. This choice permits the rear of the spacecraft to be permanently Sun-pointed, is optimal for uninterrupted observations and keeps thermal variations of the spacecraft and Earth straylight on the satellite to a minimum as the orbital plane follows, as closely as possible, the day/night terminator.

13.2.5.3 The James Webb Space Telescope: JWST

The JWST observatory consists of a passively cooled telescope, optimised for diffraction-limited performance in the near-infrared (2–5 μm) region, but with extensions to either side into the visible (0.6–2 μm) and mid- infrared (5–28 μm.). The JWST observatory includes three main elements, the Integrated Science Instrument Module (ISIM), the Optical Telescope Element (OTE) and the Spacecraft Element which comprises the spacecraft bus and the sunshield. Some of the key characteristics of JWST are:

The primary mirror will be 6.5 m in diameter and is made of 18 mirror segments of gold-coated beryllium It will have a giant shield (about 22 × 12 m) protecting the telescope and the instruments from the light of the Sun. JWST's wavelength range covered by the scientific instruments will be from about 0.6 to 28 μm (visible to the mid-infrared light), compared to the Hubble Space Telescope's 0.1–2.5 μm (ultraviolet to the near infrared)

The total observatory mass will be 6,500 kg. After launch with a European Ariane 5 launcher, JWST will operate in the Lagrangian L2 orbit, approximately 1.5 million kilometres away from the Earth. This makes its operation and pointing/stability requirements both much simpler and more efficient, in comparison with Hubble.

The science objectives of the JWST falls into four broad categories, *viz.* the observation of:

- The phase of the evolution of the Universe when the first generation of stars were formed and started to shine, thereby re-ionizing the Universe.
- The first formation of early galaxies and clusters of galaxies
- Detailed observations of star formation and the accompanying process of planetary system formation
- "Planetary systems and the origin of life" taken to mean the first high sensitivity spectroscopic observations of exoplanetary atmospheres, as well as complex molecules in the interstellar medium accompanying the star formation process

13.3 The Next Step (ESA): PLATO 2.0

The PLATO 2.0 exoplanetary research spacecraft has been selected, in February 2014, as the European Space Agency, ESA's, medium class mission for the third (M3) launch opportunity upcoming in 2024. It addresses fundamental questions such as: How do planetary systems form and evolve? Are there other systems with planets like ours, able to develop life? PLATO 2.0 consists of a spacecraft bus, equipped with 34 small (12 cm) aperture telescopes providing both a wide field-of-view and, collectively, a large photometric magnitude range. It targets bright (primarily m_v = 4–13) stars in order to detect and characterize planets down to Earth-size by photometric transits, and, whose masses can then be determined by ground-based radial-velocity follow-up measurements.

Asteroseismology will be performed for stars m_v <11.5 mag to obtain highly accurate stellar parameters, including masses and ages. The combination of bright targets and asteroseismology results in high accuracy for the bulk planet parameters: 2 %, 4–10 % and 10 % for planet radii, masses and ages, respectively. The foreseen baseline observing strategy includes two long pointings (2–3 years each) to detect and characterize, especially Earth-size, planets reaching into the habitable zone (HZ) of solar-like stars together with an additional step-and-stare phase to cover in total about 50 % of the sky during a nominal 6 year mission (during a potential extended mission of up to 8 years and longer it is possible to cover 70–80 % of the sky). During the nominal mission PLATO 2.0 will observe up to 1,000,000 stars and characterize hundreds of planets in the Earth to Super-Earth category, and thousands of planets in the Neptune to gas giant regime in orbits out to the HZ, as well as detecting many more for which partial characterization will be possible.

PLATO 2.0 will therefore provide the first large-scale catalogue of planets with radii, masses, mean densities and ages, accurate enough to carry out detailed analysis. This catalogue will include Earth-like planets at intermediate orbital distances, where surface temperatures are moderate. Coverage of this parameter range with statistical numbers of well characterized planets is unique to PLATO 2.0. First, it will provide a census for small (low-mass) planets. The results of the mission will permit to complete the knowledge of planet diversity for both low- and high-mass objects (at orbital distances < a few AU), and correlate the planet mean density as a function of orbital distance with predictions from planet formation theories. Other investigations possible to carry out include the constraining of theories for planet migration and scattering on the architecture of multiple systems, and to specify how planet and system parameters change with host star characteristics, such as type, metallicity and age. The catalogue will also allow us to study planets and planetary systems at different evolutionary phases. This will serve to identify objects which retained their primordial hydrogen atmosphere and in general the typical characteristics of planets in such low-mass, low-density range. Since PLATO 2.0 targets mainly bright stars, many of the discovered planets will be targets for future spectroscopy exploring their atmospheres. Furthermore, the mission has the potential to detect exomoons, planetary rings, binary and Trojan planets. The planetary science possible with PLATO 2.0 is complemented by its impact on stellar and galactic science via asteroseismology as well as light curves of all kinds of variable stars, together with observations of stellar clusters of different ages. This will allow us to improve stellar models and study stellar activity. A large number of well-known ages from red giant stars will probe the structure and evolution of our Galaxy. Asteroseismic ages of bright stars for different phases of stellar evolution allow calibrating stellar age-rotation relationships. Together with the results of ESA's Gaia mission, the results of PLATO 2.0 will provide a huge legacy to planetary, stellar and galactic science.

But most important, with its high accuracy and sensitivity, PLATO 2.0 will make the first unique observations changing exoplanetology into the next phase, where we will be able to carry out geophysics on Earth-like worlds and compare these observations with data from our own Solar System. At the same time, observations of the host stars will give us accurate ages for the exo-systems, which will enable the derivation of evolutionary diagrams like Fig. 13.4, creating an understanding of the context of our own world.

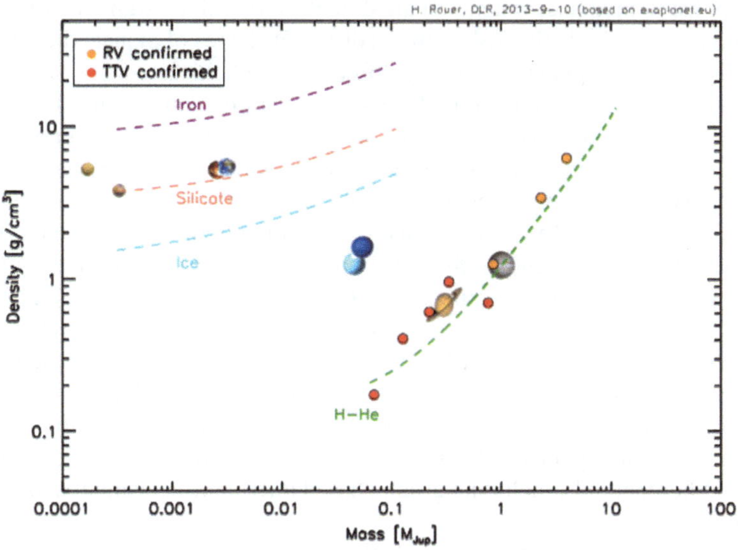

Fig. 13.4 Mean planet density versus mass with density lines for different bulk composition, for planets with orbital periods > than 50 days. For massive planets the large diversity is evident. By filling this diagram with planets of different masses (particularly low mass) and with different, precise, ages, planetary evolution could be studied (After Rauer et al. 2013)

13.4 The Next Step (NASA): WFIRST, Coronographs and Occulters

To be possibly launched in the same time frame as PLATO 2.0 or later, the Wide-Field Infrared Survey Telescope (WFIRST) is a NASA observatory designed to perform wide-field imaging and slitless spectroscopic surveys of the NIR sky. The current scenario envisages the use of an existing 2.4 m space qualified telescope (Astrophysics Focused Telescope Assets (AFTA). WFIRST-AFTA is intended to settle issues in both exoplanet and dark energy research. It is the top-ranked large space mission in the New Worlds, New Horizon, Decadal Survey of Astronomy and Astrophysics. The main instrument is a wide-field multi-filter NIR imager with optional grism spectroscopy. With the 2.4 m telescope, the possibility of a coronagraphic instrument has been added to the payload for direct imaging of exoplanets and debris disks.

With a contrast level of 10^{-9} and an inner working angle of less than 0.2 arc s, the 2.4 m telescope of WFIRST, equipped with a coronagraph will perform a survey of up to 200 of the nearest stars . This survey could directly image over a dozen known (large) radial velocity planets and could discover an additional dozen previously unknown ice and gas giants. The majority of these planets will also be characterized using a spectral resolution of ≈70 spectra in the wavelength range 400–1,000 nm via an integral field spectrograph. These spectra will allow the detection of features expected due to methane, water, and alkali metals, reveal the signature of Rayleigh scattering, and easily distinguish between different classes of planets (i.e., Neptunes versus Jupiters), and planets with very different metallicities.

The core accretion theory for planetary formation, suggests most planets are much less massive than gas giants and that the critical region for understanding planet formation is just beyond the snow-line, which is the region (1.5–4 AU) where the technique of microlensing has its greatest sensitivity (Ida and Lin 2005). WFIRST could extend the current sensitivity of the microlensing method down to lower masses of about a tenth of the Earth's mass at orbital separations ranging from 0.5 AU. While Other missions such as CoRoT (Sect. 13.2.2), Kepler (Sect. 13.2.3) and the future PLATO (Sect. 13.3) are sensitive to close-in planets (CoRoT <90 days, Kepler $\approx$ twice that and PLATO $\approx$1 year) but is unable to sense the more distant ones, WFIRST is less sensitive to close-in planets, but surveys beyond 0.5 AU is better than the other missions mentioned. WFIRST's sensitivity extends out even to unbound planets, offering the possibility to constrain their numbers and masses. Other methods, including ground based microlensing, cannot approach this sensitivity.

While microlensing from space will give comprehensive data about numbers of different classes of planets, essentially all its discoveries will be to distant for proper follow-up like in the case of e.g. PLATO. The statistical sense of microlensing missions demand that the targets will be so distant that any follow-up equipment within the foreseeable future will be unable to detect the planets again.

13.5 Further Future: Darwin, TPF and New World Observatories

Following the definition and technical studies of major missions intended for the study of exoplanets capable of sustaining life as we know it on exoplanets, carried out during the period 1996–2007, very little activity has taken place. These studies included the Darwin (ESA) and Terrestrial Planet Finder (TPF, NASA), largely carried out jointly by the two organizations. The results of these studies were that several technologies exist with the capability to study the direct light from small planets in habitable orbits around stars at least out to a distance of 30–50 pc.

To overcome the 10^{10} star-planet contrast between an Earth-like planet, located near (<0.1 arc s) its host star, optical arrangements that suppress the starlight will be required. In addition to the optical design challenges associated with this goal, wavefront control will be essential to maintain the required sub-nano meter wavefront stability within a coronagraph instrument. To enable high contrast imaging, a low-vibration telescope will need to be designed, requiring technology development in vibration isolation, low-vibration reaction wheels, and active sensing/ control of vibrations. If such efforts prove to be too challenging or costly, a fallback option is an external occulter, although this approach becomes extremely challenging (occulter size, distance, amount of energy required for repointing) for large telescope diameters.The technologies included interferometers based on destructive ('nulling') interference, as well as internal and external coronographs. After 2007, some studies have been carried out, but the extensive technological work have slowed down significantly. There are some studies being carried out w.r.t coronography by NASA. This is also true for the so-called New Worlds Missions, where studies of the deployment of star shades, i.e. external occulters intended to fly many tens of thousands of km away from the telescope have been carried out (Seager, *private communication*) while in Europe one can mention the successful Swedish PRISMA formation flying experiment utilising technology developed for ESA's Darwin program, which is necessary for any free-flying interferometer.

The first objective of Darwin was to search for planets in the habitable zone of target stars. The target stars are main sequence stars of type F, G, K, or M with corresponding B-V colours, a luminosity class of IV-V or V, and visual magnitudes of less than 12. The second objective of Darwin is a detailed analysis of the planets found. This includes the determination of the physical parameters as orbit, temperature, or evolutionary status and the analysis of the atmosphere by spectroscopy if an atmosphere is present. The third, ancillary objective of Darwin is interferometric imaging. By the analysis of biomarkers in the planet's atmosphere, Darwin will be able to discern if the planet can host life similar to that known on Earth and, if indicators for life are found, it will be able to determine the evolutionary status. The Darwin mission will therefore have a profound impact on mankind's understanding of itself.

Conclusion

After the first $\approx$20 years of observational exoplanetology the field has matured nicely, with today about 1,771 planets in 1,096 systems (http://exoplanets.eu), numbers that are changed on an almost daily basis. After this first phase of discovery, it appears clear that investigators are preparing for the next stage of detailed studies, finally being able soon to provide detailed physical parameters for planets orbiting stars other than our Sun. It appears that in this process, the understanding of the individual host stars is as important as the understanding of our own Sun in the case of the Solar System. Instruments and space missions for these next stages of investigations are being actively prepared by several space agencies, and it appears that this important scientific field can look forward with some reassurance for the near and intermediate future.

Acknowledgements We gratefully acknowledge the friendly atmosphere created by the staff of the International Space Institute in Bern, Switzerland, and the ISSI team *Characterizing stellar- and exoplanetary environments* where this chapter was discussed and planned.

References

Alonso, R., Alapini, A., Aigrain, S. (2009). *A&A, 506*, 353A.

Baglin, A., & Fridlund, M. (2006). CoRoT: From stars to habitable planets. Pre-launch studies. In M. Fridlund, A. Baglin, J. Lochard, & L. Conroy (Eds.), *The CoRoT mission pre-launch status – stellar seismology and planet finding ESA SP-1306* (p. 11). Noordwijk: ESA Publications Division.

Baglin, A., & The CoRoT Exoplanet Science Team (2009). In *Transiting planets, proceedings of the international astronomical union, IAU symposium* (Vol. 253, p.71). Cambridge/New York: Cambridge University Press.

Borucki, W. J., & The Kepler Team (2010). *Astrophysical Journal, 713*, 126B.

Borucki, W. J., & The Kepler team (2013). *Science, 340*, 587.

Borucki, W. J., Koch, D., Jenkins, J. (2009). *Science, 325*, 709B.

Brown, T. M., Charbonneau, D., Gilliland, D., Noyes, R. L., & Burrows A. (2001). *Astrophysical Journal, 552*, 699.

Butler, R. P., & Marcy, G. W. (1996). *Astrophysical Journal, 464*, 153.

Cabrera, J., Csizmadia, S., Lehman, H., Dvorak, R., Gandolfi, D., Rauer, H., Erikson, A., Dreyer, C., Eigmüller, Ph., & Hatzes, A. (2014). *Astrophysical Journal, 781*, 18.

Campbell, B., Walker, G. A. H., & Yang, S. (1988). *Astrophysical Journal, 331*, 902.

Charbonneau, D., Brown, T. M., Latham, D. W., & Mayor, M. (2000). *Astrophysical Journal, 529*, 45.

Christensen-Dalsgaard, J., Kjeldsen, H., Brown, T., Gilliland R. L., Arentoft, T., Frandsen, S., Quirion, P.-O., Borucki, W. J., Koch, D., & Jenkins, J. M. (2010). *Astrophysical Journal, 713*, 164.

Deming, D., Seager, S., Richardson, L. J., & Harrington, J. (2005). *Nature, 434*, 740.

Demory, B.-O., de Wit, J., Lewis, N., Fortney, J., Zsom, A, Seager, S., Knutson, H., Heng, K., Madhusudhan, N., Gillon, M., Barclay, T., Désert, J.-M., Parmentier, V., & Cowan, N. B. (2013). *Astrophysical Journal, 776*, 25.

Demory, B.-O., Gillon, M., Deming, D., Valencia, D., Seager, S., Benneke, B., Lovis, C., Cubillos, P., Harrington, J., Stevenson, K. B., Mayor, M., Pepe, F., Queloz, D., Ségransan, D., & Udry, S. (2011). *Astronomy and Astrophysics, 533*, 114.

Dragomir, D., & The MOST Team (2013a). 2013arXiv1302.3321D

Dragomir, D., & The MOST Team (2013b). *Astrophysical Journal, 772*, 2.

Fridlund, M., Cabrera, J., Csizmadia, S., Lehman, H., Dvorak, R., Gandolfi, D., Rauer, H., Erikson, A., Dreyer, C., Eigmueller, P., & Hatzes, A. (2014). *American Astronomical Society Meeting, 223*, 348.22.

Gray, D. F., & Hatzes A. P. (1997). *Astrophysical Journal, 464*, 412.

Hatzes, A.P., Dvorak, R., Wuchterl, G. (2010). *A&A, 520*, 93H.

Huitson, C. M., Sing, D. K., Pont, F., Fortney, J. J., Burrows, A. S., Wilson, P. A., Ballester, G. E., Nikolov, N., Gibson, N. P., Deming, D., Aigrain, S., Evans, T. M., Henry, G. W., Lecavelier des Etangs, A., Showman, A. P., Vidal-Madjar, A., & Zahnle, K. (2013). *Monthly Notices of the Royal Astronomical Society, 434*, 3252.

Ida, S., & Lin, D. N. C. (2005). In *Protostars and planets V* (LPI contribution No 1286, pp.8141). Houston: Lunar and Planetary Institute.

Knutson, H. A., Lewis, N., Fortney, J. J., Burrows, A., Showman, A. P., Cowan, N. B., Agol, E., Aigrain, S., Charbonneau, D., Deming, D., Dêsert, J.-M., Henry, G. W., Langton, J. & Laughlin, G. (2012). *Astrophysical Journal, 754*, 22.

Latham, D. W., Stefanik, R. P., Mazeh, T., Mayor, M., & Burki, G. (1989). *Nature, 339*, 38.

Léger, A., Rouan, D., Schneider, J. (2009). *A&A, 506*, 287L.

Lissauer, J. J., Marcy, G. W., Bryson, S. T., Rowe, J. F., Jontof-Hutter, D., Agol, E., Borucki, W. J., Carter, J. A., Ford, E. B., Gilliland, R. L., Kolbl, R., Star, K. M., Steffen, J. H., & Torres, G. (2014). *Astrophysical Journal, 784*, 44.

Marcy, G. W., & Butler R. P. (1996). *Astrophysical Journal, 464*, 147.

Marcy, G. W., & The Kepler Team (2014). *Astrophysical Journal Supplement, 210*, 20.
Mayor, M., & Queloz, D. (1995). *Nature, 378*, 355.
McArthur, B. E., Benedict, G. F., Barnes, R., Martioli, E., Korzennik, S., Nelan, E., & Butler, R. P. (2010). *Astrophysical Journal, 715*, 1203.
Nikolov, N., Sing, S. K., Pont, F., Burrows A. S., Fortney, J. J., Ballester, G. E., Evans, T. M., Huitson, C. M., Wakeford, H. R., Wilson, P. A., Aigrain, S., Deming, D., Gibson, N. P., Henry, G. W., Knutson, H., Lecavelier des Etangs, A., Showman, A. P., Vidal-Madjar, A., & Zahnle, K. (2014). *Monthly Notices of the Royal Astronomical Society, 437*, 46.
Queloz, D., Bouchy, F., Moutou, C. (2009). *A&A, 506*, 303Q.
Rauer, H., & The PLATO Team (2013). 2013arXiv1310.0696R.
Rowe, J. F., Bryson, S. T., Marcy, G. W., Lissauer, J. J., Jontof-Hutter, D., Mullally, F., Gilliland, R. L., Issacson, H., Ford, E., Howell, S. B., W. J., Haas, M., Huber, D., Steffen, J. H., Thompson, S. E., Quintana, E., Barclay, T., Still, M., Fortney, J., Gautier, T. N., III, Hunter, R., Caldwell, D. A., Ciardi, D. R., Devore, E., Cochran, W., Jenkins, J., Agol, E., Carter, J. A., & Geary, J. (2014). *Astrophysical Journal, 784*, 45.
Snellen, I., de Mooij, E. J. W., Albrecht, S. (2009). *Nature, 459*, 543S.
Struve, O. (1952). *The Observatory, 12*, 3.
Vidal-Madjar, A., Huitson, C. M., Bourrier, V., Desert, J.-M., Ballester, G., Lecavelier des Etangs, A., Sing, D. K., Ehrenreich, D., Ferlet, R., Hêbrard, G., & McConnell, J. C. (2013). *Astronomy and Astrophysics, 560*, 54.
Wagner, F. W., Sohl, F., Hussmann, H., Grott, M., & Rauer, H. (2011). *Icarus, 214*, 366.
Wolszczan, A., & Frail, D. A. (1992). *Nature, 355*, 145.

Chapter 14
The World Space Observatory–UV Project as a Tool for Exoplanet Science

Boris M. Shustov, Mikhail E. Sachkov, Dmitry V. Bisikalo, and Ana-Ines Gómez de Castro

Abstract During last three decades, astronomers have had practically continuous access to the 100–300 nm spectral range that is unreachable with ground-based instruments but where astrophysical processes can be efficiently studied with unprecedented capability. The successful International Ultraviolet Explorer (IUE) observatory, Russian ASTRON mission and successor instruments such as the COS and STIS spectrographs on-board the Hubble Space Telescope (HST) demonstrate the major impact that observations in the UV wavelength range have had on modern astronomy. Many exoplanetary studies have been performed in the UV domain, both Far-UV and Near-UV. This spectral region contains many resonance lines of common elements, including $Ly\ \alpha$, which provided a unique possibility to study physical and chemical properties of planetary atmospheres. Future access to space-based observatories is expected to be very limited. For the next decade, the post-HST era, the World Space Observatory UltraViolet (WSO–UV) will be the only large telescope class mission for UV observations, both spectroscopic and imaging. In its potential, the WSO–UV mission is similar to the HST, but all the observing time will be available for UV astronomy. In this chapter, we briefly outline the WSO–UV mission model, instrumentation description, science management plan as well as some of the key science issues that WSO–UV will address during its lifetime. This information should help exoplanet researchers to start to prepare their future observations with WSO–UV.

B.M. Shustov (✉) • M.E. Sachkov • D.V. Bisikalo
Russian Academy of Sciences, Institute of Astronomy, 48 Pyatnitskaya str., 119017 Moscow, Russian Federation
e-mail: bshustov@inasan.ru; msachkov@inasan.ru; bisikalo@inasan.ru

A.-I. Gómez de Castro
Faculty of Mathematics, Universidad Complutense de Madrid (AEGORA), Plaza de Ciencias 3, 28040 Madrid, Spain
e-mail: anai_gomez@mat.ucm.es

H. Lammer, M. Khodachenko (eds.), *Characterizing Stellar and Exoplanetary Environments*, Astrophysics and Space Science Library 411,
DOI 10.1007/978-3-319-09749-7_14

14.1 Introduction: UV Exoplanet Astronomy After HST

Spectroscopy during planetary transits is a powerful tool to probe exoplanet atmospheres. The chemical composition and temperature profile can be derived, providing clues on the dynamics and chemical processes in the atmosphere.

As it was shown in previous chapters, space observatories provide the major input to the exoplanet science (see Chap. 13 (Fridlund et al. 2014)). Observations in the UV domain open new highlights in exoplanet studies (see e.g., Haswell et al. (2012), Fossati et al. (2014)). Currently, only the STIS and COS instruments at HST provide the opportunity to make extremely high S/N transit observations in far-UV (FUV) and near-UV (NUV), capable of detecting and scrutinizing atmospheric constituents. NASA's Hubble Space Telescope, that was launched in 1990, is still going strong, and agency officials said in January 2013 that they plan to operate it until its instruments finally give out, potentially till 2018. The James Webb Space Telescope (JWST), the Next Generation Space Telescope with 6.5-m main aperture is to be launched in 2018 and it is optimized for operation in near- and mid-infrared. Telescope for Habitable Exoplanets and Interstellar/Intergalactic Astronomy (THEIA), a NASA-proposed 4-m optical/ultraviolet space telescope that would succeed the Hubble Space Telescope and complement JWST can be realized after 2025. THEIA is supposed to use a 40-m occulter to block starlight so as to directly image exoplanets. NASA has approved the Transiting Exoplanet Survey Satellite (TESS) mission. TESS's primary goal would be to identify terrestrial planets orbiting nearby stars. TESS's team estimates that the mission will be able to detect as many as 2,700 planets, including several hundred Earth-size worlds. The TESS mission is now scheduled for launch in 2017. The CHaracterising ExOPlanets Satellite (CHEOPS) project was selected in October 2012 from among 26 proposals as the first S-class ("small") space mission in ESA's Science Programme. Scheduled for launch in 2017, this mission aims to bring an optical Ritchey-Chrétien telescope with an aperture of 30 cm, mounted on a standard small satellite platform, into a sun-synchronous orbit of about 800 km altitude. This will be the photometric mission.

For the nearest future the World Space Observatory – Ultraviolet project will be the only 2-m class observatory fully devoted to UV observations, both photometric and spectroscopic, that can be used for exoplanet studies. In this chapter we describe the WSO–UV project with its general objectives and main features, the details and status of instrumentation, WSO–UV ground segment and science management plan as well as prospects of exoplanet observations with this observatory.

14.2 The WSO–UV Mission

The main instrument of the WSO–UV is 1.7 m aperture telescope with instrumentation designed to carry out high resolution spectroscopy, long-slit low resolution spectroscopy and direct sky imaging (Sachkov et al. 2014a).

Fig. 14.1 The T-170M telescope scheme

The T-170M telescope (see Fig. 14.1) is designed as a powerful concentrator of radiation for spectroscopy and direct images at 115–310 nm (Sachkov et al. 2014a). It is a Ritchey-Chrétien reflective optical design with a focal length of 17 m. The telescope provides an accessible field of view of 30 arc min on the telescope focal surface (Shustov et al. 2009). The T-170M telescope has inherited the successful

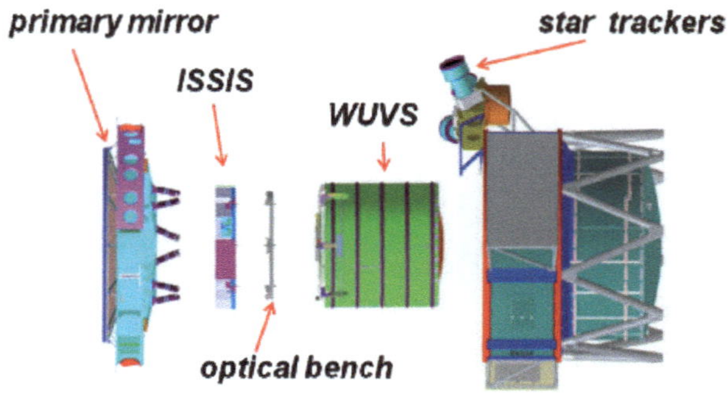

Fig. 14.2 WSO–UV instrument compartment

experience gained during the Soviet ASTRON project. The space telescope is under the responsibility of Russia. WSO–UV has been developed as a multipurpose, observatory-type mission (Shustov et al. 2011) carrying instrumentation for UV imaging and spectroscopy (see Fig. 14.2).

14.3 WUVS: WSO–UV Spectrographs

The WSO–UV spectrographs (WUVS) (Werner et al. 2009; Reutlinger et al. 2011) is a set of three instruments (see Fig. 14.3)

- Far UV high resolution spectrograph (VUVES) that will carry out echellé spectroscopy with resolution about 50,000 in the 115–176 nm range.
- Near UV high resolution spectrograph (UVES) to carry out echellé spectroscopy with resolution about 50,000 in the 174–310 nm range.
- Long Slit Spectrograph (LSS) that will provide low resolution (R = 1,000), long slit spectroscopy in the 115–305 range. The spatial resolution will be 0.5 arc s.

The WUVS instrument is under the responsibility of Russia. The main characteristics of the instrument are presented in the Table 14.1.

14.4 Comparison of WSO–UV and HST Spectrograph Efficiency

According to WUVS current design all spectrographs will be equipped with CCD detectors instead of previously proposed MCPs (Kappelmann et al. 2006). Their characteristics will allow detailed spectral observations and analyses of objects up to $V = 17^m$ (Klochkova et al. 2009). The prospects of WSO–UV for spectroscopic studies are discussed in a paper (Sachkov 2010). Efficiency of WUVS channels can be considered as a product of the relevant parameters ("efficiencies") of all the elements of their optical paths. These parameters are: reflectance of $Al + MgF_2$

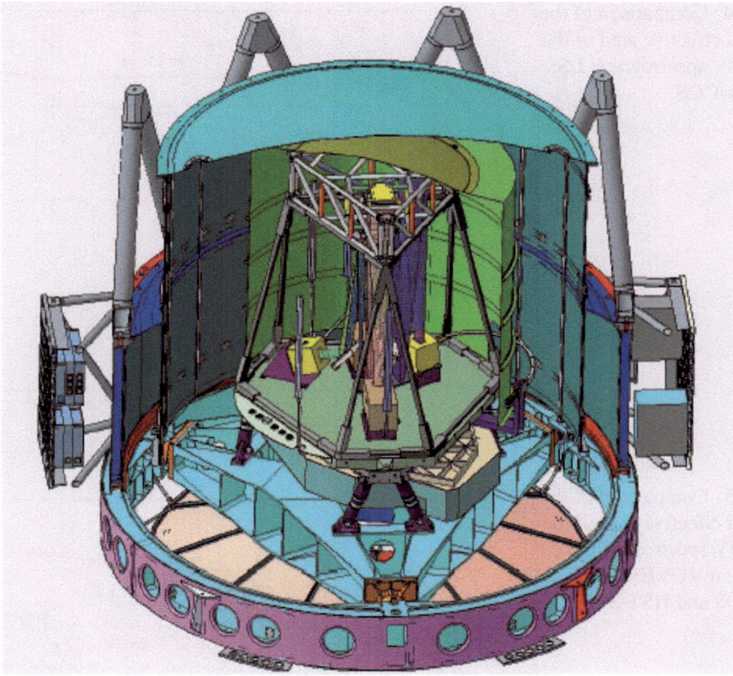

Fig. 14.3 WSO-UV spectrographs (WUVS)

Table 14.1 WUVS main characteristics

Parameter	Far UV (VUVES)	Near UV (UVES)	LSS
Wavelength range, nm	115–176	174–310	115–305
Spectral resolution	50,000	50,000	1,000
Detectors	CCD	CCD	CCD
Exposure time, s	1–3,600	1–3,600	1–3,600
Dark current, e^{-}/ pixel/ hour	3	3	3
Effective area, cm^2	250 (at 130 nm)	1,000 (at 175 nm)	500 (at 150 nm)
	300 (at 176 nm)	1,000 (at 250 nm)	2,000 (at 220 nm)

coating, transmission of MgF_2 window for the UV detectors, efficiency of gratings and CCD's sensitivity (quantum yield). The data for the plot are not theoretical values but measured ones. We compare the effective area of WUVS echellé spectrographs and LSS with COS/HST and STIS/HST (see Figs. 14.4 and 14.5). The spectral resolution provided by VUVES and UVES channels of WUVS is similar to that offered by STIS (HST) at medium resolution with its echelle gratings E140 and E230, but it is higher than the maximum resolution provided by HST COS ($R = 20{,}000$). It should be pointed out that we compare WUVS's effective area with HST initial parameters before their degradation during on orbit operations. WSO–UV spectrographs have have a sensitivity in the $Ly\alpha$ wavelength region.

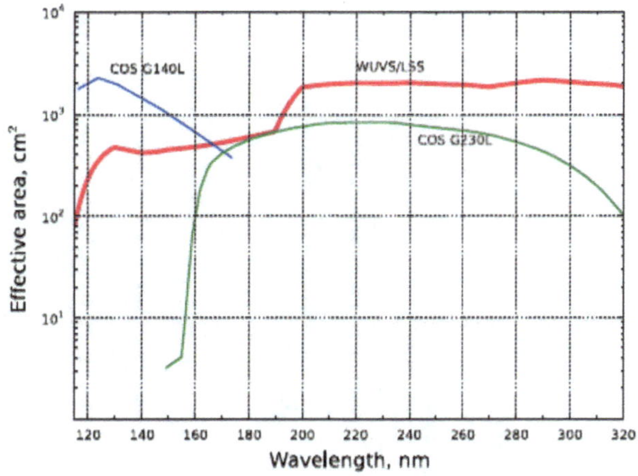

Fig. 14.4 Comparison of the expected effective area of the WSO–UV spectrograph LSS and HST/COS

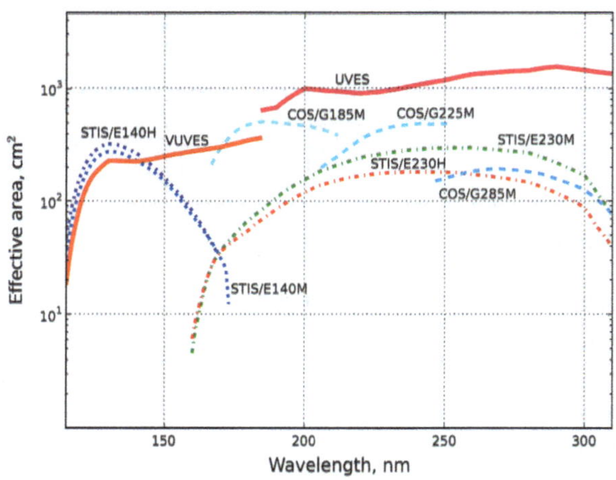

Fig. 14.5 Comparison of the expected effective area of the WSO–UV spectrographs, UVES and VUVES, and HST/COS and HST/STIS

14.5 ISSIS: Imaging and Slitless Spectroscopy Instrument for Surveys

The WSO–UV Imaging and Slitless Spectroscopy Instrument for Surveys (ISSIS) is a multipurpose instrument with a mode selector wheel that permits imaging and slitless spectroscopy in the 115–320 nm spectral range (Gómez de Castro et al. 2013). The instrument is equipped with two MCP detectors, with CsI and CsTe photocathods for FUV and NUV observations, respectively. The resolution in the slitless spectroscopy mode is about 500 and the spatial resolution is less than 0.1 arc s. ISSIS will be the first UV imager located in such a high altitude orbit. This

Fig. 14.6 WSO-UV imagers (ISSIS)

Table 14.2 ISSIS main characteristics

Parameter	Far UV channel	Near UV channel
Wavelength range, nm	115–175	185–320
Field of view, arcsec	70 × 75	70 × 75
Scale, arcsec/pix	0.037	0.037
Detectors	CsI MCP	CsTe MCP

has the advantage of being above the geocoronal emission and thus dimishing the UV background significantly. The current design was approved in the Preliminary Design Review (PDR) in June 2012 (see Fig. 14.6). According to the international agreement between Federal Space Agency (Roscosmos, Russia) and Ministry of Industry, Energetics and Tourism (Spain) the construction of the ISSIS instrument is the responsibility of Spain. Main characteristics of the ISSIS instrument are presented in Table 14.2. A crucial issue in the studies of the interaction between exoplanets and their host stars is the impact of the stellar magnetic activity both, in the measurement process and in the atmospheric evolution. ISSIS instrument on board WSO-UV is equipped with photon counting detectors that will allow to track the stellar activity evolution while transit occurs and its interaction with the exoplanet atmosphere.

14.6 WSO–UV Orbit

One of the main advantages of WSO–UV that allow to compensate the factor of 2 difference in the collecting surface between HST and WSO–UV is the more efficient geosynchronous orbit (Boyarchuk et al. 2013). It has an inclination of $51.^{o}6$ (see Fig. 14.7). Earth occultation periods will be short and the orbital period will allow long term monitoring and rapid access to targets of opportunity. Geosynchronous orbit was chosen based mainly on launcher capabilities, residence time in the Earth Radiation Belts, continuous visibility zones, minimum duration of the Earth shadow periods, stability of the orbit and available technical equipment of the Space and Ground Segments for radio communication. WSO–UV will use the Russian NAVIGATOR platform (Sachkov 2007), that was designed in Lavochkin Science & Technology Association (Russia) as a unified unit for several missions including Radioastron (successfully launched 2011) and Spektrum-Roentgen-Gamma (launch scheduled 2015) and WSO–UV. The platform is also used for commercial satellites, which have successfully proved the concept. The platform weighs 1,300 kg and has a payload mass of 1,600 kg. With assistance through the fine guidance system the pointing stabilization is about 0.03 arc s. The bus provides 300 W power for all instruments and a data download up to 4 Mbit/s. WSO–UV will be launched from Baikonur (Kazakhstan) with a Proton rocket. Main characteristics of the platform are presented in Table 14.3.

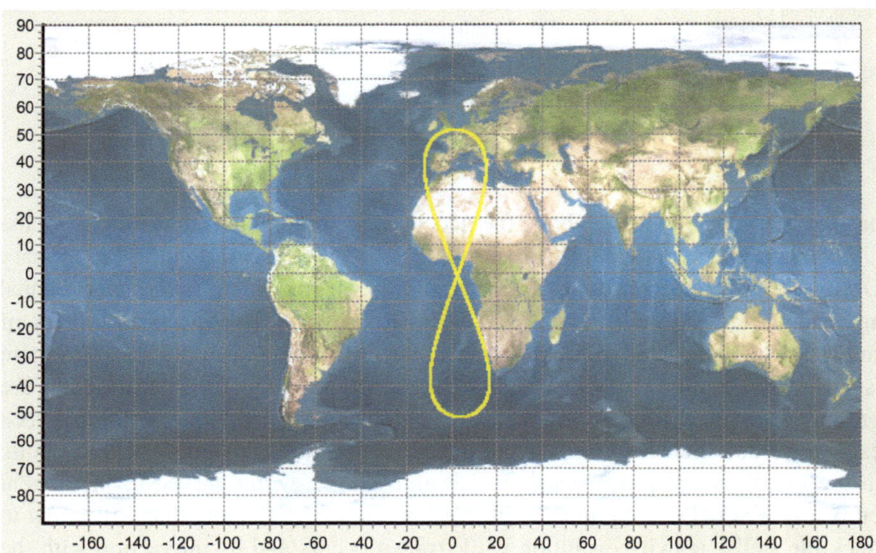

Fig. 14.7 Trace of the WSO–UV orbit

Table 14.3 NAVIGATOR platform

Parameter	Specification
Life time	> 5 years
Spacecraft mass with propellant	2,900 Kg
Payload mass	1,600 Kg
The power consumption for payload	750 W
Data transmission rate	2 Mbit s^{-1}
Stabilization and pointing accuracy using the Fine	
Guidance System data	0,1 arc s (3 σ)

14.7 WSO–UV Science Management Plan

WSO–UV will work as a targeted space observatory with a core program, an open program for scientific projects from the world-wide community and national (funding bodies) programs for the project partners (Malkov et al. 2011). The Core Program (CP) of scientific observations with WSO–UV is defined to allow conduction of high impact or legacy scientific projects that deserve large amounts of observing time. The projects are selected on the basis of their scientific excellence. The Open Program (OP) consists of astronomical observations obtained with the WSO–UV by astronomers who may or may not belong to the WSO–UV international consortium. The Funding Bodies Program (FBP) is the guaranteed time granted to each one of the national bodies funding the WSO–UV project. A Time Allocation Committee (TAC), appointed by the Agencies funding the project, which will select the scientific programs for the CP and the OP. The membership of the OP TAC will be renewed every 2 years. The CP TAC will be the same as the OP TAC selected for the first 2 years of the WSO–UV mission. National Time Allocation Committees will select the scientific programs for the FBP.

The time for astronomical observations will be distributed according to the following scheme:

- During the first and second years: 50 % of the total time will be granted to the CP, 48 % to the FBP and 2 % to the Director Discretionary Time (DDT). CP shall be completed within the first 2 years of the mission;
- For the following years: 58 % for the FBP, 40 % OT, 2 % DDT.

Currently, as it is decided by WSO–UV scientific committee, the main topics of the Core Program, i.e. the science drivers of the WSO–UV observatory, are (Gómez de Castro et al. 2009):

- The determination of the diffuse baryonic content in the Universe and its chemical evolution. The main topics will be the investigation of baryonic content in warm and hot Inter Galactic Matter, of damped Lyman-α systems, the role of starbursts and the formation of galaxies.

- The physics of accretion and outflows: stars, black holes, and all those objects dominated by accretion mechanisms. The efficiency and time scales of the phenomena will be studied, together with the role of the radiation pressure and the disk instabilities.
- The study of the formation and evolution of the Milky Way. The Milky Way history could be tracked through observations complementary to those obtained by the GAIA mission.
- The investigation of the extrasolar planetary atmospheres and astrochemistry in presence of strong UV radiation fields.

Because of the strong interest of astronomical community to such observations it was decided to include extrasolar planetary atmospheres investigations into the WSO–UV Core Program.

14.8 WSO–UV Status 2014

The project has entered in Phase C (Sachkov et al. 2014b). The telescope T-170M passed its vibration and static as well as transportation tests in 2012 (see Fig. 14.8). The WSO-UV team pays the special attention to the optics quality. It is diffraction – limited. The microscopic roughness (microroughness) parameter of both primary and secondary mirrors is about 0.75 nm (rms). Deposition of coatings onto mirrors of the T-170M telescope are carried out in a new facility specially created in Russia for WSO–UV project and equipped with a 2.6-m diameter Denton deposition vacuum chamber. The obvious choice for a coating with high reflectance above 115 nm is Al protected with MgF_2. The chamber has been designed with a size that would enable uniform coating of 2-m class mirrors. Ion Beam Assisted Deposition technic

Fig. 14.8 The T-170M telescope model for vibration tests at the Lavochkin Science & Technology Association (Lavochkin Science & Technology Association)

is used. Installation and testing of the Ground Segment first version, prepared by Spanish team, was carried out in Moscow during the period 2012–2013. All the evaluated tests were successful. Both main instruments, spectrographs and cameras, passed successfully their Preliminary Design Review Phase. Structural-thermal models of both main instruments, WUVS and ISSIS, are delivered to Moscow (Lavochkin Science & Technology Association) where they should be attached to the relevant T-170M telescope model. Thermal-vacuum tests are scheduled for 2014. The Payload as well as the Space Craft will be completed at the beginning of 2017 with a foreseen launch in late 2017 – early 2018.

14.9 WSO–UV Ground Segment

The WSO–UV Ground Segment (GS) is composed of all the infrastructure and facilities involved in the preparation and execution of the WSO–UV mission operations, which typically encompass real-time monitoring and control of the spacecraft, telescope and instruments as well as reception, processing and storage of the scientific data. There will be two complete GS systems: the Russian one will be located in Moscow (Lavochkin Science & Technology Association and Institute of Astronomy of the RAS), and the Spanish one will be sited at Madrid. The satellite operations will be shared between both Ground Control Centers, transferring the mission control from one center to the other on a regular basis.

The science operations system and a fraction of the mission operations system are part of the Spanish contribution to the WSO–UV. The Remote Proposal System, the Science Data Processing System, the Science Archive and the Scheduling systems are defined by the international science team composed of Spanish and Russian Science Support Teams based at the Universidad Complutense de Madrid (UCM) and Institute of Astronomy of the Russian Academy of Science (INASAN). The Science Team is part of the man power of the GS, and is responsible of laying the foundation of and supervising all the operations related to the mission primary users: the scientists. At mission level, the Science Team constitutes the core of the future WSO–UV international observatory. One of the main challenges of GS development is the management of shared information between both centers and the alignment of all the operational data (telemetry, telecommand and planning) according to the operational shifts (Lozano et al. 2010).

Conclusion

As it was discussed in various chapters of this book, the Instrumentation of the WSO–UV project is very important and helpful for exoplanet studies and the characterization of the exoplanet-stellar environment. The UV domain is suitable for detecting atoms that show resonance lines in this spectral window.

(continued)

However, there are several difficulties that are at the origin of the major uncertainties on any estimations of exoplanetary atmosphere properties

- The relative faintness of the UV stellar emissions;
- The variability of the sources;
- Signal contamination by both the sky background signal (at some spectral lines) and the instrument response.

While the first difficulty can be resolved by focusing on a few close-by and UV-bright stars, the signal variability from both the source and the instrument is a real problem that should be addressed to build a reliable diagnostic to extract an accurate description of the upper atmosphere and of the interaction region between the exoplanet and the impinging wind from its host star.

It should also be mentioned that WSO–UV can be used not only for standard exoplanet observations and characterizations but also for observations of biomarkers. Biomarkers like ozone have very strong transitions in the ultraviolet. These are electronic molecular transitions, hence several orders of magnitude stronger than the vibrational or rotational transitions observed in the infrared or radio range. The spectral resolving power required to detect biomarkers in the atmosphere of exoplanets is not a crucial capability. A resolution of $R \sim 10{,}000$ is adequate to these investigations, and even $R \leq 1{,}000$ could be enough to detect the broad band signatures of many molecules. The presence of biomarkers and other constituents in the atmospheres can be searched by WSO-UV high resolution spectrographs for about 100 exoplanets orbiting K, G and F-type main sequence stars.

Finally we conclude that the World Space Observatory–Ultraviolet is the solution to the problem of future access to UV spectroscopy. The project has entered in Phase C with a foreseen launch in late 2017 – early 2018. WSO-UV is ideally placed in time, after HST, together with CHEOPS and TESS and just before the JWST operations. During the period 2018–2028 it would be the main instrument for UV exoplanet research.

Current information on the WSO–UV project can be found at the official web site: http://wso-uv.org.

Acknowledgements D. Bisikalo acknowledge the support by the International Space Science Institute (ISSI) in Bern, Switzerland and the ISSI team *Characterizing stellar- and exoplanetary environments*. The authors also acknowledge the support by the RFBR projects 12-02-00047 and 14-02-00215.

References

Boyarchuk, A. A., Shustov, B. M., Moisheev, A. A., & Sachkov, M. E. (2013). *Solar System Research, 47*, 499.

Fossati, L., Haswell, C. A., Linsky, J. L., & Kislyakova, K. G. (2014). In H. Lammer & M. L. Khodachenko (Eds.), *Characterizing stellar and exoplanetary environments* (pp. 59). Heidelberg/New York: Springer.

Fridlund, M., Rauer, H., & Anders, E. (2014). In H. Lammer & M. L. Khodachenko (Eds.), *Characterizing stellar and exoplanetary environments* (pp. 253). Heidelberg/New York: Springer.

Gómez de Castro, A. I., Pagano, I., Sachkov, M., Lecavelier Des Étangs, A., Piotto, G., González, R., Shustov, B (2009). In M. Chavez, E. Bertone, D. Rosa-Gonzalez, & L. H. Rodrigez-Merino (Eds.), *New quests in stellar astrophysics. II. Ultraviolet properties of evolved stellar populations* (p. 319). Berlin: Springer.

Gómez de Castro, A. I., Sestito, P., Sanchez Doreste, N., López-Martínez, F., Seijas, J., Rodríguez, P., Gómez, M., Lozano, J. M., Shustov, B., Sachkov, M., Moisheev, A. (2013). In J. C., Lara, L. M. Quilis, & J. Gorgas (Eds.), *Proceedings of the X scientific meeting of the Spanish astronomical society (SEA)*, Valencia (Highlights of spanish astrophysics VII), p. 820.

Haswell, C. A., Fossati, L., Ayres, T., France, K., Froning, C. S., Holmes, S., Kolb, U. C., Busuttil, R., Street, R. A., Hebb, L., Collier Cameron, A., Enoch, B., Burwitz, V., Rodriguez, J., West, R. G., Pollacco, D., Wheatley, P. J., & Carter, A. (2012). *Astrophysical Journal, 760*(79), 23.

Kappelmann, N., Barnstedt, J., Gringel, W., Werner, K., Becker-Ross, H., Florek, S., Graue, R., Kampf, D., Reutlinger, A., Neumann, C., Shustov, B., Sachkov, M., Panchuk, V., Yushkin, M., Moisheev, A., & Skripunov, E. (2006). In M. Turner & M. G. Hasinger (Eds.), *Space telescopes and instrumentation II: Ultraviolet to gamma ray* (Proceeding of the SPIE 6266, id. 62660X).

Klochkova, V., Panchuk, V., Sachkov, M., & Yushkin, M. (2009). In M. Chavez, M. E. Bertone, D. Rosa-Gonzalez, & L. H. Rodrigez-Merino (Eds.), *New quests in stellar astrophysics. II. Ultraviolet properties of evolved stellar populations* (p. 337). Berlin: Springer.

Lozano, J. M., & The WSO-UV Team (2010). *Space Optics, 213993*, 693.

Malkov, O., Sachkov, M., Shustov, B., Kaigorodov, P., Yáñez, F. J., & Gómez de Castro, A. I. (2011). *Astrophysics and Space Science, 335*, 323.

Reutlinger, A., Sachkov, M., Gál, C., Brandt, C., Haberler, P., Zuknik, K.-H., Sedlmaier, T., Shustov, B., Moisheev, A., Kappelmann, N., Barnstedt, J., & Werner, K. (2011). *Astrophysics and Space Science, 335*, 311.

Sachkov, M. (2007). *AIP Conference Proceedings, 938*, 148.

Sachkov, M. (2010). *Astrophysics and Space Science, 329*, 261.

Sachkov, M., Shustov, B., Savanov, I., & Gómez de Castro, A. I. (2014a). *Astronomische Nachrichten 335*, 46.

Sachkov, M., Shustov, B., & Gómez de Castro, A. I. (2014b). *Advance Space Research, 53*, 990.

Shustov, B., Sachkov, M., Gómez de Castro, A.I., Huang, M., Werner, K., Kappelmann, N., & Pagano, I. (2009). *Astrophysics and Space Science, 320*, 187.

Shustov, B., Sachkov, M., Gómez de Castro, A.I., Werner, K., Kappelmann, N., & Moisheev, A. (2011). *Astrophysics and Space Science, 335*, 273.

Werner, K., Shustov, B., Sachkov, M., Gómez de Castro, A. I., Huang, M., Kappelmann, N., & Zhao, G. (2009). *AIP Conference Proceedings, 1135*, 314.

Chapter 15
Ground-Based Exoplanet Projects

Eike W. Guenther

Abstract A large number instruments have specifically been developed to detect, and study extrasolar planets. The number of new, ground-based instruments is already so large that it is not possible discussing them adequately in just one article. I will therefor just outline what the new concepts are, and mention only a few examples that demonstrate what the new instruments will offer to us.

15.1 Introduction: Ground-Based Exoplanet Research

Many different types instruments are being used to detect and study extrasolar planets. It is interesting to note that the multi-purpose instruments that had initially being used for exoplanet research, have given way to instruments that were specifically for this purpose. Because the book recently published by Perryman (2011) gives a comprehensive overview of the methods, and instruments used, I will only concentrate the new concepts. The methods, and the historic background will only be mentioned when it is absolutely necessary for the understanding the new ideas. The number of new projects, and concepts is in fact so large that discussing all of them in an adequate manner will be far beyond the scope of this article. In order to demonstrate what these new instruments will offer to us, and how large the gain will be, I will thus mention only a few examples that stand for many similar instruments. Since the techniques used in radio astronomy were covered extensively in Chap. 11 (Griessmeier 2014), and space instrumentation was discussed in Chap. 13 (Fridlund et al. 2014) and Chap. 14 (Shustov et al. 2014), this chapter will focus only on ground-based, optical, near-infrared (NIR) and mid-infrared (MIR) instruments.

E.W. Guenther (✉)
Thüringer Landessternwarte Tautenburg, Sternwarte 5, 07778 Tautenburg, Germany
e-mail: guenther@tls-tautenburg.de

H. Lammer, M. Khodachenko (eds.), *Characterizing Stellar and Exoplanetary Environments*, Astrophysics and Space Science Library 411,
DOI 10.1007/978-3-319-09749-7_15

15.2 Radial Velocity Measurements

The first extrasolar planet orbiting a main sequence star, and most of the extrasolar planets known today, were discovered using the radial-velocity (RV) technique. This method will also play an important role in the future. The RV-method is, for example, essential for studying the planet population, or for determining the masses of the planets discovered by methods, like transits. A small draw-back of this method is that it only allows to determine the minimum mass of a planet, or $v\,sin(\,i\,)$. In principle, the minimum mass can only be converted into the true mass, if the inclination of the orbit (i) is known. However, for a large sample of planets, we can simply assume that the inclination angles will be randomly oriented. This allows converting the average $m\,sin(\,i\,)$-values into average masses of the planets. Statistically, the mass of a planet is $m_{true} = 1.27 \times m\,sin(\,i\,)$.

Today, RV-measurements are so sensitive that the RV-jitter induced by stellar activity is often much larger than the errors of the measurements themselves. Thus, planets of very small masses can only be detected, if they are orbiting inactive stars. Because the RV-method is more sensitive for planets in short-period orbits, planets of very small mass can only be detected if they orbit inactive stars at small orbital distances. Stellar activity is not a show-stopper, it just makes the analysis more complicated. If the diagnostics of stellar activity are carefully analyzed, it is possible to correct the RV-measurements for these effects (e.g., Hatzes et al. 2011 and Lagrange et al. 2013). It is thus not only important to achieve the highest possible accuracy of the measurements, but it is equally important to have an instrument that allows to make use of the most important diagnostics for stellar activity. For example, the spectral range of the spectrograph should include the CaII H and K lines. Other diagnostics are the equivalent width, and the asymmetry of photospheric lines. To make use of these, the spectrograph has to have a sufficient resolution, and the spectrum should contain many photospheric lines that are not too much affected by telluric lines (or by lines from an absorption cell, if a cell is used).

In order to distinguish stellar activity from RV-signals of orbiting planets, a new concept is to obtain RV-measurements at different wavelength bands. If the amplitude of the variations is different in different wavelength bands, the signal is presumably not due to a planet but, for example, due to stellar spots. This argument was used by Huélamo et al. (2008) who showed that the periodic RV-variation in TW Hya were most likely not due to a planet. The reason for this effect is that spots are cooler than the photosphere and thus have a different spectral-energy distribution and different line-strength than spectral-lines compared the normal photosphere. For example, if we observe a Sun-like star with cool pots, the amplitudes will in general be larger in the optical than in the NIR, because the brightness difference between a spot and the photosphere is larger in the optical than in the NIR. Crockett et al. (2012) showed that at least for T Tauri stars, the RV-amplitude caused by stellar spots is a factor 2–3 smaller at 2.3 μm than at 0.67 μm. However, if we observe the TiO-lines, which are particularly strong in a spot, the amplitudes should be larger. However, obtaining the same RV-precision as an optical spectrograph in the

NIR is quite difficult. Once such instruments will become standard, high-resolution IR-spectroscopy will play an important role for confirming planets, particularly those orbiting active stars.

In order to show in which direction the new developments will go, I will discuss the two methods how RV-measurements are done today, in the next tow sections. The methods fall into two categories: One is the absorption cell method, the other the emission line method.

15.2.1 The Absorption Cell Method

In this method an absorption cell is placed in front of the slit of the spectrograph, which creates a dense forest of absorption lines in the spectrum of the star. Since the lines from the cell are recorded simultaneously with the spectrum of the star, both sets of lines have the same instrumental wavelength-shifts. Any instrumental wavelength-shift can thus be removed from the stellar spectrum by using the lines from the cell as reference.

Using an HF-absorption cell Walker et al. (1992) observed 16 stars and found periodic RV-variations for γ Cephei. Although they detected RV-variations in this star, they were not convinced that these were due to a planet. This was only shown much later by Hatzes et al. (2003). Because Iodine is a less dangerous molecule, and produces a forest of lines, I_2-cells were used ever since the 1990s.

The RV is determined by modeling the observed spectrum using very-high spectral resolution spectra of the cell (e.g. iodine) and a stellar spectrum taken without the cell (template). The observed spectra taken with the cell are then modeled by iteratively convolving the template and the spectrum of the cell with the instrumental profile (IP) of the instrument, and combing them. The high-resolution spectrum of the iodine is usually obtained with an Fourier-Transform spectrograph (FTS). The high-resolution spectrum of the star (template) is often obtained by deconvolving a spectrum of that star with a very high signal-to-noise (S/N) (Marcy and Butler 1992).

The absorption cell method has the advantage that iodine-cells are very cheap, and that high-precision RV-measurements can be obtained even if the spectrograph is not stabilized, because the instrumental wavelength-shift are removed. The instrumental requirements are thus relatively low. Therefor, almost any high-resolution spectrograph can be used as a high-precision RV-machine just by adding an absorption cell to it. Absorption cells are therefor foreseen for the high-resolution, optical spectrographs planed for the 30-m TMT (www.tmt.org), and the 24.5-m GMT (www.gmto.org).

However, the method has the disadvantage that the accuracy of the RV-measurements depends on the quality of the template, and the accuracy with which the PSF is modeled. A spectrograph with a highly asymmetric PSF should thus be avoided. For obtaining a high quality template, it is usually required to take a spectrum of that star without the cell with a long exposure time. Thus, a relatively large amount of observing time is required before even the first RV-measurement is obtained. The need for templates also implies that stars with variable spectra are not really suitable for this method. Another draw-back of this method is that the cell absorbs some light from the star. How much depends on the exact cell that is being used but 30 % are quite typical. Another disadvantage is that iodine has only lines in the wavelength range between 500 and 630 nm.

If all instrumental effects are removed, the accuracy is given by the signal-to-noise (S/N), the resolving power (R), and the wavelength coverage (B in Å) in the form Hatzes and Cochran (1992)

$$\sigma \approx 1.45\times 10^{9}\cdot (S/N)^{-1}\cdot R^{-1}\cdot B^{-1/2}(ms^{-1})(\text{Hatzes and Cochran} 1992) \qquad (15.1)$$

The wavelength coverage thus directly affects the accuracy of the RV-measurements.

In order to obtain RV-measurements in the near infrared (NIR), other molecules than iodine have been used. As mention above, RV-variations caused by spots are usually smaller in the infrared than in the optical, which allows to discriminate variations caused by planets from other effects. NIR observations also have the advantage that later type objects are usually brighter in this wavelength regime. Thus, infrared RV-measurements are particularly useful for active and late type stars. Using an Ammonia ($^{14}NH_3$) gas cell on the NIR spectrograph CRIRES (CRyogenic high-resolution InfraRed Echelle Spectrograph, VLT) in the wavelength-range from 2292 to 2350 nm Bean et al. (2010) demonstrated that an accuracy of up to 3 m/s can be achieved (Fig. 15.1). CRIRES is currently being upgraded to operate as a cross-dispersed Echelle spectrograph. It will receive new absorption cells and new detectors. The upgraded instrument, CRIRES+, will have a much larger wavelength coverage and therefor an significantly increased efficiency. It will be very well suited for high precision RV-work in the NIR. A very promising new type of infrared cell containing a mixture of different gases, has recently been developed by Valdivielso et al. (2010). This cell produces a dense grid of lines in the whole H and K-band. This is a huge improvement compared to the cells that have been used up to now.

15.2.2 The Emission-Line Method

Another possibility for obtaining high-precision RV-measurements is to use a light source that produces a large number of well determined emission lines as a wavelength reference. Naively, one may think that one could simply use a slit spectrograph and inject light from an emission line source into the slit. However, if the star moves in the slit, for example due to seeing, the position of the stellar lines on the detector would change but the positions of the emission lines would not. In a typically Echelle spectrograph this effect is often larger than than 100 m/s. The easiest way to overcome this problem is to use optical fibres to bring the light from the telescope to the spectrograph. If the fibres are additionally equipped with so-called image scramblers, any motion of the star at the telescope focus does not

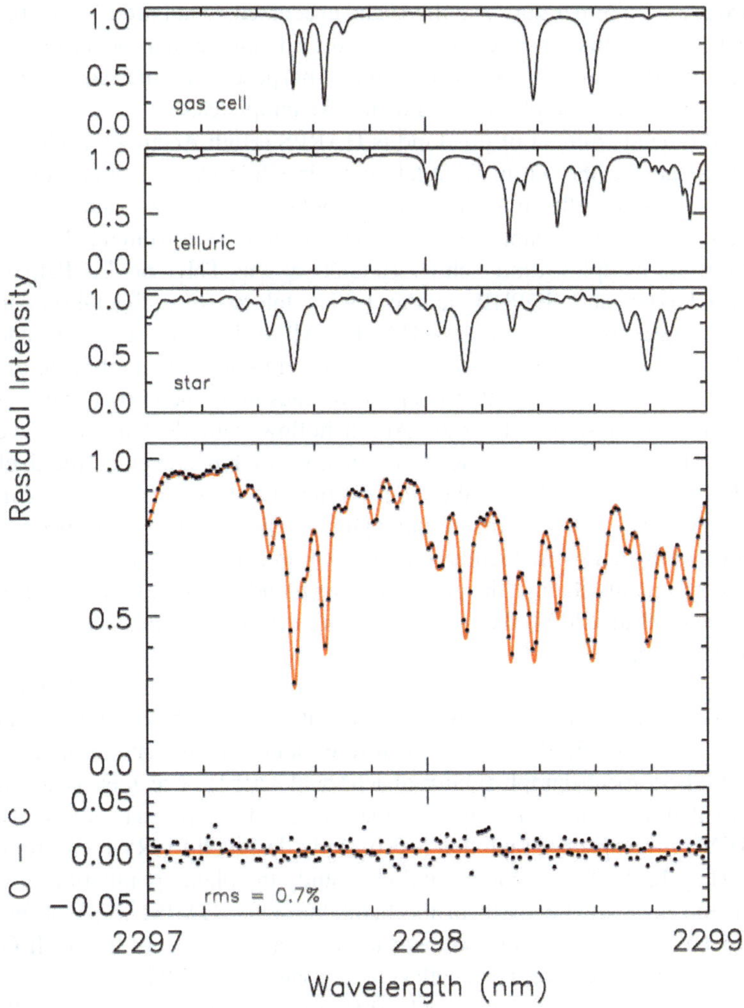

Fig. 15.1 Example for the components needed to model the observed IR-spectrum. The observed spectrum (*lower panel*) is modeled with three components: the ammonia cell, the telluric lines, and the stellar spectrum (From Bean et al. 2010)

effect the position of the light at the entrance of the spectrograph. A second fibre is then used to inject the light from the calibration source into the spectrograph. A new development is the use of none-circular fibres (for example octagonal or hexagonal ones). Such fibres produce a flat output, so that any motion of the star at the input does not affect the RV-measurements (Avila 2012). Because the light from the emission line source is carried in a different fibre than the stellar light, one has to ensure that there is no differential motion between the light coming from the star and the calibration source. The spectrograph thus has to be very stable. Ideally, such a spectrograph is temperature-stabilized and evacuated. Compared to the absorption cell method, the emission-line method has the advantage that the spectrum of the star is undisturbed and can be used for many purposes. Another advantage is that there is no need to spend observing time to take templates.

A prototype instrument of this kind is HARPS (High Accuracy Radial velocity Planet Searcher) at the 3.6-m ESO telescope in Chile (Mayor et al. 2003). Since HARPS is also one of the most productive exoplanet-discovery machines ever built, most new instruments adapt the basic concept of this instrument. For example, HARPS-N (https://plone.unige.ch/HARPS-N) at the TNG on La Palma Island (Canary Islands), and SOPHIE at the 1.93 m telescope at the Observatoire de Haute-Provence, are quite similar to HARPS. All of them are being upgraded with non-spherical fibres. The great advantage of the emission line method compared to the absorption line method is that lamps have lines in a much broader wavelength-range than cells. Using a Thorium-Argon hollow cathode lamp as wavelength reference, HARPS routinely achieves an accuracy of 1 m/s, even on relatively faint stars. However, one problem of the lamps is that the emission lines can shift when the lamp ages. The reason is that at the typical resolution of Echelle spectrographs, many lines of the lamps are blends consisting of several components. If the intensity of these components change, the position of the blend changes. However, as pointed out by Mayor et al. (2009) the aging effect is far more significant for Argon lines than for Thorium lines.

One way to overcome the aging effects is to use many lamps. For example, seven hollow cathode lamps are used in the Calar Alto high-Resolution search for M dwarfs with Exo-Earths with Near-infrared and optical Echelle Spectrographs (CARMENES) spectrograph (Quirrenbach et al. 2012). The idea is to calibrate the lamp that is used every night against a set of lamps that are rarely used. CARMENES is a spectrograph that is especially designed for RV-measurements of M-stars (Fig. 15.2). The scientific goal is to study the planet population of M-stars down to the mass of the Earth in the habitable zone. CARMENES observations will also be important to find suitable targets for instruments like CHEOPS, or NGST. Other NIR spectrographs with similar goals are HPF (Habitable-zone Planet Finder) (Mahadevan et al. 2012) and SPIRou (SpectroPolarimétre Infra-Rouge a near infrared spectro-polarimeter Barrick et al. 2012). CARMENES covers the wavelength region from 0.5 to 1.0 μm in the optical and 1.0–1.7 μm in the NIR. Thorium-Neon lamps are used for the optical and Uranium-Neon lamps for the infrared arm. Because of its spectral range, which covers the region of the spectrum were M stars are brightest, and because of its high sensitivity, it is expected that there will be a significant gain in detecting planets orbiting around M-stars compared to previous instruments.

Hollow cathode lamps are not the only possible calibration source. A Fabry-Pérot can be made in such a way that it produces a dense grid of equally spaced spectral reference lines of roughly equal intensity. Therefore, a Fabry-Pérot can also be used as a calibration source. If the Fabry-Pérot is sufficiently stable, it would be possible to reach higher accuracy than with the hollow cathode lamps. A study by Wildi et al. (2011) shows that, using modern technology, it is possible to achieve

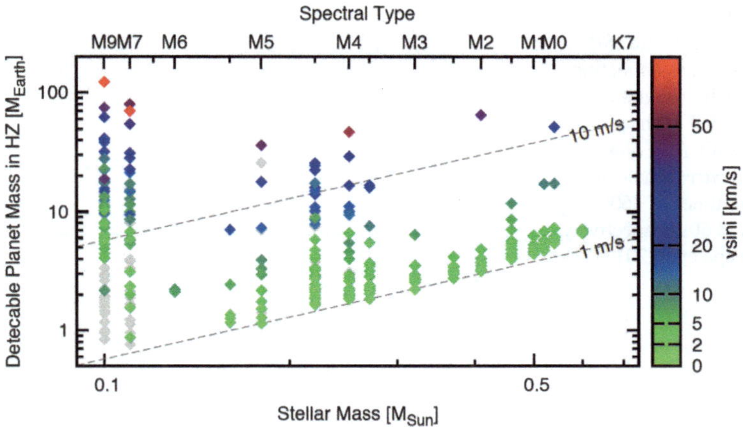

Fig. 15.2 Expected sensitivity of CARMENES for detecting planets in the habitable zone of M-stars. The *grey diamonds* are for $S/N = 150$ in J; the *coloured diamonds* account for the S/N-limitation of faint stars that will be observed in the survey ($S/N < 150$ for $J < 9$ mag) (From Quirrenbach et al. 2012)

a stability of 0.1 m/s during a night with a Fabry-Pérot. Because the Fabry-Pérot can be referenced against the lamps during daytime, RV-measurements with a long-term accuracy of 0.1 m/s are possible. In the past few years, a laser frequency comb (LFC) has also been developed as an alternative calibration source. Since the LFC produces a sequence of equally spaced emission lines, and since the combs can now be made in such a way that they cover a large wavelength-range, such a device seems to be ideal. Using an LFC, it should be possible to achieve an accuracy of the order of 0.01 m/s (Murphy et al. 2007; Wilken et al. 2010; Phillips et al. 2012; Ycas et al. 2012; Molaro et al. 2013).

ESPRESSO (Echelle SPectrograph for Rocky Exoplanets and Stable Spectroscopic Observations) (Pepe et al. 2010, 2013) is an instrument which represents the next step in the evolution of high-precision spectrographs. ESPRESSO is an optical spectrograph that covers the wavelength range is from 380 to 780 nm. The instrument will be installed at the combined Coudé laboratory of the VLT. ESPRESSO can either be used with one, two, or four unit telescopes (UTs) of the VLT. Using an LFC as a wavelength reference source, and none-spherical fibres, it is expected that ESPRESSO will achieve an RV-accuracy of 0.1 m/s with an exposure time of 20 min on a 9th magnitude star using just one UT (Fig. 15.3). This implies that it will be possible to determine the mass of planets down to the mass of the Earth in the habitable zone of solar like stars. A stand-alone RV-program with ESPRESSO will thus allow to study the planet population of solar-like stars in a yet unexplored mass distance regime. ESPRESSO will be a very important source for targets to be observed with CHEOPS. Given that the main science targets of PLATO will be solar-like stars of 9th–11th magnitude, ESPRESSO will allow to measure the mass of rocky planets in the habitable zone that PLATO will detect. For the first time

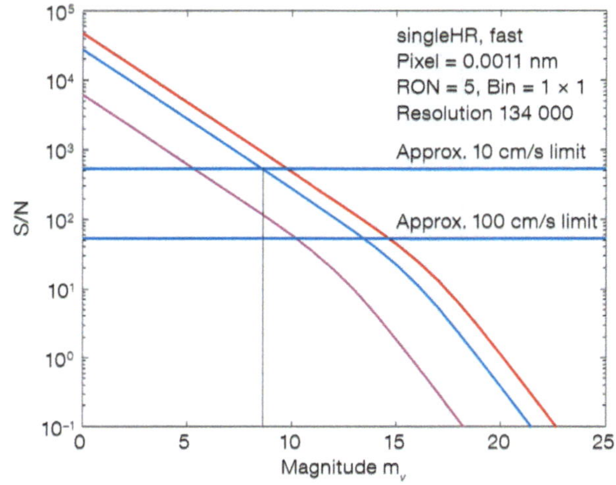

Fig. 15.3 Achievable signal-to-noise ratio (S/N ratio) plotted as a function stellar visible magnitude for the single UT with ESPRESSO. *Red, blue* and *magenta curves* indicate exposure times of 3600s, 1200s and 60s, respectively (From Pepe et al. 2013)

it will be possible to measure the density of planets having the same mass as the Earth in the habitable zone. These observations will allow us to find out whether these planets have thick gaseous hydrogen envelopes, or relatively thin atmospheres like our Earth. It will then be possible to answer the question whether the small sized planets that Kepler has detected in great numbers are rocky and potentially habitable planets, or planets with rocky cores that have thick hydrogen atmospheres and are thus inhabitable. If ESPRESSO uses all four UTs, it will be able to reach an accuracy of 1 m/s for stars as faint as 15th magnitude in 20 min, 2–3 magnitudes fainter than HARPS.

15.2.3 Photometric Observations of Transits

Many transiting planets have been discovered by dedicated ground-based telescopes. A first generation instrument was OGLE. The OGLE-experiment was originally designed for detecting microlensing events (see below) but was also used to detect transiting planets. Although eight transiting extrasolar planets were detected with OGLE, the disadvantage of this experiment was that the stars were all quite faint, typically 16th–17th magnitude in V. This made the followup observations challenging. The second generation instruments, Super-WASP, HAT, TrES, KELT, XO, thus observed much brighter stars. The typically brightnesses of stars in these surveys are 9th–13th magnitude in V. All of them use wide-angle cameras of short-focal length with huge field-of views, and typically detect hot Jupiters. Super-WASP-North is located on the island of La Palma, and Super-WASP-South at the South African Astronomical Observatory in South Africa. Each instrument consists of eight wide-angle cameras of 11.1 cm aperture and a field of view of $7.8 \times 7.8^\circ$. Up to now Super-WASP has discovered 87 planets with radii in the range from 0.7 to 1.9 R_{Jup}, and with orbital periods between 0.8 and 8 days.

The next generation of surveys aims in detecting smaller planets. The way to do this is to increase the photometric accuracy, and to observe smaller stars where planets of a given size lead to deeper transits. One of the new projects is the Next-Generation Transit Survey (NGST). NGTS is a wide-field photometric survey designed to discover transiting exoplanets with a size between 1.5 and 6 M_{Earth} around relatively bright stars ($V < 13$) (http://www.ngtransits.org/). A number of steps are taken to increase the photometric accuracy, compared to the previous experiments. Because the main targets are K, and early M-stars, which are brighter in the red than in the blue part of the spectrum, NGTS will operate in the wavelength range between 600 and 900 nm. The experiment will be built close to Cerro Paranal (Chile), a very dry site of superb photometric quality. NGTS consists of 12 robotic 20 cm f/2.8 telescopes on independent equatorial mounts with a field of view of 3° (Wheatley et al. 2013). A prototype has been tested on La Palama were it delivered a photometric accuracy of better than 0.1 % The current plan foresees to observe about 40,000 stars. This should lead to the discovery of about 100 transiting planets.

The discovery of transiting planets of M stars is also the aim of the MEarth[1] project. MEarth is observing 2,000 individual M-stars using 8 robotic telescopes of 40 cm aperture. GJ1214b, a planet with 2.7 R_{Earth} has been found in this survey (Charbonneau et al. 2009).

Observations of very shallow transits require an extremely high photometric accuracy and are thus often considered to be the domain of space telescopes. However, ground-based telescope can also make a contribution to this field. The giant telescopes that are now being planed, the European Extremely Large Telescope (E-ELT), planed telescope of 39 m aperture (http://www.eso.org/public/teles-instr/e-elt), the TMT (planed Thirty Meter Telescope, www.tmt.org), and the GMT (Giant Magellan Telescope, planed telescope of 24.5 m aperture, www.gmto.org) can certainly contribute to this field. The giant telescopes not only collect more photons but the telescope aperture (a) also acts like a low-pass filter for the scintillation of the Earth's atmosphere (σI). According to Young (1967) $\sigma I \sim a^{-2/3}$. The scintillation thus is also reduced in a giant telescope.

A new idea is to place a telescope in Antarctica, were the dry, clear atmosphere allows to obtain a very high photometric accuracy. Antarctica also has the advantage that long uninterrupted time-series can be obtained during the 3 winter month when it is dark. A pilot project is Antarctic Search for Transiting Exoplanets (ASTEP). It is an optical 40 cm f/4.7 Newton telescope with a field of view of $1 \times 1^\circ$, which is located at Concordia Base at Dome C (Dome C: 75° 06′ 01″ South, 123° 19′ 27″ East, elevation 3,233 m) (Daban et al. 2010). The photometric accuracy achieved with ASTEP is 0.3 (0.7) mmag, or $3\,10^{-4}$ ($7\,10^{-4}$) per hour under good (typical) conditions in the visible. ASTEP has detected the secondary transit of WASP-19b which is only $3.9 \pm 0.8\,10^{-4}$ deep. Dome A is further south and much higher than

[1] http://www.cfa.harvard.edu/MEarth/Welcome.html

Dome C (Dome A: 80° 03″ 22′ South, 77° 22′ 26″ East, elevation 4,084 m). A study of the photometric conditions at this site by Wang et al. (2013) shows that the extinction due to clouds is less than 0.1 (0.4) mag during 45 % (75 %) of the dark time. Both sites are thus very promising for exoplanet research.

15.2.4 Spectroscopic Observations of Transits

Time-resolved spectroscopy has proven to be a very powerful tool for studying transiting extrasolar planets. Time resolved spectroscopy was for example used to confirm the planet WASP-33 (Collier Cameron et al. 2010). During the primary transit the planet crosses the disk of the star. In the optical regime, the rear side of a transiting planet can be considered as totally black. The situation thus is quite similar to the observation of a dark spot in the case of Doppler-Imaging. Like a dark spot, a transiting planet thus causes a hump in the profile of each line in the stellar spectrum of the star. The size of the hump depends on the size of the planet, and how the hump moves across the line profile depends on where the planet crosses the stellar disk. A stationary hump in the middle of the line profile indicates a planet that orbits over the poles. A prograde planet will cause a hump to move from blue to red, and a retrograde one one that moves in the opposite direction. Time-resolved spectroscopy works particularly well for rapidly rotating stars. However, even for slowly rotating stars, we can still measure the Doppler-shift of the spectral-lines (Rossiter-McLaughlin effect).

If a transiting planet has an atmosphere, its structure and composition can be studied by observing the planet during the primary, and secondary transit, as well as out of transit.[2] Since a review has recently published by Tinetti et al. (2013), I will just give a few illustrating examples here. For example, from the analysis of high-resolution NIR-spectra obtained with CRIRES@VLT at different orbital phases of HD 209458b, Snellen et al. (2010) concluded that strong winds are flowing from the irradiated dayside to the non-irradiated night-side of this planet. In another study Rodler et al. (2013) detected CO-absorption in the atmosphere of the hot Jupiter HD 189733b by analyzing high-resolution NIR spectra.

The prospects for such observations is very high, once the giant telescopes like the E-ELT, TMT, and GMT become available. Simulations show that the E-ELT instrument METIS and the proposed instrument HIRES should allow to study the detailed chemical composition of the full population of exoplanets down to Earth-sized planets. METIS is the third instrument foreseen to build for the E-ELT. It is a mid-infrared imager and spectrograph which will offer imaging and medium-resolution spectroscopy in the wavelength range from 3 to 14 μm, and high-resolution integral field spectroscopy from 3 to 5.3 μm. HIRES is a proposed

[2]The same can also be done by observing the transit photometrically in one, or several colours, but this can be taken as just a special case of low-resolution spectroscopy.

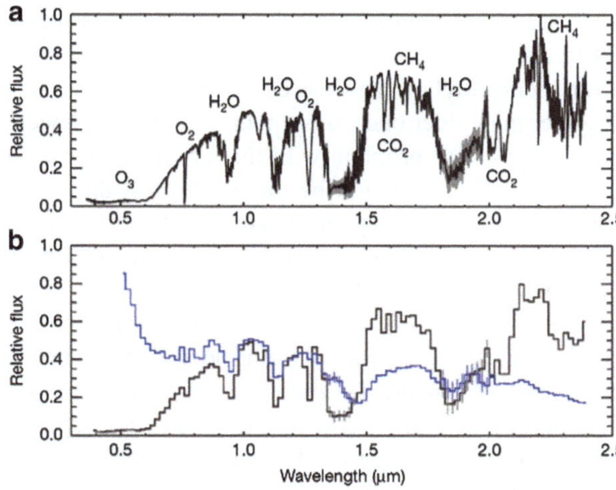

Fig. 15.4 (**a**) The transmission spectrum of the Earth's atmosphere, with some of the major atmospheric constituents marked. (**b**) Comparison between the Earth's transmission (*black*) and reflection (*blue*) spectra. (From Pallé et al. 2009)

high-resolution optical spectrograph for the E-ELT. According to Birkby et al. (2013), an instrument like HIRES might have the potential to detect the oxygen biomarker at 0.76 μm in the atmospheres of Earth-like planets orbiting in the habitable zone of M-dwarfs. However, the detailed analysis by Rodler and López-Morales (2014) shows that the detection of O_2 in the atmosphere of an Earth twin will only be feasible with the ELTs if the planet is orbiting a bright new by ($d \leq 8$ pc) M-dwarf with a spectral type later than M3. In order to find out what the potential of transmission spectroscopy for terrestrial planets is, Pallé et al. (2009) obtained a transmission spectrum of the Earth atmosphere by using a lunar eclipse. The transmission spectrum shows some biologically relevant atmospheric features which are weak in the reflection spectrum (such as ozone, molecular oxygen, water, carbon dioxide and methane). Molecular nitrogen (N_2), which is missing in the reflection spectrum, can also be seen in transmission spectrum (Fig. 15.4).

15.3 Direct Imaging and Interferometry

The technical challenge of the direct imaging technique is not so much the brightness of extrasolar planets at near- and mid-IR wavelengths (NIR, MIR), but the contrast between planets and stars. An instrument that aims in detecting planets in this way has to overcome a huge brightness contrast. It is thus not surprising that it is easier to detect planets at larger, rather than smaller separation. This means that the planets, which have so far been directly imaged, all have long orbital periods. The long orbital periods makes it very difficult to measure their masses dynamically. Quite often the masses are thus estimated from evolutionary tracks. This can be problematic, because evolutionary tracks published by different authors often lead to different masses. This is particularly the case of young planets. The ages of the stars are also quite often poorly determined. On the other hand, young planets are particularly interesting because we can learn more about the formation of planets from them. Anyhow, because at NIR and MIR wavelengths young planets are much brighter than older ones, most planets that were detected in this way have relatively young ages. The problem with that are not only the errors of the masses, the real problem is that it should be the aim to test the models by comparing them with observations, not to use the models to determine the masses of planets. The goal of the future thus must be to detect planets, determine their masses, and then to compare the results with the models.

A new approach is to measure the masses and radii of the planets by determining the log (g) and T_{eff} -values from the spectra of the planets, and then to calculate the masses from these values (e.g., Currie et al. 2013). This approach has the advantage that it also works for planets at large separation, and it even works for the so-called free-floating planets. The ongoing discussion of the connection between brown dwarfs, free-floating planets, and planets in orbit around star has recently been summarized by Pinfield et al. (2013) in the form that "the disk instability can lead to a range of substellar mass objects, including masses reasonable close to that of Jupiter, with some of these objects being ejected into the field". It is thus very well possible that at least some free-floating planets formed in the same way as some planets in orbit around stars. It would thus be very interesting to find out whether there is a difference between free-floating planets, and planets orbiting stars, or not.

To illustrate what has been achieved with the currently available high-contrast imagers (e.g. HiCIAO/SUBARU, NaCo/VLT, NICI/Gemini, OSIRIS Keck, PISCES/LBT, or Project1640/Palomar), I will discuss the three illustrating examples: HR8799 b,c,d,e (Fig. 15.5), βPic b, and GJ504.

HR8799 is an A5V star with a debris disk. It has 1.56 M_{Sun}, an age of 20–50 Mrys, and a distance of 39.4 ± 1.0 pc. The inclination of the disk is 26 ± 3°, as derived from Herschel observations (Matthews et al. 2014). The planets HR8799 b, c, d have a distances of 24, 38, and 68 AU, respectively. Because the system is viewed almost face-on, and because the primary is an A-star, determining the masses of the planets by means of RV-measurements is (almost) impossible. Their masses thus have been estimated from evolutionary tracks. Assuming an age of 60 (30) Myrs, their masses are 7 (5), 10 (7) and 10 (7) M_{Jup}. The fourth planet,

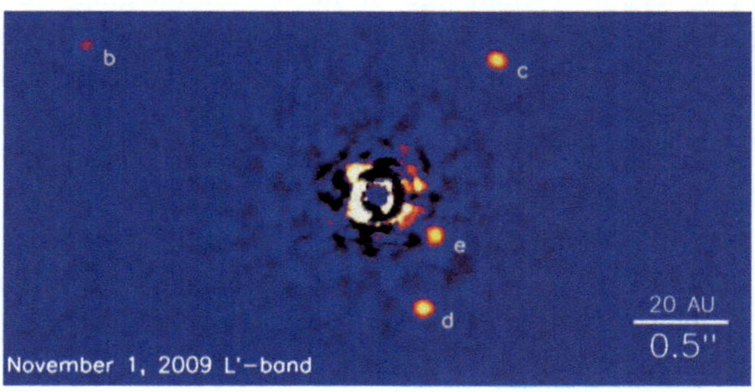

Fig. 15.5 Image of HR8799 taken with the Keck II telescope with the Angular Differential Imaging technique (ADI) in the L'-band (3.776 μm) showing four planets of this system (From Marois et al. 2010)

HR8799e, has a projected distance of about 15 AU, and a mass of about 9 M_{Jup} (Marois et al. 2008, 2010). β Pic is an A6V star with a debris disk which has a mass of about 1.8 M_{sun}. This star is at a distance of 19.3 ± 0.2 pc, and it has an age of about 12^{+8}_{-4} *Mrys*. A planet was discovered using the AO-imager NaCo on the VLT (Lagrange et al. 2009; Currie et al. 2013; Absil et al. 2013). It has an orbital period of about 20 years. By modeling the spectrum of the planet Currie et al. (2013) derived a $T_{eff} = 1575 - 1650\,K$ and $log(g) = 3.8 \pm 0.2$ for the planet. Form these values they obtain a mass of $7^{+3}_{-4}\,M_{Jup}$, and a radius of $\approx 1.65 \pm 0.06 R_{Jup}$.

GJ504 is a G0V-star at a distance of 17.56 ± 0.08 pc. It has an age of 160^{+350}_{-60} *Mrys*. A planet of $4.0^{+4.5}_{-1.0} M_{Jup}$ is at a projected distance of 43.5 AU from the star (Kuzuhara et al. 2013). With a temperature of $510^{+30}_{-20}\,K$, GJ 504 b is also significantly cooler than the other planets. The contrast between the planet and the star is only $3\,10^{-7}$ in the H-band (Kuzuhara et al. 2013).

These three examples show us that, although contrast of $3\,10^{-7}$ have been achieved, the planets discovered by the direct imaging technique are very massive, relatively young, and have orbital periods of 20 years, or more. The aim of future instruments thus is reach a higher contrast-level between the star and the planet. This will allow us to detect fainter planets at smaller distances from the stars. One problem is that the quasi-static speckles of AO-systems can be easily confused with a planet. It is thus of key importance to remove these. One possibility is to make use of the image rotation in an alt-azimuth telescope. If we do not use the image derotator, the image of the planet would start to move over the detector, whereas the static speckles would stay at the same position (angular differential imaging (ADI)). By subtracting images taken at different times, the static "speckles" are removed. Other methods are to take images inside and outside of strong molecular bands in the atmosphere of planets, or to use polarization. An important part of such an instrument is a coronagraph. A coronagraph blocks the light from the star but not the light from the planet. A large number of different concepts have been studied. A new concept is the achromatic four-quadrant phase mask coronagraph (A4Q) (Mawet et al. 2006). Such a device will be part of the new planet imager at the VLT, SPHERE (Beuzit et al. 2008; Vigan et al. 2012). SPHERE will also use an adaptive-optics system that is optimized for planet detection.

The next generation of ground-based telescopes will dramatically enlarge the discovery space. It might even become possible to to detect old, giant planets in reflected light and possibly even some rocky planets. The instrument study EPICS for the E-ELT shows that such an instrument could reach a contrast level of 10^{-9} at separations of 0.1 arcsec (Kasper et al. 2010). The same study show that with this instrument it would be possible to detect about 40 planets in the mass range of 10–40 M_{Earth}, and hundreds of planets of higher mass. A study for the high-contrast imager PFI (Planet Formation Imager), which is proposed for the TMT, shows that such an instrument could reach a contrast-level of 10^{-8} at separations of 0.1 arcsec (Macintosh et al. 2006).

Observations with interferometers may also have the potential to detect extrasolar planets. Particularly interesting is the concept of a Nulling Interferometry, in which the light from the star is suppressed by destructive interference. Such instruments have, or are being developed for the Keck and the LBT interferometers (Serabyn et al. 2012; Hinz et al. 2003; Angle and Woolf 1997).

15.4 Astrometry, Polarization, Microlensing

After the launch of Gaia, ground-based astrometric observations naturally has to focus on measurements that Gaia can not do. For example, observations of stars that are either too bright, or too faint for Gaia. Up to now the role of ground based astrometric measurements for exoplanet research was somewhat limited, as no extrasolar planet has so far been detected using this method. However, the accuracy that has been achieved is already impressive. Using just the standard imaging cameras FORS1/2 at the VLT, Lazorenko et al. (2009) achieved a precision of 50 μas for 17–19 mag stars, and Muterspaugh et al. (2005) one of 26 μas. In the 1990s, the Mark III optical interferometer already achieved an accuracy of about $6-10$ *mas* (Shao et al. 1990; Hummel et al. 1994). It is hoped that future interferometers reach a precision of 30 μas (Woillez et al. 2010; Eisenhauer et al. 2011; Sahlmann et al. 2013).

Polarization can also be used to detect and characterize planets. For example, the fractional polarization for the Earth is 24.6 % in the B, 19.1 % in the V, 13.5 % in R, and 8.3 % I (Bazzon et al. 2013). A highly sensitive polarimeter, called Zurich IMaging POLarimeter, or ZIMPOL for short, is also part of the SPHERE instrument (Schmid et al. 2006). ZIMPOL has demonstrated a polarimetric sensitivity of 10^{-5} (Milli et al. 2013). This means that planets with $M \leq 25\ M_{Earth}$ would be in reach of this instrument (Milli et al. 2013). Sterzik et al. (2012) obtained spectro-polarimetric observation of the Earth-shine which clearly show the prospects of this technique, as the signatures of oxygen, ozone and water were seen.

Gravitational microlensing has proven to be a very powerful tool to detect and study extrasolar planets. An interesting aspect of this method is that the results are complementary to the other ones, because microlensing allows to detect small-mass planets orbiting at large orbital distance from the host stars. A review on this subject has recently been published by Gaudi (2012).

Conclusion

In recent years, the instrumental development has shown a clear trend away from the multi-purpose towards instruments towards instruments dedicated to a very special purpose. It is interesting to note the radial velocity measurements and transit observation, which have played an important role in the past, will continue to play an important role in the future. The direct imaging technique is now being developed to become a standard tool that allows not only to detect very massive planets at very large distances but in

(continued)

the near future perhaps low-mass planets with orbital periods short enough that their masses can be determined dynamically. Polarization measurements and microlensing will also continue to play an important role in the future while astrometric observations with ground-based telescopes will be limited to those case that Gaia can not do.

Acknowledgements I am thankful to José A. Caballero and Roman Follert for their suggestions to the article. Moreover, I acknowledge the support by the International Space Science Institute (ISSI) in Bern, Switzerland and the ISSI team *Characterizing stellar- and exoplanetary environments*.

References

Avila, G. (2012). *Proceedings of the SPIE, 8446*, article id. 84469L, 6pp.

Absil, O., Milli, J., Mawet, D., Lagrange, A.-M., Girard, J., Chauvin, G., Boccaletti, A., Delacroix, C., & Surdej, J. (2013). *Astronomy and Astrophysics, 559*, L12.

Angel, J. R. P., & Woolf, N. J. (1997). *Astrophysical Journal, 475*, 373.

Barrick, G. A., Vermeulen, T., Baratchart, S., & The SPIRou team (2012). *Proceedings of the SPIE, 8451*, article id. 84513J, 15pp.

Bazzon, A., Schmid, H.M., & Gisler, D. (2013). *Astronomy and Astrophysics, 556*, A117.

Bean, J.L., Seifahrt, A., Hartman, H., Nilsson, H., Wiedemann, G., Reiners, A., Dreizler, S., & Henry, T.J. (2010). *Astrophysical Journal, 713*, 410.

Beuzit, J.-L., Feldt, M., Dohlen, K., Mouillet, D., Puget, P., Wildi, F., Abe, L., Antichi, J., Baruffolo, A., Baudoz, P., Boccaletti, A., Carbillet, M., Charton, J., Claudi, R., Downing, M., Fabron, Ch., Feautrier, Ph., Fedrigo, E., Fusco, Th., Gach, J.-L., Gratton, R., Henning, Th., Hubin, N., Joos, F., Kasper, M., Langlois, M., Lenzen, R., Moutou, C., Pavlov, A., Petit, C., Pragt, J., Rabou, P., Rigal, F., Roelfsema, R., Rousset, G., Saisse, M., Schmid, H.-M., Stadler, E., Thalmann, Ch., Turatto, M., Udry, St., Vakili, F., & Waters, R. (2008). *Proceedings of the SPIE, 7014*, article id. 701418, 12pp.

Birkby, J., de Kok, R., Brogi, M., Schwarz, H., Albrecht, S., de Mooij, E., & Snellen, I. (2013). *The Messenger, 154*, 57.

Charbonneau, D., Berta, Z. K., Irwin, J., Burke, Ch. J., Nutzman, Ph., Buchhave, L. A., Lovis, Ch., Bonfils, X., Latham, D. W., Udry, St., Murray-Clay, R. A., Holman, M.J., Falco, E. E., Winn, J. N., Queloz, D., Pepe, F., Mayor, M., Delfosse, X., & Forveille, Th. (2009). *Nature 462*, 891.

Cameron, A. C., Guenther, E., Smalley, B., McDonald, I., Hebb, L., Andersen, J., Augusteijn, Th., Barros, S. C. C., Brown, D. J. A., Cochran, W. D., Endl, M., Fossey, S. J., Hartmann, M., Maxted, P. F. L., Pollacco, D., Skillen, I., Telting, J., Waldmann, I. P., & West, R. G. (2010). *Monthly Notices of the Royal Astronomical Society, 407*, 507.

Crockett, C. J., Mahmud, N. I., Prato, L., Johns-Krull, Ch. M., Jaffe, D. T., Hartigan, P. M., & Beichman, Ch. A. (2012). *Astrophysical Journal, 761*, 164.

Currie, Th., Burrows, A., Madhusudhan, N., Fukagawa, M., Girard, J.H., Dawson, R., Murray-Clay, R., Kenyon, S., Kuchner, M., Matsumura, S., Jayawardhana, R., Chambers, J., & Bromley, B., (2013). *Astrophysical Journal, 776*, 15.

Daban, J.-B., Gouvret, C., Guillot, T., Agabi, A., Crouzet, N., Rivet, J.-P., Mekarnia, D., Abe, L., Bondoux, E., Fanteï-Caujolle, Y., Fressin, F., Schmider, F.-X., Valbousquet, F., Blanc, P.-E., Le van Suu, A., Rauer, H., Erikson, A., Pont, F., & Aigrain, S. (2010). *Proceedings of the SPIE, 7733*, article id. 77334T, 9pp.

Eisenhauer, F., Perrin, G., Brandner, W., Straubmeier, C., Perraut, K., Amorim, A., Schöller, M., Gillessen, S., Kervella, P., Benisty, M., Araujo-Hauck, C., Jocou, L., Lima, J., Jakob, G., Haug, M., Clénet, Y., Henning, T., Eckart, A., Berger, J.-P., Garcia, P., Abuter, R., Kellner, S., Paumard, T., Hippler, S., Fischer, S., Moulin, T., Villate, J., Avila, G., Gräter, A., Lacour, S., Huber, A., Wiest, M., Nolot, A., Carvas, P., Dorn, R., Pfuhl, O., Gendron, E., Kendrew, S., Yazici, S., Anton, S., Jung, Y., Thiel, M., Choquet, E., Klein, R., Teixeira, P., Gitton, P., Moch, D., Vincent, F., Kudryavtseva, N., Ströbele, S., Sturm, S., Fédou, P., Lenzen, R., Jolley, P., Kister, C., Lapeyrére, V., Naranjo, V., Lucuix, C., Hofmann, R., Chapron, F., Neumann, U., Mehrgan, L., Hans, O., Rousset, G., Ramos, J., Suarez, M., Lederer, R., Reess, J.-M., Rohloff, R.-R., Haguenauer, P., Bartko, H., Sevin, A., Wagner, K., Lizon, J.-L., Rabien, S., Collin, C., Finger, G., Davies, R., Rouan, D., Wittkowski, M., Dodds-Eden, K., Ziegler, D., Cassaing, F., Bonnet, H., Casali, M., Genzel, R., & Lena, P. (2011). *The Messenger, 143*, 16.

Fridlund, M., Rauer, H., & Erikson, A. (2014). H. Lammer & M. L. Khodachenko (Eds.), *Characterizing stellar and exoplanetary environments* (pp. 253). Heidelberg/New York: Springer.

Gaudi, B. S. (2012). *Annual Review of Astronomy and Astrophysics, 50*, 411.

Grießmeier, J.-M. (2014). H. Lammer & M. L. Khodachenko (Eds.), *Characterizing stellar and exoplanetary environments* (pp. 213). Heidelberg/New York: Springer.

Hatzes, A.P., & Cochran, W.D. (1992). M.-H. Ulrich (Ed.). *ESO workshop on high resolution spectroscopy with the VLT. European southern observatory conference and workshop proceedings*, Garching b. München (Vol. 40, p. 275)

Hatzes, A. P., Cochran, W. D., Endl, M., McArthur, B., Paulson, D. B., Walker, G. A. H., Campbell, B., & Yang, S. (2003). *Astrophysical Journal, 599*, 1383.

Hatzes, A. P., Fridlund, M., Nachmani, G., Mazeh, T., Valencia, D., Hébrard, G., Carone, L., Pätzold, M., Udry, St., Bouchy, F., Deleuil, M., Moutou, C., Barge, P., Bordé, P., Deeg, H., Tingley, B., Dvorak, R., Gandolfi, D., Ferraz-Mello, S., Wuchterl, G., Guenther, E., Guillot, T., Rauer, H., Erikson, A., Cabrera, J., Csizmadia, S., Léger, A., Lammer, H., Weingrill, J., Queloz, D., Alonso, R., Rouan, D., & Schneider, J. (2011). *Astrophysical Journal, 743*, 75.

Hinz, P. M., Angel, J. R. P., McCarthy, D. W., Jr., Hoffman, W. F., & Peng, C. Y. (2003). *Proceedings of the SPIE, 4838*, 108.

Huélamo, N., Figueira, P., Bonfils, X., Santos, N. C., Pepe, F., Gillon, M., Azevedo, R., Barman, T., Fernández, M., di Folco, E., Guenther, E. W., Lovis, C., Melo, C. H. F., Queloz, D., & Udry, S. (2008). *Astronomy and Astrophysics, 489*, L9.

Hummel, C. A., Mozurkewich, D., Elias, N. M., II, Quirrenbach, A., Buscher, D. F., Armstrong, J. T., Johnston, K. J., Simon, R. S., Hutter, D. J. (1994). *Astronomical Journal, 108*, 326.

Kasper, M., Beuzit, J.-L., Verinaud, Ch. & The EPICS team (2010). *Proceedings of the SPIE, 7735*, article id. 77352E, 9pp.

Kuzuhara, M., Tamura, M., Kudo, T., Janson, M., Kandori, R., Brandt, T. D., Thalmann, C., Spiegel, D., Biller, B., Carson, J., Hori, Y., Suzuki, R., Burrows, A., Henning, T., Turner, E. L., McElwain, M. W., Moro-Martín, A., Suenaga, T., Takahashi, Y. H., Kwon, J., Lucas, P., Abe, L., Brandner, W., Egner, S., Feldt, M., Fujiwara, H., Goto, M., Grady, C. A., Guyon, O., Hashimoto, J., Hayano, Y., Hayashi, M., Hayashi, S. S., Hodapp, K. W., Ishii, M., Iye, M., Knapp, G. R., Matsuo, T., Mayama, S., Miyama, S., Morino, J.-I., Nishikawa, J., Nishimura, T., Kotani, T., Kusakabe, N., Pyo, T.-S., Serabyn, E., Suto, H., Takami, M., Takato, N., Terada, H., Tomono, D., Watanabe, M., Wisniewski, J. P., Yamada, T., Takami, H., Usuda, T. (2013). *Astrophysical Journal, 774*, 11.

Lagrange, A.-M., Gratadour, D., Chauvin, G., Fusco, T., Ehrenreich, D., Mouillet, D., Rousset, G., Rouan, D., Allard, F., Gendron, É., Charton, J., Mugnier, L., Rabou, P., Montri, J., & Lacombe, F. (2009). *Astronomy and Astrophysics, 493*, L21.

Lagrange, A.-M., Meunier, N., Chauvin, G., Sterzik, M., Galland, F., Lo Curto, G., Rameau, J., & Sosnowska, D. (2013). *Astronomy and Astrophysics, 559*, A83.

Lazorenko, P. F., Mayor, M., Dominik, M., Pepe, F., Segransan, D., & Udry, S. (2009). *Astronomy and Astrophysics, 505*, 903.

Mahadevan, S., Ramsey, L., Bender, C., & The Habitable-zone Planet Finder (HPF) team (2012). *Proceedings of the SPIE, 8446*, article id. 84461S, 14pp.

Marcy, G. W., & Butler, P. R. (1992). *Publications of the Astronomical Society of the Pacific. 104*, 270.
Marois, C., Macintosh, B., Barman, T., Zuckerman, B., Song, I., Patience, J., Lafreniére, D., & Doyon, R. (2008). *Science, 322*, 1348.
Marois, C., Zuckerman, B., Konopacky, Q. M., Macintosh, B., & Barman, T. (2010). *Nature, 468*, 1080.
Matthews, B., Kennedy, G., Sibthorpe, B., Booth, M., Wyatt, M., Broekhoven-Fiene, H., Macintosh, B., & Marois, Ch. (2014). *Astrophysical Journal, 780*, 97.
Mawet D., Riaud, P., Baudrand, J., Baudoz, P., Boccaletti, A., Dupuis, O., & Rouan, D. (2006). *Astronomy and Astrophysics, 448*, 801.
Mayor, M., Udry, S., Lovis, C., Pepe, F., Queloz, D., Benz, W., Bertaux, J.-L., Bouchy, F., Mordasini, C., & Segransan, D. (2009). *Astronomy and Astrophysics, 493*, 639.
Mayor, M., Pepe, F., Queloz, D., & The HARPS team (2003). *The Messenger, 114*, 20.
Macintosh, B., Troy, M., Doyon, R., Graham, J., Baker, K., Bauman, B., Marois, Ch., Palmer, D., Phillion, D., Poyneer, L., Crossfield, I., Dumont, Ph., Levine, B. M., Shao, M., Serabyn, G., Shelton, Ch., Vasisht, G., Wallace, J. K., Lavigne, J.-F., Valee, Ph., Rowlands, N., Tam, K., Hackett, D. (2006). *Proceedings of the SPIE, 6272*, article id. 62720N.
Milli, J., Mouillet, D., Mawet, D., Schmid, H. M., Bazzon, A., Girard, J. H., Dohlen, K., & Roelfsema, R. (2013). *Astronomy and Astrophysics, 556*, A64.
Molaro, P., Esposito, M., Monai, S., Lo Curto, G., González Hernández, J. I., Hänsch, T. W., Holzwarth, R., Manescau, A., Pasquini, L., Probst, R. A., Rebolo, R., Steinmetz, T., Udem, Th., & Wilken, T. (2013). *Astronomy and Astrophysics, 560*, A61.
Murphy, M. T., Udem, T., Holzwarth, R., Sizmann, A., Pasquini, L., Araujo-Hauck, C., Dekker, H., D'Odorico, S., Fischer, M., Hänsch, T. W., & Manescau, A. (2007). *Monthly Notices of the Royal Astronomical Society, 380*, 839.
Muterspaugh, M. W., Lane, B. F., Konacki, M., Burke, Bernard F., Colavita, M. M., Kulkarni, S. R., & Shao, M. (2005). *Astrophysical Journal, 130*, 2866.
Pallé, E., Zapatero Osorio, M. R., Barrena, R., Montañés-Rodríguez, P., & Martín, E. L. (2009). *Nature, 459*, 814.
Pepe, F. A., Cristiani, S., Rebolo Lopez, R., & The ESPRESSO team (2010). *Proceedings of the SPIE, 7735*. article id. 77350F.
Pepe, F. A., Cristiani, S., Rebolo, R., Santos, N. C., Dekker, H., & The ESPRESSO team (2013). *The Messenger, 152*, 6.
Perryman, M. (2011). *The exoplanet handbook*. Cambridge: Cambridge university press.
Phillips, D., Glenday, A., Li, C., Cramer, C., Furesz, G., Chang, G., Benedick, A.J., Chen, L.-J., Kärtner, F.X., Korzennik, S., Sasselov, D., Szentgyorgyi, A., Walsworth, R. L. (2012). *Optics Express, 20*, 13711.
Pinfield, D. J., Beaulieu, J.-P., Burgasser, A. J., Delorme, P., Gizis, J., & Konopacky, Q. (2013). *Memorie della Societa Astronomica Italiana, 84*(4), 1154.
Quirrenbach, A., Amado, P.J., Seifert, W., & The CARMENES team (2012). *Proceedings of the SPIE, 8446*, article id. 84460R, 13pp.
Rodler, F., Kürster, M., Barnes, J. R. (2013). *Monthly Notices of the Royal Astronomical Society, 432*, 1980.
Rodler, F., & López-Morales, M. (2014). *Astrophysical Journal, 781*, 54.
Sahlmann, J., Henning, T., Queloz, D., & The PRIMA team (2013). *Astronomy and Astrophysics, 551*, A52.
Schmid, H. M., Beuzit, J.-L., Feldt, M., Gisler, D., Gratton, R., Henning, Th., Joos, F., Kasper, M., Lenzen, R., Mouillet, D., Moutou, C., Quirrenbach, A., Stam, D. M., Thalmann, C., Tinbergen, J., Verinaud, C., Waters, R., & Wolstencroft, R. (2006). C. Aime & F. Vakili (Eds.), *IAU Colloq.-200: direct imaging of exoplanets: science & techniques, IAU colloq* (Vol. 200, p. 165).
Serabyn, E., Mennesson, B., Colavita, M. M., Koresko, C., & Kuchner, M. J. (2012). *Astrophysical Journal, 748*, 55.

Shao, M., Colavita, M. M., Hines, B. E., Hershey, J. L., Hughes, J. A., Hutter, D. J., Kaplan, G. H., Johnston, K. J., Mozurkewich, D., Simon, R. S., & Pan, X. P. (1990). *Astronomical Journal, 100*, 1701.

Shustov, B. M., Sachkov, M. E., Bisikalo, D. V., & Gómez de Castro, A.-I. (2014). H. Lammer & M. L. Khodachenko (Eds.), *Characterizing stellar and exoplanetary environments* (pp. 275). Heidelberg/New York: Springer.

Sterzik, M. F., Bagnulo, S., & Pallé, E. (2012). *Nature, 483*, 64.

Snellen, I. A. G., de Kok, R. J., de Mooij, E. J. W., & Albrecht, S. (2010). *Nature, 465*, 1049.

Tinetti, G., Encrenaz, T., & Coustenis, A. (2013). *Astronomy and Astrophysics Review, 21*, 63.

Valdivielso, L., Esparza, P., Martín, E. L., Maukonen, D., & Peale, R. E. (2010). *Astrophysical Journal, 715*, 1366.

Vigan, A., Langlois, M., Martinez, P., Le Mignant, D., Dohlen, K., Moutou, C., Gry, C., Madec, F. (2012). *Proceedings of the SPIE, 8446*, article id. 844699, 11pp.

Walker, G. A. H., Bohlender, D., Walker, A. R., Irwin, A. W., Yang, St. L. S., Larson, A. (1992). *Astrophysical Journal. 396*, L91.

Wang, L., Macri, L. M., Wang, L., Ashley, M. C. B., Cui, X., Feng, L.-L., Gong, X., Lawrence, J. S., Liu, Q., Luong-Van, D., Pennypacker, C. R., Shang, Z., Storey, J. W. V., Yang, H., Yang, J., Yuan, X., York, D. G., Zhou, X., Zhu, Z., & Zhu, Z. (2013). *Astronomical Journal, 146*, 139.

Wheatley, P. J., Pollacco, D. L., Queloz, D., Rauer, H., Watson, Ch. A., West, R. G., Chazelas, B., Louden, T. M., Walker, S., Bannister, N., Bento, J., Burleigh, M., Cabrera, J., Eigmüller, Ph., Erikson, A., Genolet, L., Goad, M., Grange, A., Jordán, A., Lawrie, K., McCormac, J., & Neveu, M. (2013). *European Physical Journal Web of Conferences, 47*, 13002.

Wildi, F., Pepe, F., Chazelas, B., Lo Curto, G., & Lovis, C. (2011). *Proceedings of the SPIE, 8151*, article id. 81511F.

Wilken, T., Lovis, C., Manescau, A., Steinmetz, T., Pasquini, L., Lo Curto, G., Hänsch, Th. W., Holzwarth, R., & Udem, Th. (2010). *Proceedings of the SPIE, 7735*, articl id. 77350T.

Woillez, J., Akeson, R., Colavita, M., & The ASTRA team (2010). *Proceedings of the SPIE, 7734*, article id. 773412, 10pp.

Ycas, G., Quinlan, F., Diddams, S. Osterman, St., Mahadevan, S., Redman, St., Terrien, R., Ramsey, L., Bender, Ch. F., Botzer, B., & Sigurdsson, St. (2012). *Optics Express, 20*, 6631.

Young, A. (1967). *Astrophysical Journal, 72*, 747.

Index

H. Lammer, M. Khodachenko (eds.), *Characterizing Stellar and Exoplanetary Environments*, Astrophysics and Space Science Library 411,
DOI 10.1007/978-3-319-09749-7

MIX
Papier aus verantwortungsvollen Quellen
Paper from responsible sources
FSC® C105338

If you have any concerns about our products,
you can contact us on
ProductSafety@springernature.com

In case Publisher is established outside the EU,
the EU authorized representative is:
Springer Nature Customer Service Center GmbH
Europaplatz 3, 69115 Heidelberg, Germany

Printed by Libri Plureos GmbH
in Hamburg, Germany